Edexcel AS Biology

Ann Fullick

STUDENTS' BOOK

This book also includes

Active Book

A PEARSON COMPANY

CONTENTS

Unit 1 Lifestyle, transport, genes and health

TOPIC 1
Lifestyle, health and risk

TOPIC 2
Genes and health

Unit 2 Development, plants and the environment

How to use this book

This book contains a number of great features that will help you find your way around your AS Biology course and support your learning.

Introductory pages

Each topic has two introductory pages to help you identify how the main text is arranged to cover all that you need to learn. The left-hand page gives a brief summary of the topic, linking the content to three key areas of How Science Works: *What are the theories? What is the evidence? What are the implications?*

The right-hand page of the introduction consists of a topic map that shows you how all the required content of the Edexcel specification for that topic is covered in the chapters, and how that content all interlinks. Links to other topics are also shown, including where previous knowledge is built on within the topic.

Main text

The main part of the book covers all you need to learn for your course. The text is supported by many diagrams and photographs that will help you understand the concepts you need to learn.

Key terms in the text are shown in bold type. These terms are defined in the interactive glossary that can be found on the ActiveBooK CD-ROM using the 'search glossary' feature.

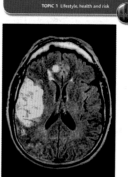

Topic 3 **The voice of the genome**

Introductory pages

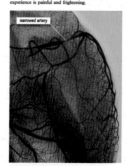

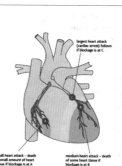

Main text

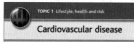

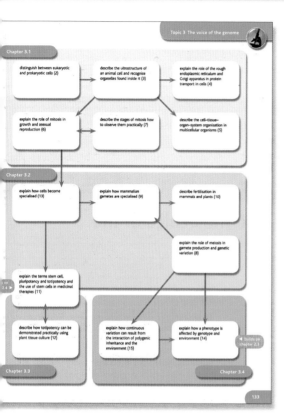

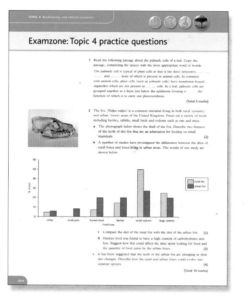

Examzone page

Examzone pages

At the end of each topic you will find two pages of exam questions from past papers. You can use these questions to test how fully you have understood the topic, as well as to help you practise for your exams.

HSW boxes

How Science Works is a key feature of your course. The many HSW boxes within the text will help you cover all the new aspects of How Science Works that you need. These include how scientists investigate ideas and develop theories, how to evaluate data and the design of studies to test their validity and reliability, and how science affects the real world including informing decisions that need to be taken by individuals and society.

Practical boxes

Your course contains a number of core practicals that you may be tested on. These boxes indicate links to core practical work. Your teacher will give you opportunities to cover these investigations.

Question boxes

At the end of each section of text you will find a box containing questions that cover what you have just learnt. You can use these questions to help you check whether you have understood what you have just read, and whether there is anything that you need to look at again.

The contents list shows you that there are two units and four topics in the book, matching the Edexcel AS specification for biology. Page numbering in the contents list, and in the index at the back of the book, will help you find what you are looking for.

How to use your ActiveBook

The ActiveBook is an electronic copy of the book, which you can use on a compatible computer. The CD-ROM will only play while the disc is in the computer. The ActiveBook has these features:

Find Resources

Click on this tab to see menus which list all the electronic files on the ActiveBook.

Student Book tab

Click this tab at the top of the screen to access the electronic version of the book.

Key words

Click on any of the words in **bold** to see a box with the word and what it means. Click 'play' to listen to someone read it out for you to help you pronounce it.

Interactive view

Click this button to see all the icons on the page that link to electronic files, such as documents and spreadsheets. You have access to all of the features that are useful for you to use at home on your own. If you don't want to see these links you can return to **Book view**.

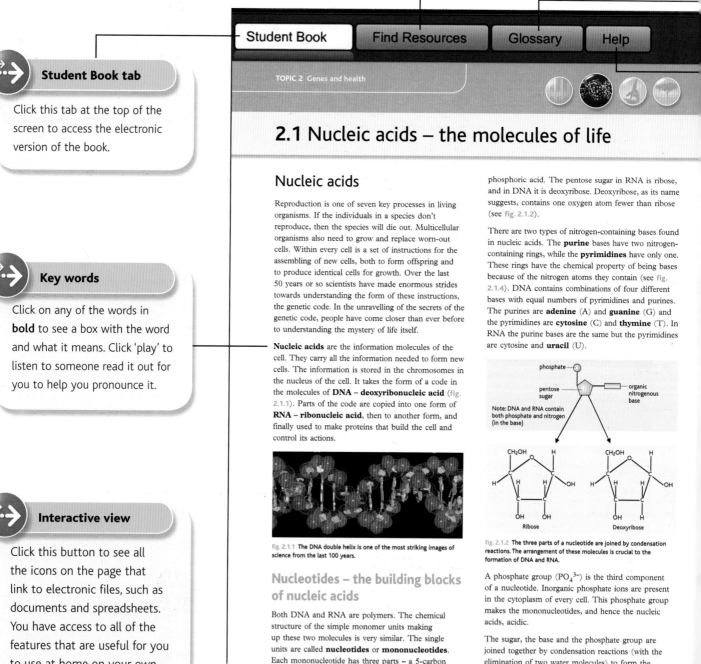

Student Book | **Find Resources** | **Glossary** | **Help**

TOPIC 2 Genes and health

2.1 Nucleic acids – the molecules of life

Nucleic acids

Reproduction is one of seven key processes in living organisms. If the individuals in a species don't reproduce, then the species will die out. Multicellular organisms also need to grow and replace worn-out cells. Within every cell is a set of instructions for the assembling of new cells, both to form offspring and to produce identical cells for growth. Over the last 50 years or so scientists have made enormous strides towards understanding the form of these instructions, the genetic code. In the unravelling of the secrets of the genetic code, people have come closer than ever before to understanding the mystery of life itself.

Nucleic acids are the information molecules of the cell. They carry all the information needed to form new cells. The information is stored in the chromosomes in the nucleus of the cell. It takes the form of a code in the molecules of **DNA – deoxyribonucleic acid** (fig. 2.1.1). Parts of the code are copied into one form of **RNA – ribonucleic acid**, then to another form, and finally used to make proteins that build the cell and control its actions.

fig. 2.1.1 The DNA double helix is one of the most striking images of science from the last 100 years.

Nucleotides – the building blocks of nucleic acids

Both DNA and RNA are polymers. The chemical structure of the simple monomer units making up these two molecules is very similar. The single units are called **nucleotides** or **mononucleotides**. Each mononucleotide has three parts – a 5-carbon or **pentose sugar**, a nitrogen-containing base and

phosphoric acid. The pentose sugar in RNA is ribose, and in DNA it is deoxyribose. Deoxyribose, as its name suggests, contains one oxygen atom fewer than ribose (see fig. 2.1.2).

There are two types of nitrogen-containing bases found in nucleic acids. The **purine** bases have two nitrogen-containing rings, while the **pyrimidines** have only one. These rings have the chemical property of being bases because of the nitrogen atoms they contain (see fig. 2.1.4). DNA contains combinations of four different bases with equal numbers of pyrimidines and purines. The purines are **adenine** (A) and **guanine** (G) and the pyrimidines are **cytosine** (C) and **thymine** (T). In RNA the purine bases are the same but the pyrimidines are cytosine and **uracil** (U).

fig. 2.1.2 The three parts of a nucleotide are joined by condensation reactions. The arrangement of these molecules is crucial to the formation of DNA and RNA.

A phosphate group (PO_4^{3-}) is the third component of a nucleotide. Inorganic phosphate ions are present in the cytoplasm of every cell. This phosphate group makes the mononucleotides, and hence the nucleic acids, acidic.

The sugar, the base and the phosphate group are joined together by condensation reactions (with the elimination of two water molecules) to form the nucleotide (see fig. 2.1.2).

74

☑ Page turn   ☑ Interactive

Glossary

Click this tab to see all of the key words and what they mean. Click 'play' to listen to someone read them out to help you pronounce them.

Help

Click on this tab at any time to search for help on how to use the ActiveBook.

Building the polynucleotides

Mononucleotides are themselves linked together by condensation reactions to form polynucleotide strands (nucleic acids) which can be millions of units long. The sugar of one nucleotide bonds to the phosphate group of the next nucleotide so polynucleotides always have a hydroxyl group at one end and a phosphate group at the other.

To form DNA, nucleotides containing the bases C, G, A and T join together. RNA is made up of long chains of nucleotides containing C, G, A and U. Knowledge of how these units join together, and the three-dimensional structures that are produced in DNA in particular, is the basis of our understanding of molecular genetics.

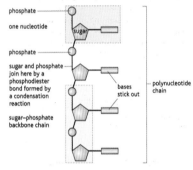

fig. 2.1.3 **A polynucleotide strand like this makes up the basic structure of both DNA and RNA.**

RNA molecules form single polynucleotide strands which may be folded into complex shapes or remain as long thread-like molecules. A DNA molecule is made up of two polynucleotide strands twisted around each other. The sugars and phosphates form the backbone of the molecule. Pointing inwards from this 'spine' are the bases, which pair up in specific ways. A purine always pairs with a pyrimidine – in DNA adenine pairs with thymine and cytosine with guanine. This results in the famous DNA **double helix**, a massive molecule that resembles a spiral staircase.

The two strands of the double helix are held together by hydrogen bonds between the **complementary base pairs** (fig. 2.1.4). There are ten of these pairs for each complete twist of the helix. The two strands are known as the 5' (5 prime) and 3' (3 prime) strand, named

according to the number of the carbon atom in the pentose sugar to which the phosphate group is attached in the first nucleotide of the chain. As you will see, these features of the structure of DNA and RNA are crucial to the way the molecules work within cells.

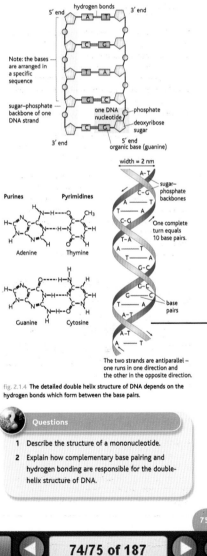

fig. 2.1.4 **The detailed double helix structure of DNA depends on the hydrogen bonds which form between the base pairs.**

Questions

1 Describe the structure of a mononucleotide.

2 Explain how complementary base pairing and hydrogen bonding are responsible for the double-helix structure of DNA.

75

Zoom feature

Just click on a section of the page and it will magnify so that you can read it easily on screen. This also means that you can look closely at photos and diagrams.

Topic 1 Lifestyle, health and risk

This topic deals with one of the most crucial areas of your life – your health. It looks at how the way that people live – their lifestyle – might affect their health. It focuses in particular on the effect of diet, level of exercise and habits such as smoking on the well-being of the heart and circulatory (cardiovascular) system.

What are the theories?

The cardiovascular system plays a vital role in your body. This topic looks at how the different elements of the system work together and how your lifestyle and genetic inheritance might affect its health. It looks at what happens when things go wrong with your cardiovascular system and how they might be prevented or cured.

Diet is a crucial part of your well-being studied in this topic. What impact does diet have on your health? What are the effects of being overweight or underweight? The topic also looks at how food is used in the body. You will learn something about the biochemistry of food – how foods such as carbohydrates and fats are built up and broken down, and the different roles they play in your body.

What is the evidence?

All the information we have about diet, illness and health comes from scientific research and you will be looking at how this is carried out. You will also be considering how to evaluate the results and conclusions from scientific studies, including those which produce conflicting evidence about the same problem. There will be opportunities to carry out your own investigations, for example on the levels of vitamins in food and the effects of different substances on the heart rate.

What are the implications?

Science does not exist in a moral vacuum. In this topic you will begin to explore some ethical issues, for example using animals in medical research and what is involved in investigating human health.

Finally there is the question of how to use all this information in everyday life. What are the risks of getting cardiovascular disease? How do you assess them and make decisions about your lifestyle based on them? How does the way people estimate risk affect their lifestyle choices?

The map opposite shows you all the knowledge and skills you need to have by the end of this topic. The colour in each box shows which chapter they are covered in and the numbers refer to the Edexcel specification.

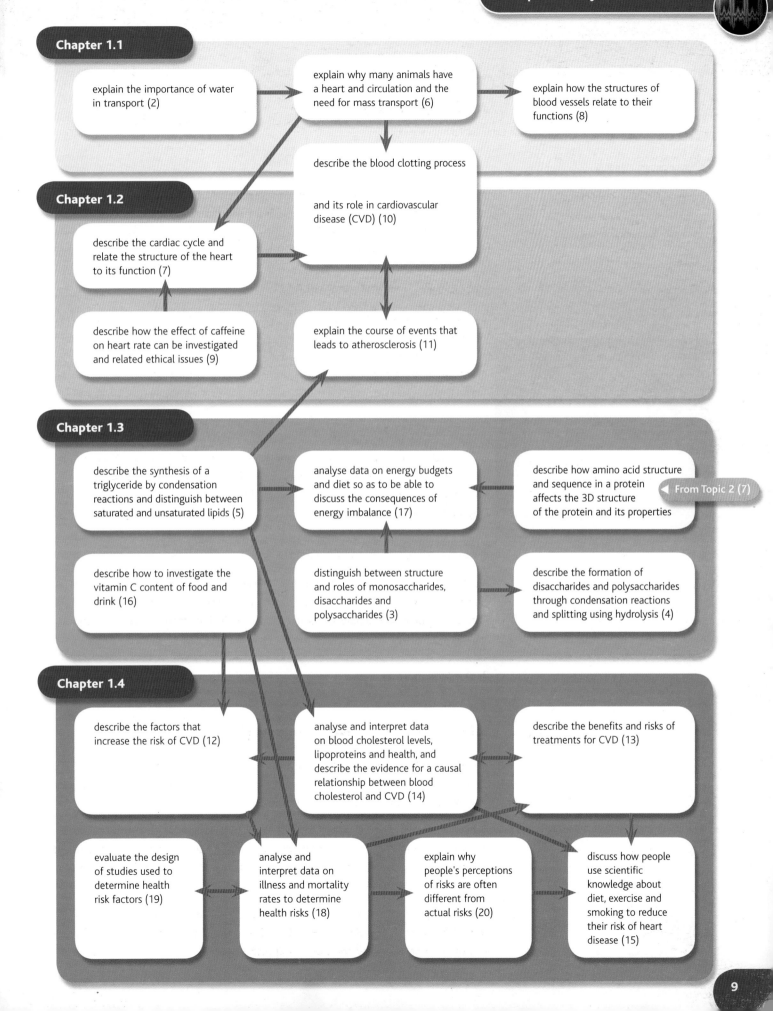

Chapter 1.1

explain the importance of water in transport (2)

explain why many animals have a heart and circulation and the need for mass transport (6)

explain how the structures of blood vessels relate to their functions (8)

describe the blood clotting process and its role in cardiovascular disease (CVD) (10)

Chapter 1.2

describe the cardiac cycle and relate the structure of the heart to its function (7)

describe how the effect of caffeine on heart rate can be investigated and related ethical issues (9)

explain the course of events that leads to atherosclerosis (11)

Chapter 1.3

describe the synthesis of a triglyceride by condensation reactions and distinguish between saturated and unsaturated lipids (5)

analyse data on energy budgets and diet so as to be able to discuss the consequences of energy imbalance (17)

describe how amino acid structure and sequence in a protein affects the 3D structure of the protein and its properties

◀ From Topic 2 (7)

describe how to investigate the vitamin C content of food and drink (16)

distinguish between structure and roles of monosaccharides, disaccharides and polysaccharides (3)

describe the formation of disaccharides and polysaccharides through condensation reactions and splitting using hydrolysis (4)

Chapter 1.4

describe the factors that increase the risk of CVD (12)

analyse and interpret data on blood cholesterol levels, lipoproteins and health, and describe the evidence for a causal relationship between blood cholesterol and CVD (14)

describe the benefits and risks of treatments for CVD (13)

evaluate the design of studies used to determine health risk factors (19)

analyse and interpret data on illness and mortality rates to determine health risks (18)

explain why people's perceptions of risks are often different from actual risks (20)

discuss how people use scientific knowledge about diet, exercise and smoking to reduce their risk of heart disease (15)

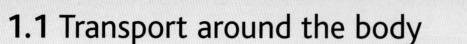

1.1 Transport around the body

Transport systems

Within any living organism, substances need to be moved from one place to another. Cells require a supply of chemicals, such as glucose and oxygen for cellular respiration. These must be transported from outside the organism into the cells. Respiration supplies energy for the other reactions of life but it also produces the toxic waste product carbon dioxide. This and other waste products need to be removed from the cells and the body before they damage them.

Transport in simple organisms

One of the main ways substances move in and out of cells is by **diffusion**. Diffusion is the free movement of particles in a liquid or a gas down a **concentration gradient** from an area where they are at a relatively high concentration to an area where they are at a relatively low concentration.

For a unicellular organism such as an *Amoeba*, nutrients and oxygen can diffuse directly into the cell from its external environment, and waste substances can diffuse directly out. This works well because the surface area of the *Amoeba*'s membrane in contact with the outside is very large relative to the volume of the inside of its cell. That is, its **surface area to volume ratio** is large. Because the organism is just one cell, substances do not need to travel from cell to cell inside it (see **fig. 1.1.1**).

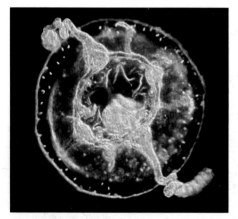

fig. 1.1.1 The surface area to volume ratio of this tiny jellyfish larva is relatively large and so simple diffusion can supply its transport needs.

 HSW Modelling organisms

The surface area to volume ratio of an organism largely determines whether diffusion alone will allow substances to move in and out of all of the cells. However, it isn't easy to calculate the surface area to volume ratio of organisms such as elephants, people and oak trees. It's tricky even for an *Amoeba* because of its irregular shape. So scientists use models to help show what happens in the real situation. A simple cube makes surface area to volume calculations easy. The bigger the organism gets, the smaller the surface area to volume ratio becomes. The distance from outside the organism to the inside gets longer, and there is less surface for substances to enter through. So it takes longer for substances to diffuse in.

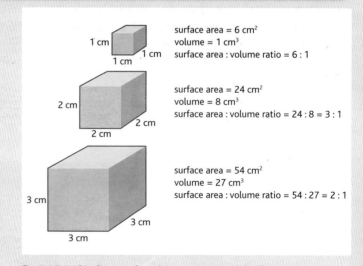

surface area = 6 cm²
volume = 1 cm³
surface area : volume ratio = 6 : 1

surface area = 24 cm²
volume = 8 cm³
surface area : volume ratio = 24 : 8 = 3 : 1

surface area = 54 cm²
volume = 27 cm³
surface area : volume ratio = 54 : 27 = 2 : 1

fig. 1.1.2 In this diagram the cubes represent models of organisms.

Transport in large organisms

In contrast to unicellular organisms such as an *Amoeba*, larger organisms are made up of billions of cells, often organised into specialised tissues and organs. Substances need to travel long distances from the outside to reach the cytoplasm of all the cells. Nutrients and oxygen would eventually reach the inner cells of the body by simple diffusion, but not fast enough to sustain the processes of life.

Complex organisms have evolved specialised systems to get food and oxygen into their bodies (in humans these systems are the gut and the lungs) and to remove waste (the gut, lungs, skin and kidneys). They also have an internal transport system which carries substances to every cell in the body, delivering oxygen and nutrients and taking away waste quickly so that cells can carry out their reactions efficiently. In large complex organisms such as humans, chemicals made in a cell in one part of the body – eg a **hormone** such as insulin or adrenaline – may have an effect on a different type of cell elsewhere in the body. So substances made internally need to be moved around the body as well.

In humans this transport system is the heart and circulatory system and the blood which flows through it. This is an example of a **mass transport system** – substances are transported in the flow of a fluid with a mechanism for moving it around the body. All large complex organisms have some form of mass transport system. Substances are delivered over short distances from the mass transport system to individual cells deep in the body by processes such as diffusion, **osmosis** (the movement of water along a concentration gradient through a partially permeable membrane) and **active transport** (in which energy is used to move substances against a concentration gradient).

Features of mass transport systems

Mass transport makes an effective transport system. Most mass transport systems have certain features in common. They have:

- a system of vessels that carry substances – these are usually tubes, sometimes following a very specific route, sometimes widespread and branching

- a way of making sure that substances are moved in the right direction, eg nutrients in and waste out

- a means of moving materials fast enough to supply the needs of the organism – this may involve mechanical methods (eg the pumping of the heart) or ways of maintaining a concentration gradient so that substances move quickly from one place to another (eg using active transport)

- a suitable transport medium.

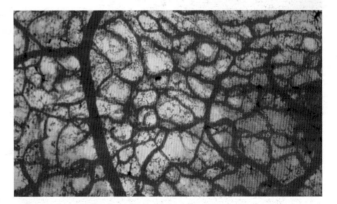

fig. 1.1.3 **The human transport system puts every cell within easy diffusion distance of a blood capillary.**

In this unit you will be looking at how the human cardiovascular system fulfils its transport functions – and what happens when it goes wrong.

 Questions

1 Explain why large animals cannot take in all the substances they need from outside the body through their skin.

2 In humans oxygen enters the body and carbon dioxide leaves it through the lungs. The lungs are made of thousands of tiny air sacs surrounded by blood vessels. How does this help the two gases to diffuse quickly into and out of the blood?

Water in living organisms

Water is the medium in which all the reactions take place in living cells. Without it substances could not move around the body. Water is one of the reactants in the process of photosynthesis, on which all life depends. And water is a major habitat – it supports more life than any other part of the planet. Understanding the properties of water will help you understand many key systems in living organisms, including transport systems.

Each water molecule is slightly polarised. This means it has a very slightly negative part – the oxygen atom – and very slightly positive parts – the hydrogen atoms. This separation of charge is called a **dipole**, and the tiny charges are represented as δ+ and δ–. One of the most important results of this charge separation is that water molecules form **hydrogen bonds**. The slightly negative oxygen atom of one water molecule will attract the slightly positive hydrogen atoms of other water molecules in a weak **electrostatic attraction** called a hydrogen bond. This means that the molecules of water 'stick together' more than you might otherwise expect, because although each individual hydrogen bond is weak, there are a great many of them (as shown in **fig. 1.1.6**). Water has relatively high melting and boiling points compared with other substances that have molecules of a similar size – it takes more energy to overcome the attractive forces of all the hydrogen bonds.

fig. 1.1.4 Between 60 and 70% of your body is water. Understanding this special chemical will show you why drinking plenty is vital to the health of all your body systems.

The chemistry of water

The importance of water to biological systems is due to the basic chemistry of its molecules. The simple chemical formula of water is H_2O, which tells us that two atoms of hydrogen are joined to one atom of oxygen to make up each water molecule (see **fig. 1.1.5**).

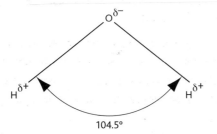

fig. 1.1.5 A model of a water molecule.

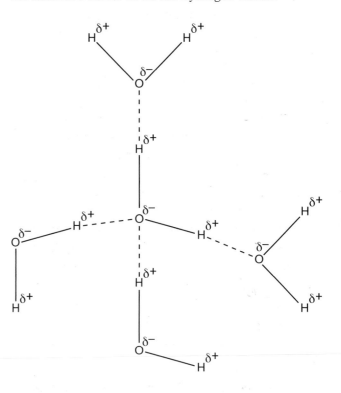

fig. 1.1.6 Water molecules form hydrogen bonds which hold them together.

Why is water important?

The properties of water make it very important in biological systems for several reasons.

1 **Water is an unusual and excellent solvent.** Many other substances will dissolve in it. The fact that the water molecule has a dipole means that many **ionic substances** like sodium chloride (salt), which are made up of positive and negative ions, will dissolve in it. The positive and negative ions separate and become surrounded by water molecules which keep them in solution. Polar substances (compounds with covalent bonds but with small charges on different parts of the molecule) will not usually dissolve in organic solvents such as ethanol, but they will dissolve in water. However, water can also carry many non-polar substances – chemicals that do not form ions. They may form **colloids**, with solute particles larger than the solvent particles. The solute particles are spread through the water but do not separate out. Because the chemical reactions within cells occur in water (in aqueous solution) the ability of water to act as a solvent is vitally important for the processes of life.

Some substances that do not dissolve in water are important in the body. Insoluble particles form **emulsions** (tiny droplets of one liquid suspended in another liquid) or **suspensions** (a solid mixed with a liquid, in which the particles will separate out if the mixture is not constantly moved or stirred). Blood is a suspension of cells and platelets in **plasma**. Fats may be transported in the blood as emulsions of tiny fat droplets (oils form emulsions in a similar way – see **fig. 1.1.7**). Alternatively they may be broken down (eg in digestion) into smaller, soluble molecules to be transported around the body. Then they are built back into insoluble molecules where these are needed, for example in the cell membrane or to form energy storage molecules.

2 **Water has one of the highest known surface tensions.** Surface tension is a property of liquids, when they behave as if the surface is covered by a thin elastic skin. Water has a high surface tension because water molecules form hydrogen bonds which tend to pull them down and together. There is no such attraction between the different

fig. 1.1.7 If oil and water are shaken together hard, the tiny oil droplets form an emulsion. The lymph found in the villi of the small intestine looks milky white like this because of all the tiny fat droplets from digestion emulsified in the lymph.

molecules where water and air meet. So the water layer holds together forming a thin skin of surface tension. Surface tension is of great importance in plant transport systems, and also affects life at the surface of ponds, lakes and other water masses.

3 **The water molecule is amphoteric.** This means that it can act both as an **acid** (it forms H^+ ions and is a proton donor) and as a **base** (it forms OH^- ions and is a proton acceptor).

This ability of water molecules to both donate and receive protons makes it an ideal medium for the biochemical reactions occurring in cells. It acts as a **buffer**, helping to prevent reactions in progress from changing the pH inside the cell. Any excess H^+ or OH^- ions are mopped up.

Questions

1 How are hydrogen bonds formed between water molecules and what effect do they have on the properties of water?

2 Plasma is a solution, cytoplasm is a colloid and blood is a suspension. Explain how the properties of solutions, colloids and suspensions adapt these biological materials to carry out their functions in the body.

The role of blood

In humans the mass transport system is the **cardiovascular system**. This is made up of a series of vessels with a pump (the heart) to move blood through the vessels. The blood is the transport medium and its passage through the vessels is called the **circulation**. The system delivers the materials needed by the cells of the body, and carries away the waste products of their metabolism. It also carries out other functions as well, such as:

- carrying hormones (chemical messages) from one part of the body to another

- forming part of the defence system of the body

- distributing heat.

Chapter 1.2 focuses on the human cardiovascular system. Here you are going to take a closer look at the transport medium – blood.

The components of blood

Your blood is a complex mixture carrying a wide variety of cells and substances to all areas of your body (see **fig. 1.1.8**).

Blood component	Main features
Plasma 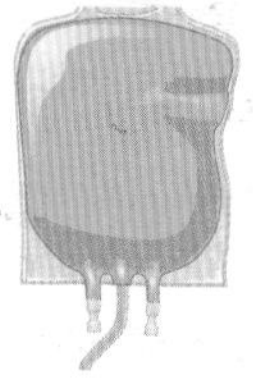	This straw-coloured liquid is the main component of the blood, and it consists largely of water. Plasma contains a wide range of dissolved substances to be transported and also **fibrinogen**, which is a soluble substance vital for the clotting of the blood.
Erythrocytes (red blood cells) 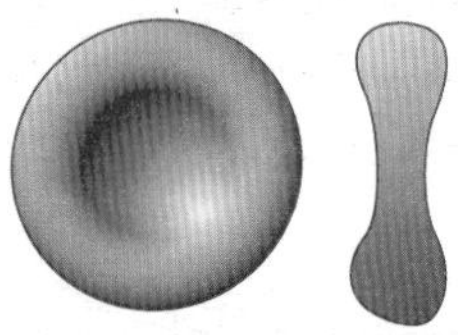	These cells look like biconcave discs. There are approximately 5 million per mm^3 of blood (4–5 million per mm^3 in women, 5–6 million per mm^3 blood in men). They contain **haemoglobin**, a red pigment which carries oxygen and gives them their colour. The scientific name for red blood cells is erythrocytes and they are formed in the red bone marrow of the short bones such as the ribs. Mature erythrocytes do not contain a nucleus and have a limited life of about 120 days.
Leucocytes (white blood cells) 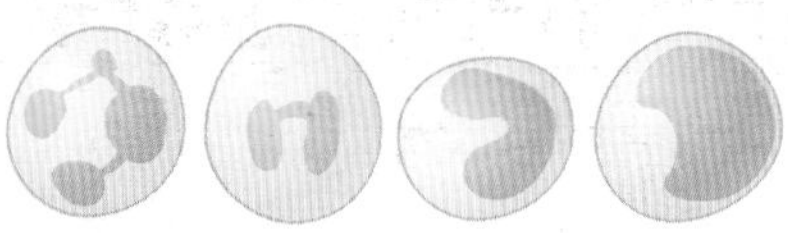	These are much larger than erythrocytes, but can also squeeze through tiny blood vessels as they can change their shape. There are around 4 000–11 000 per mm^3 of blood. They all contain a nucleus and have colourless cytoplasm. There are several types. Most are formed in the white bone marrow of the long bones such as the humerus in the arm and the femur in the leg, although one type, called **lymphocytes**, are formed in the lymph glands and spleen. Their main function is to defend the body against infection.
Platelets 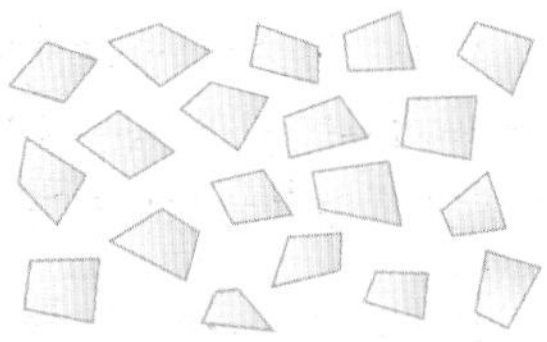	Platelets are tiny fragments of large cells called **megakaryocytes** which are found in the bone marrow. There are about 150 000–400 000 platelets per mm^3 of blood. They are involved in the clotting of the blood.

fig. 1.1.8 The main components of blood. Only the plasma and erythrocytes have a transport function.

The main functions of blood

Your blood carries out a wide variety of functions. The plasma:

- transports digested food products (eg glucose and amino acids) from the small intestine to all the parts of the body where they are needed either for immediate use or storage
- transports food molecules from storage areas to the cells that need them
- transports excretory products (eg carbon dioxide and urea) from cells to the organs such as the lungs or kidneys that excrete them from the body
- transports chemical messages (hormones) from where they are made to where they cause changes in the body
- helps to maintain a steady body temperature by carrying heat around the system from deep-seated organs (eg the gut) or very active tissues (eg leg muscles in someone running)
- acts as a buffer to pH changes.

If a blood vessel is broken, the fibrinogen in the plasma together with the platelets clots the blood. The clot seals the blood vessel and so prevents excessive blood loss. If the broken blood vessel is on the surface of the skin the clot also prevents the entry of disease-causing bacteria or viruses (**pathogens**) which could cause infection. The clot dries to form a scab which protects the new skin growing beneath it and falls off once the damage is fully healed.

The red blood cells transport oxygen from the lungs to all the cells. They are well adapted for transporting oxygen. The shape of the cells – a biconcave disc – means that they have a large surface area to volume ratio, so oxygen can diffuse into and out of them rapidly. Having no nucleus leaves all the space inside the cells for the haemoglobin molecules that carry the oxygen. In fact, each red blood cell contains around 250 million molecules of haemoglobin and can carry 1 000 million molecules of oxygen! Haemoglobin also carries some of the carbon dioxide produced in respiration back to the lungs. The rest is transported in the plasma.

The white blood cells defend against disease in two main ways:

- Some types make **antibodies** which destroy pathogens or **antitoxins** which neutralise the poisons (toxins) made by pathogens. Once the body has encountered a pathogen, these white cells can make antibodies to this pathogen very quickly if it invades again. This is the basis of the body's immunity to diseases.

- Some white blood cells engulf and digest pathogens in a process known as **phagocytosis**.

Substances move between the plasma or red blood cells and the body cells by diffusion or active transport. The tiniest blood vessels have walls only one cell thick and substances pass easily across these into other cells. Every cell in the body is close to one of these small vessels.

HSW Finding out about blood

Knowledge about the cells of the blood and how blood transports substances around the body only appeared as scientists developed the right tools.

- Microscopes have become very powerful, so detailed images of the components of the blood and the smallest blood vessels are now available.

- Radio-opaque dyes in combination with X-rays and magnetic resonance imaging (**MRI scans**) show more detailed internal pictures of the heart and circulation than ever before. They allow doctors and scientists to track the blood as it transports substances around the body.

Questions

1 Why does blood look red although plasma is yellow?

2 Explain the role of the blood transport system in the defence against infection.

3 Red blood cells are unusual in not having a nucleus. Explain how this is an adaptation for their role in carrying oxygen, and why they have a limited life.

Transporting oxygen and carbon dioxide

As you have seen, the erythrocytes are adapted for transporting oxygen. The blood/transport system also carries away the carbon dioxide produced during respiration by cells.

Oxygen

The haemoglobin molecule packed in the red blood cells transport oxygen. Each haemoglobin molecule can pick up four molecules of oxygen. The concentration of oxygen in the red blood cells when the blood enters the lungs is relatively low. Oxygen moves into the red blood cells from the air in the lungs by diffusion. Because the oxygen is picked up and bound to the haemoglobin, the free oxygen concentration in the cytoplasm of the red blood cells stays low. This maintains a steep concentration gradient from the air in the lungs to the red blood cells, so more and more oxygen diffuses in and is loaded onto the haemoglobin.

In the body tissues the oxygen levels are relatively low. The concentration of oxygen in the cytoplasm of the red blood cells is higher than in the surrounding tissue. As a result oxygen moves out into the body cells by diffusion down its concentration gradient. The haemoglobin molecules give up some of their oxygen. When you are at rest or exercising gently, only about 25% of the oxygen carried by the haemoglobin is released into your cells (see **fig. 1.1.9**). There is another 75% reserve in the transport system for when you are very active.

Carbon dioxide

Waste carbon dioxide diffuses from the respiring cells of the body tissues into the blood along a concentration gradient. The reaction of the carbon dioxide with water is crucial. When carbon dioxide is dissolved in the blood it reacts slowly with the water to form carbonic acid, H_2CO_3. The carbonic acid separates to form the ions H^+ and HCO_3^-.

$$CO_2 + H_2O \rightleftharpoons H_2CO_3 \rightleftharpoons HCO_3^- + H^+$$

About 5% of the carbon dioxide is carried in solution in the plasma. A further 10–20% combines with haemoglobin molecules to form **carbaminohaemoglobin**. Most of the carbon dioxide is transported in the cytoplasm of the red blood cells as hydrogencarbonate ions. The enzyme **carbonic anhydrase** controls the rate of the reaction between carbon dioxide and water to form carbonic acid. In the body tissues there is a high concentration of carbon dioxide in the blood so carbonic anhydrase catalyses the formation of carbonic acid.

In the lungs the carbon dioxide concentration is low so carbonic anhydrase catalyses the reverse reaction and free carbon dioxide diffuses out of the blood and into the lungs (**fig. 1.1.10**).

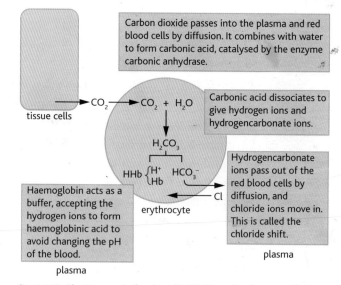

Carbon dioxide passes into the plasma and red blood cells by diffusion. It combines with water to form carbonic acid, catalysed by the enzyme carbonic anhydrase.

Carbonic acid dissociates to give hydrogen ions and hydrogencarbonate ions.

Hydrogencarbonate ions pass out of the red blood cells by diffusion, and chloride ions move in. This is called the chloride shift.

Haemoglobin acts as a buffer, accepting the hydrogen ions to form haemoglobinic acid to avoid changing the pH of the blood.

tissue cells

erythrocyte

plasma

fig. 1.1.10 The transport of carbon dioxide from the tissues to the lungs depends on the reaction of carbon dioxide with water, controlled by an enzyme in the red blood cells.

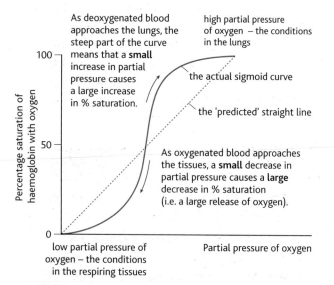

As deoxygenated blood approaches the lungs, the steep part of the curve means that a **small** increase in partial pressure causes a large increase in % saturation.

high partial pressure of oxygen – the conditions in the lungs

the actual sigmoid curve

the 'predicted' straight line

As oxygenated blood approaches the tissues, a **small** decrease in partial pressure causes a **large** decrease in % saturation (i.e. a large release of oxygen).

Percentage saturation of haemoglobin with oxygen

low partial pressure of oxygen – the conditions in the respiring tissues

Partial pressure of oxygen

fig. 1.1.9 Oxygen dissociation curve for human haemoglobin.

The blood-clotting mechanism

You have a limited amount of blood. In theory, a minor cut could endanger life as the torn blood vessels allow blood to escape.

First, and most immediately, your blood volume will fall. If you lose too much blood you will die.

Second, pathogens can get into your body through an open wound.

In normal circumstances your body has a damage-limitation system in the clotting mechanism of the blood. This mechanism seals up damaged blood vessels to minimise blood loss and prevent pathogens getting in.

Plasma, blood cells and platelets flow from a cut vessel. Contact between the platelets and components of the tissue (eg collagen fibres in the skin) causes the platelets to break open in large numbers. They release several substances, of which two are particularly important.

1 **Serotonin** causes the smooth muscle of the blood vessel to contract. This narrows the blood vessels, cutting off the blood flow to the damaged area.

2 **Thromboplastin** is an enzyme which sets in progress a cascade of events that leads to the formation of a clot (see **fig. 1.1.11**).

   • Thromboplastin catalyses the conversion of a large protein called **prothrombin** found in the plasma into another enzyme called **thrombin**. This happens on a large scale at the site of a wound. Calcium ions need to be present in the blood at the right concentration for this reaction to happen.

   • Thrombin acts on another plasma protein called **fibrinogen**, converting it to a substance called **fibrin**. This forms a mesh of fibres.

   • More platelets and blood cells pouring from the wound get trapped in the fibrin mesh. This forms a clot.

   • Special proteins in the structure of the platelets contract, making the clot tighter and tougher.

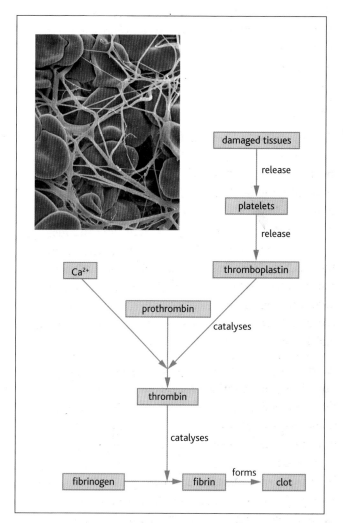

fig. 1.1.11 The cascade of events that results in a life-saving clot. This seals the blood vessels and protects the delicate new skin underneath.

In a cascade system such as clot formation, a relatively small event is amplified through a series of steps. However, sometimes the body's clotting mechanism is triggered in the wrong place, and this can lead to serious problems in the blood vessels. A clot in the vessels that supply your heart with blood can cause a heart attack, as you will see in chapter 1.2.

Questions

1 Explain the role of diffusion in the transport of both oxygen and carbon dioxide around the body.

2 In the genetic condition haemophilia, part of the normal cascade mechanism for blood clotting is missing. What problems do you think this leads to?

Blood circulation

Circulation systems

Many animals have a circulatory system in which a heart pumps blood around the body. Insects have an open circulatory system, with the blood circulating in large open spaces. However, most larger animals – including the vertebrates – have a closed circulatory system with the blood contained within tubes. The blood makes a continuous journey out to the most distant parts of the body and back to the heart.

Animals such as fish have a **single circulation**. The heart pumps deoxygenated blood to the gills, where the blood takes in oxygen and becomes oxygenated. The blood then travels on around the rest of the body of the fish, giving up oxygen to the body cells before returning to the heart.

Birds and mammals need far more oxygen than fish. Not only do they have to move around without the support of water, but they also maintain a constant body temperature which is usually higher than their surroundings. This takes a lot of energy, so their cells need plenty of oxygen and food and produce a lot of waste products that need to be removed quickly. For this reason birds and mammals possess the most complex type of transport system, known as a **double circulation** because it involves two circulatory systems.

The **systemic circulation** carries **oxygenated** (oxygen-rich) blood from the heart to the cells of the body where the oxygen is used, and carries the **deoxygenated** blood (blood that has given up its oxygen to the body's cells) back to the heart.

The **pulmonary circulation** carries deoxygenated blood from the heart to the lungs to be oxygenated, and carries the oxygenated blood back to the heart (see **fig. 1.1.12**).

These separate circulatory systems make sure that the oxygenated and deoxygenated blood cannot mix, so the tissues receive as much oxygen as possible. Another big advantage is that the fully oxygenated blood can be delivered quickly to the body tissues at high pressure. The blood going through the tiny blood vessels in the lungs is at relatively low pressure so it does not damage the vessels and allows gas exchange to take place. If this oxygenated blood at low pressure then went straight

into the big vessels that carry it around the body it would move very slowly. Because it returns to the heart, the oxygenated blood can be pumped hard and sent around the body at high pressure. This means it reaches all the tiny capillaries between the body cells quickly, supplying oxygen for an active way of life.

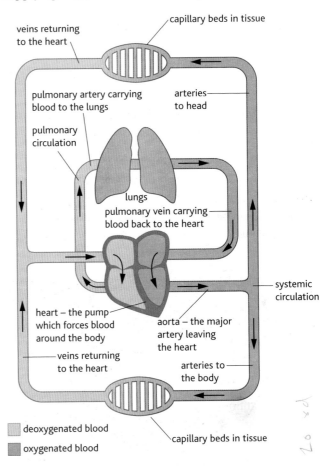

fig. 1.1.12 A double circulation sends blood at high pressure carrying lots of oxygen to the active cells of the body.

The blood vessels

The blood vessels that make up the circulatory system can be thought of as the biological equivalent of a road transport system. The **arteries** and **veins** are like the wide 'motorways' carrying heavy traffic while the narrow town streets resemble the vast branching and spreading **capillary** network. In this capillary network substances carried by the blood are exchanged with cells in the same way as goods are uploaded from factories or offloaded to shops and homes. The structures of the different types of blood vessel closely reflect their functions in your body.

We shall look at the structure and the function of the types of blood vessel separately. However, you should remember that the vessels do not exist as separate structures – they are all interlinked within the whole circulatory system.

Arteries

Arteries carry blood *away* from the heart towards the cells of the body. The structure of an artery is shown in **fig. 1.1.13**. Almost all arteries carry oxygenated blood. The only exceptions are:

* the pulmonary artery, carrying deoxygenated blood from the heart to the lungs

* during pregnancy, the umbilical artery carries deoxygenated blood from the fetus to the placenta.

The arteries leaving the heart branch off in every direction, and the diameter of the **lumen** (the central space inside the blood vessel) gets smaller the further it is from the heart. The very smallest branches of the arterial system, furthest from the heart, are the **arterioles**.

Blood is pumped out from the heart in a regular rhythm, about 70 times a minute. Each heartbeat sends a high-pressure surge of blood into the arteries. The major arteries close to the heart must withstand these pressure surges. Their walls contain a lot of elastic fibres so they can stretch to accommodate the greater volume of blood without being damaged (see **fig. 1.1.14**). Between surges the elastic fibres return to their original length, squeezing the blood and so moving it along in a continuous flow. The pulse you can feel in an artery is the effect of the surge each time the heart beats. The blood pressure in all arteries is relatively high, but it falls in arteries further away from the heart (the **peripheral** arteries).

In the peripheral arteries the muscle fibres in the vessel walls contract or relax to change the size of the lumen, controlling the blood flow. The smaller the lumen, the harder it is for blood to flow through the vessel. This controls the amount of blood that flows into an organ, so regulating the activity of the organ.

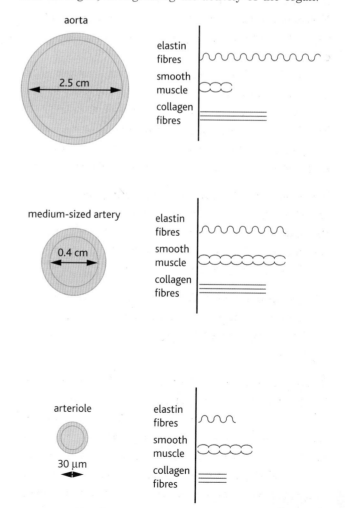

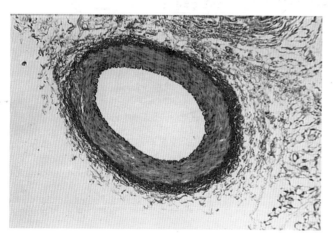

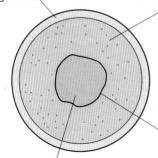

external layer of tough tissue

The middle layers of the artery wall contain elastic fibres and smooth muscle, arteries nearest the heart have more elastic fibres, those further from the heart have a greater proportion of muscle tissue.

Smooth lining allows easiest possible flow of blood.

Lumen is small when artery **unstretched** by flow of blood from heart.

fig. 1.1.13 The structure of an artery enables it to cope with the surging of the blood as the heart pumps.

fig. 1.1.14 The relative proportions of different tissues in different arteries. Collagen gives general strength and flexibility to both arteries and veins.

Capillaries

Arterioles feed into networks of capillaries. These are minute vessels that spread throughout the tissues of the body. The capillary network links the arterioles and the venules. Capillaries branch between cells – no cell is far from a capillary, so substances can diffuse between cells and the blood quickly. Also, because the diameter of each individual capillary is small, the blood travels relatively slowly through them, again giving more opportunity for diffusion to occur (**fig. 1.1.15**). The smallest capillary is no wider than a single red blood cell.

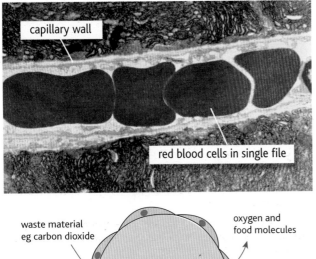

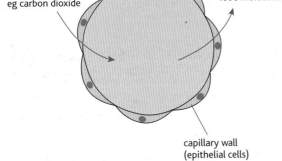

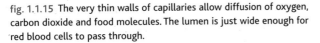

fig. 1.1.15 The very thin walls of capillaries allow diffusion of oxygen, carbon dioxide and food molecules. The lumen is just wide enough for red blood cells to pass through.

Capillaries have a very simple structure well suited to their function. Their walls are very thin, containing no elastic fibres, smooth muscle or collagen. This helps them fit between individual cells and also allows rapid diffusion of substances between the blood and the cells. The walls consist of just one very thin cell. Oxygen and other molecules quickly diffuse out of the blood in the capillaries into the nearby body cells, and carbon dioxide and other waste molecules diffuse in.

Blood entering the capillary network from the arteries is oxygenated. By the time it leaves, it carries less oxygen and more carbon dioxide.

Veins

Veins carry blood back towards the heart. Most veins carry deoxygenated blood. The exceptions are:

*   the pulmonary vein which carries oxygen-rich blood from the lungs back to the heart for circulation around the body

*   during pregnancy, the umbilical vein carries oxygenated blood from the placenta into the fetus.

Tiny **venules** lead from the capillary network, merging into larger and larger vessels leading back to the heart (**fig. 1.1.16**). Eventually only two veins carry the returning blood to the heart – the **inferior vena cava** from the lower parts of the body and the **superior vena cava** from the upper parts of the body.

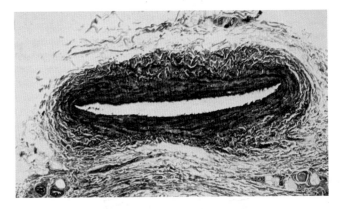

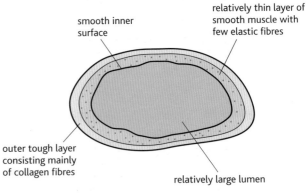

fig. 1.1.16 Veins do not have to withstand the high pressures of the arterial system and this is reflected in their structure.

Veins can hold a large volume of blood – in fact more than half of the body's blood is in the veins at any one time. They act as a blood reservoir. The blood pressure in the veins is relatively low – the pressure surges from the heart are eliminated as the blood passes through the capillary beds. This blood at low pressure must be returned to the heart to be oxygenated again and recirculated. There are two main ways in which this is achieved:

• At frequent intervals throughout the venous system there are one-way valves. These are called **semilunar valves** because of their half-moon shape. They are formed from infoldings of the inner wall of the vein. Blood can pass through towards the heart, but if it starts to flow backwards the valves close, preventing this (see **fig. 1.1.17**).

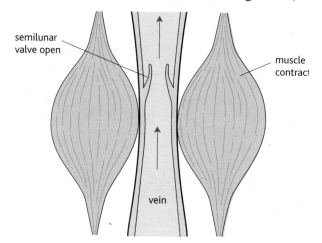

semilunar valve open

muscle contract

vein

Blood moving in the direction of the heart forces the valve open, allowing the blood to flow through.

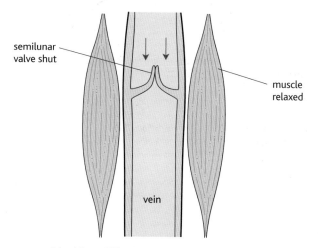

semilunar valve shut

muscle relaxed

vein

A backflow of blood will close the valve, ensuring that blood cannot flow away from the heart.

fig. 1.1.17 Valves in the veins makes sure that blood flows in only one direction – towards the heart. The contraction of large muscles encourages blood flow through the veins.

• Many of the larger veins are situated between the large muscle blocks of the body, particularly in the arms and legs. When the muscles contract during physical activity they squeeze these veins. With the valves keeping blood travelling in one direction, this squeezing helps to return the blood to the heart.

So, the main types of blood vessels – the arteries, veins and capillaries – have very different characteristics. These affect the way the blood flows through the body, and the roles they play in the body. Some of these differences are summarised in **fig. 1.1.18**.

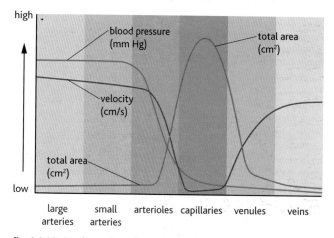

high

blood pressure (mm Hg)

total area (cm²)

velocity (cm/s)

total area (cm²)

low

large arteries | small arteries | arterioles | capillaries | venules | veins

fig. 1.1.18 Graphs to show the surface area of each type of blood vessel in your body, along with the velocity and pressure of the blood travelling in them.

Questions

1 In fish, the blood cannot be supplied to the body tissues at high pressure. Why not – and why does this not matter?

2 Why is a double circulation ideal for an active animal which maintains its own body temperature independently of the environment?

3 Why are valves important in veins but unnecessary in arteries?

4 Look at the graph in **fig 1.1.18**. Explain carefully what the graph shows, and the links between the different lines on the graph.

HSW New ways of treating cancer?

An understanding of the role of the blood and the blood vessels has been increasingly important in the effort to overcome cancer.

Cancers involve the uncontrolled growth of abnormal cells, often resulting in a large mass of cells known as a **tumour**.

The formation of new blood vessels anywhere in the body is known as **angiogenesis**. Cancer cells seem able to stimulate angiogenesis to supply blood to the fast-growing cells of a tumour. The cancer cells produce small **activator molecules** which set off a series of reactions which stimulate the surrounding healthy epithelial cells to produce a new, rich network of blood vessels (**fig. 1.1.19**).

Understanding of the way cancer cells work suggested two possible new ways of treating the disease – cutting off the blood supply to a tumour or blocking the activator molecules so angiogenesis will not take place. Either way the cancer cells will not be able to grow.

fig. 1.1.19 In most of your body tissues a rich blood supply is a good thing. Cutting off the blood supply to a tumour is one approach to treating cancer.

Blocking the blood vessels

In the US a group of scientists led by Dr Erkki Ruoslahti has developed a system based on nanoparticles (very tiny particles) which they hope will treat a variety of cancers. They identified a **peptide** molecule which binds to breast cancer tumours in mice. Small blood clots are found on the endothelium of the blood vessels of tumours, but not in normal blood vessels. Dr Ruoslahti's team showed that their peptide molecule binds to these blood clots, using a genetic variant of mice that don't produce fibrinogen so lack these blood clots. The peptide didn't bind to tumours in the mutant mice but did in the normal ones. The team then linked the peptide to nanoparticles. Once these particles have bound to the blood vessel walls, more clotting takes place and more nanoparticles bind. The clots that form block the tumour blood vessels, starving the cancer cells of oxygen and nutrients. At present the particles block only about 20% of the blood vessels, which isn't enough to stop the tumour growing, but the team are working on improving this. They also hope to produce nanoparticles that carry anticancer drugs right into the tumour, so that they deliver a two-pronged attack. Much of this research uses 'knockout' mice which have been bred to develop cancer (see below).

Blocking the receptors

In the UK, 32 000 men are diagnosed with prostate cancer each year. This cancer often spreads to other tissues and organs such as the bones. As a result 10 000 men die of prostate cancer each year. Some promising research by scientists led by Dr Isaiah Fidler at the University of Texas suggests a new way of stopping the tumours spreading. They worked on mice infected with a drug-resistant strain of prostate cancer. The drug they tried was Glivec (also known as Gleevec), already used to treat leukaemia. Some of the mice were given a mixture of traditional drug treatment and Glivec. Others were given no treatment. The mice were given a mixture of drugs because the human patients would be given the traditional treatment as well. Even if the tumour is drug-resistant (as in these mice) the old treatment might give some benefit. Evidence from the treatment of other cancers suggests a combination of Glivec with other drugs gives the best results. The results are shown in **table 1.1.1**.

Glivec appears to block receptors in the epithelial cells of the blood capillaries which normally respond to growth signals produced by the cancer cells. With these receptors blocked the blood vessels do not grow and branch, and soon the tumour cells stop growing and start to die. Researchers are now trying to find the best combinations of drugs and will then start human trials.

	Tumours present in bone	Median tumour mass	Spread to lymph nodes?
Glivec combination treatment	In 4 out of 18 mice	0.1 g	In 3 cases
Control group (no treatment)	In all 19 mice	1.3 g	In all 19 cases

table 1.1.1 These results were published by the Texan team in the *Journal of the National Cancer Institute*.

Mice as models – ethical questions

In 2007 the British scientist Sir Martin Evans was one of three people awarded the Nobel Prize for Medicine for his work on gene technology and 'knockout' mice. Using animals in research is very expensive and time-consuming. Knockout mice (**fig 1.1.20**) are mice with specific genes silenced or replaced, so that they develop cancer or other diseases. This minimises the number of mice needed in research to work out what happens in a disease and to try out different experimental treatments. Much of our recent knowledge of diseases and how to treat them has come from work using knockout mice.

Professor Steve Brown, head of the Mammalian Genetics Unit of the Medical Research Council, said the research was a 'grand, groundbreaking achievement'. He commented: 'Without this toolkit, we would be considerably hampered'.

Many scientists regard knockout mice as part of the essential toolkit of medical research. Many non-scientists are equally happy with the use of such animals in medical research, feeling that potential benefits to human health outweigh issues of animal welfare. However, other people find the idea of animals being described as 'toolkits' disconcerting. They have ethical concerns about deliberately changing the genetic code of an animal to ensure it gets a disease. These are questions that science cannot answer. Scientists can present their evidence, but society has to decide what is acceptable in research and what is not.

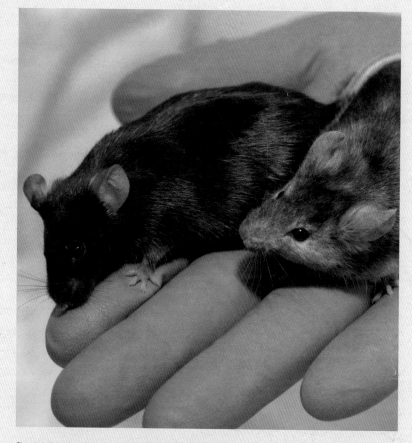

fig. 1.1.20 Knockout mice have revolutionised research into disease.

Questions

1 How did the use of mutant mice enable Dr Erkki Ruoslahti and his team to identify the binding site of the new chemical they hope will lead to a novel cancer treatment?

2 What are the main advantages of using 'knockout' mice in cancer research? And what, in your opinion, are the ethical arguments for and against the use of such organisms in scientific and medical research?

3 Explain why cancer cells need a greater blood supply than normal cells.

1.2 The heart and health

The human heart

In most animal transport systems the heart is the organ that moves the blood around the body. In mammals it is a complex, four-chambered muscular bag, which sits in the chest protected by the ribs and sternum. In an average lifetime your heart will beat about 3 000 000 000 (3×10^9) times and will pump over 200 million litres of blood! In this chapter you will look at how a healthy heart works and how it can go wrong, as well as some other diseases of the cardiovascular system.

The structure of the heart

The human heart (**fig. 1.2.1**) is not a single pump but two, joined together and working in perfect synchrony. The right side of the heart receives blood from the body and pumps it to the lungs. The left side of the heart receives blood from the lungs and pumps it to the body. The blood in each side of the heart does not mix with the blood from the other side. The heart is made of a unique type of muscle known as **cardiac muscle** which has special properties – it can carry on contracting regularly without resting or getting fatigued.

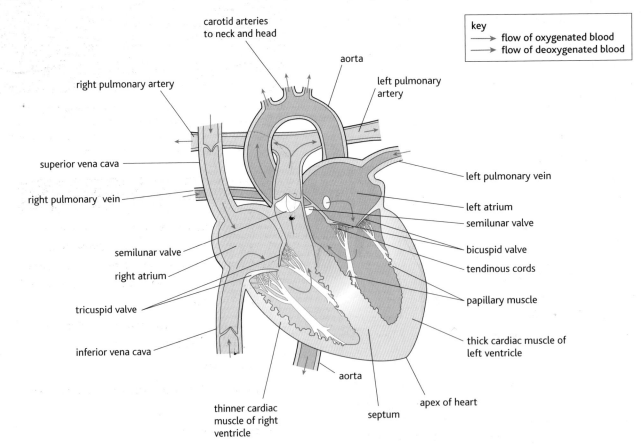

fig. 1.2.1 The structure of the human heart.

- The **inferior vena cava** collects blood (low in oxygen, high in carbon dioxide) from the lower parts of the body, while the **superior vena cava** receives blood from the head, neck, arms and chest. This blood is delivered to the right atrium.

- The **right atrium** receives the blood from the great veins. As it fills with blood, the pressure builds up and opens the tricuspid valve so the right ventricle starts to fill with blood too. When the atrium is full it contracts, forcing the blood into the ventricle. The atrium has thin muscular walls because it receives blood at low pressure from the two venae cavae and needs to exert relatively little pressure to move the blood into the ventricle. One-way semilunar valves (like the valves in veins described in chapter 1.1) at the entrance to the atrium stop a backflow of blood into the veins.

- The **tricuspid valve** is made up of three flaps and is also known as an **atrioventricular valve** as it separates an atrium and a ventricle. The valve allows blood to pass from the atrium to the ventricle, but not in the other direction. The tough **tendinous cords** (also known as valve tendons or heartstrings) make sure the valves are not turned inside out by the great pressure exerted when the ventricles contract.

- The **right ventricle** is filled with blood under some pressure when the right atrium contracts, then the ventricle contracts. Its muscular walls produce the pressure needed to force blood out of the heart into the **pulmonary arteries**. This carries the deoxygenated blood to the capillary beds of the lungs. **Semilunar valves** (like those in veins) prevent the blood flowing back from the artery into the ventricle.

- The blood returns from the lungs to the left side of the heart in the **pulmonary veins**. The blood is at relatively low pressure after passing through the extensive capillaries of the lungs. The blood returns to the **left atrium**, another thin-walled chamber which performs the same function as the right atrium. It contracts to force blood into the **left ventricle**. Backflow is prevented by the **bicuspid valve**.

- As the left atrium contracts, the left ventricle is filled with blood under high pressure. The left ventricle then contracts to force the blood out of the heart and into the **aorta**, the major artery of the body.

This carries blood away from the heart at even higher pressure to other major arteries that branch off from it. The muscular wall of the left side of the heart is much thicker than that of the right. The right side pumps blood to the lungs, which are relatively close to the heart. The left side has to produce sufficient force to move the blood under pressure to all the extremities of the body and overcome the elastic recoil of the arteries. Semilunar valves prevent the blood flowing back from the aorta into the ventricle.

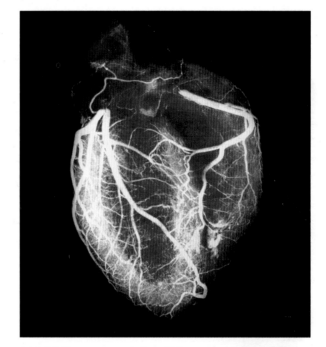

fig. 1.2.2 The coronary arteries, which you can see clearly here, carry oxygenated blood from the aorta to the heart muscle. As your heart beats continuously the muscle needs a good supply of food and oxygen, provided by the coronary arteries.

Questions

1 Describe the path of blood around the human body, identifying at which points the blood is oxygenated and where it is deoxygenated. Explain how this system efficiently supplies cells with the oxygen they need.

2 Discuss the relationship between structure and function for these parts of the heart:
   - semilunar valves
   - muscle wall (and its thickness) in the atria and ventricles
   - tendinous cords.

How your heart works

The beating of your heart produces sounds which are your **heartbeat**. The sounds are made not by the contracting of the heart muscle but by blood hitting the heart valves. The two sounds of a heartbeat are often described as 'lub-dub'. The first sound comes as the blood is forced against the atrioventricular valves when the ventricles contract. The second sound comes as a backflow of blood hits the semilunar valves in the pulmonary artery and aorta as the ventricles relax. The rate of your heartbeat shows how fast your heart is contracting.

The cardiac cycle

Your heart is continuously contracting then relaxing. The contraction of the heart is called **systole**. Systole can be divided into **atrial systole**, when the atria contract together forcing blood into the ventricles, and **ventricular systole**, when the ventricles contract. Ventricular systole happens about 0.13 s after atrial systole, and forces blood out of the ventricles into the pulmonary artery and the aorta. Between contractions the heart relaxes and the atria fill with blood. This relaxation stage is called **diastole**. One cycle of systole and diastole makes up a single heartbeat, which lasts about 0.8 s in humans. This is known as the **cardiac cycle** (**fig. 1.2.3**).

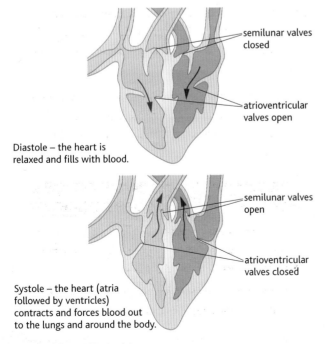

fig. 1.2.3 The cardiac cycle.

Some of the changes in the heart which cause the sounds of the heartbeat are shown in **fig. 1.2.4**. In this diagram you can see how differences in pressure in the regions of the heart cause the different valves to close. You can also see how the stages of the cardiac cycle relate to the pressure changes and the recording of the electrical activity of the heart known as an **electrocardiogram** or **ECG**.

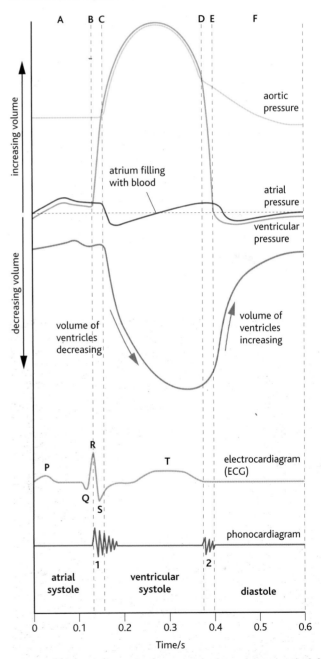

fig. 1.2.4 This graph shows the pressure and volume changes on the left side of the heart during the human cardiac cycle. You can see how these changes relate to the heart sounds and also to an ECG recording from a healthy heart.

What affects the heartbeat?

Your heart beats continuously throughout life. An average adult heart rate is about 70 beats per minute, although in small children it is much higher. The heart responds to the needs of the body – during physical exercise when the tissues need more oxygen, your heart beats faster to supply more oxygenated, glucose-carrying blood to the tissues and to remove the increased waste products. Many other things can also affect the heart rate. For example, stress or excitement can raise it, whilst rest and relaxation can lower it. Many substances that we take into our bodies affect the heart rate. These include permissible drugs such as caffeine and nicotine, as well as many illegal drugs which can cause sudden death because of the way they affect the heart.

Caffeine and the heart

The effect of caffeine on the heart rate can be studied using small pond organisms called *Daphnia*. This avoids the need for experimenting on humans or other large animals, which some people consider unethical. But how far can the results of these experiments be considered valid for all animals?

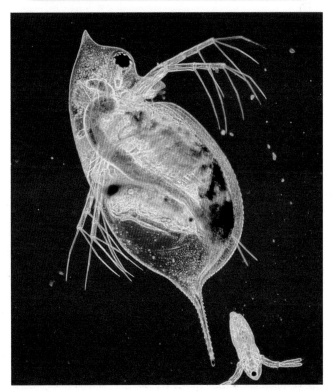

fig. 1.2.5 **It is easy to measure the heart rate of the pond organism** *Daphnia*.

Control of the heart rate

In the very early embryo, cells that are destined to become the heart begin contracting rhythmically long before the actual organ forms. We say they have **intrinsic rhythmicity**. This is retained throughout life, and an adult heart removed from the body will continue to contract as long as it is bathed in an oxygen-rich fluid that contains suitable nutrients. The intrinsic rhythm of the heart is around 60 beats per minute. It is maintained by a wave of electrical excitation similar to a nerve impulse which spreads through special tissue in the heart muscle. Because your heart has its own basic rhythm, your body's resources are not wasted on maintaining this vital but continuous activity.

The intrinsic rhythm of the heart, although very important, does not explain how your heart is able to respond to changes in your body's requirements. This sensitivity is supplied by nerves that control the heart. One nerve speeds up the heart rate and another nerve slows it down. These nerves are controlled by the **cardiovascular centre** in your brain. The cardiovascular centre responds to the variable level of carbon dioxide in your blood – this changes with the demands you make on your body, going up as you exercise. Receptors in your blood vessels respond to these changes in the carbon dioxide levels by sending signals to your cardiovascular centre, which in turn sends signals to your heart. This nervous control enables your heart to react to situations such as exercise or fear, when a faster heart rate is needed, or sleep and relaxation, when the heart needs to slow down.

Hormones also affect your heart rate. The hormone adrenaline, produced in situations when you are excited, angry, nervous or scared, speeds up the heart rate.

Questions

1 Explain how the changes in volume and pressure in the atria and ventricles, shown in **fig.1.2.4**, relate to the events of the cardiac cycle.

2 Describe how the heart rate is controlled by the body, and the factors that can affect it.

Blood pressure

Normal blood pressure

The pressure of blood in your arteries varies as the heart beats (see **fig. 1.2.6**). When the heart contracts, the blood pressure is at its highest. As the heart relaxes and refills, the blood pressure in the arteries falls. Although the inner walls of your blood vessels are mainly smooth, there is some friction between the blood and the vessel walls. This friction is called peripheral resistance and it slows down the flow of blood.

Relatively little pressure is lost as the blood passes through your arteries because they are wide and so offer relatively little resistance to the flow. However, the elastic recoil of the artery walls means that the pulsing of the heartbeat is gradually reduced in vessels that are further away from a direct link to the heart.

In the narrower arterioles and capillaries a greater surface area of vessel wall is in contact with the blood so the peripheral resistance increases. This slows down the flow of blood. However, the blood pressure does not go up, even though the capillaries are very narrow. This is because there are so many capillaries, and together they have a greater total cross-sectional area than that of the main arteries. So the blood pressure falls as blood travels slowly through the capillaries.

The blood pressure in your arteries varies through the day, and it is affected by changes in the blood vessels themselves as well as in the way the heart works. The smooth muscle in the arteries contracts in response to nervous and hormonal stimuli. This makes the arteries **constrict** (get narrower), which raises the blood pressure. When the smooth muscle relaxes, the arteries **dilate** (get wider) and the blood pressure falls. Constricting and dilating the arteries is one way in which the body controls local blood pressure. Changing the pressure changes the flow rate – so more blood can be delivered rapidly to exercising muscles, for example. However, permanent changes in the arteries, such as narrowing due to atherosclerosis (see page 30), can cause permanently raised blood pressure which may lead to severe health problems. Measuring blood pressure is used as an indicator of the health of both the heart and the blood vessels.

Vessel	Lumen diameter	Rate of blood flow/cm s^{-1}
artery	0.4 cm	40–10
arteriole	30.0 μm	10–0.1
capillary	8.0 μm	less than 0.1
venule	20.0 μm	less than 0.3
vein	0.5 cm	0.3–5.0

table 1.2.1 Blood flow rate in the cardiovascular system.

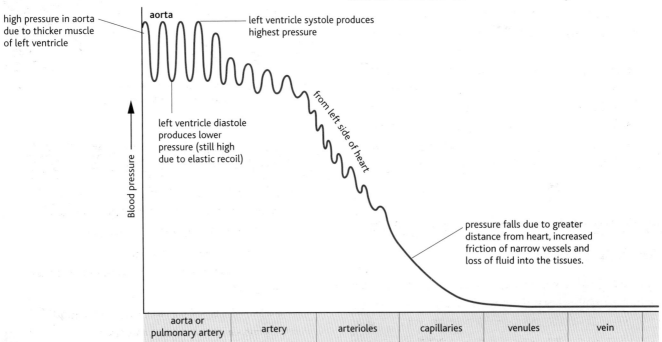

high pressure in aorta due to thicker muscle of left ventricle

aorta

left ventricle systole produces highest pressure

left ventricle diastole produces lower pressure (still high due to elastic recoil)

from left side of heart

pressure falls due to greater distance from heart, increased friction of narrow vessels and loss of fluid into the tissues.

Blood pressure

aorta or pulmonary artery | artery | arterioles | capillaries | venules | vein

fig. 1.2.6 Blood pressure changes as the blood passes through the cardiovascular system.

HSW Measuring blood pressure

Measuring blood pressure is an important diagnostic tool. There are two main ways of doing it – and arguments for and against patients measuring it for themselves.

Blood pressure is usually measured with a **sphygmomanometer**, either manually using a cuff connected to a mercury manometer and a stethoscope, or using an automatic machine. In the manual method (see **fig. 1.2.7**) the cuff is placed around the upper arm as close to the heart as possible. The cuff is then inflated until the blood supply to the lower arm is completely cut off. A stethoscope is used to listen for the sound of blood passing through the blood vessels at the elbow as air is slowly let out of the cuff. This is a cheap method that has been used for many years, but it needs a trained health professional and relies on a reading made as a column of mercury moves quickly down a scale, so it is subject to human error.

Automatic sphygmomanometers also inflate a cuff around the wrist or the arm to control blood flow. A microphone in the cuff detects the blood sounds and an automatic reading appears on a screen. Automatic machines have the advantage that they do not rely on human judgement to take a reading and they can be used by anyone in their own home. The patient can take several readings over time, and are likely to be more relaxed so the readings give a more accurate picture of blood pressure than a snapshot under stress at the doctor's.

Some machines detect blood flow at the wrist or thumb, and rely on a calculation to convert this reading to that of the blood pressure taken on the arm. This is a source of potential error. Another disadvantage is that a medically trained expert is not present to help you interpret the results. For example, because your blood pressure fluctuates naturally, one high reading may not be significant. But if you don't know that and feel anxious, your blood pressure may rise every time you measure it!

Making sense of blood pressure measurements

The first blood to get through the cuff is that under the highest pressure, when the heart is contracting strongly (systole). This is the **systolic blood pressure**. The blood sounds return to normal at the point when even the lowest pressure during diastole is sufficient to force

blood through the cuff. This gives the **diastolic blood pressure**. Blood pressure is measured in an old unit of pressure, millimetres of mercury (mm Hg), because that is what a mercury sphygmomanometer shows. Automatic machines also use mm Hg because this unit is commonly understood.

A systolic reading of 120 mm Hg and a diastolic reading of 80 mm Hg (120/80) is regarded as 'normal'. As you have seen, your blood pressure varies widely throughout the day depending on what you are doing. A doctor or nurse would only become concerned when measurements taken at rest and sustained over a period of time remain much higher or lower than 'normal'. A sustained value of over 140/90 is called **hypertension** and a sustained value of around 90/60 or lower is called **hypotension**. A weakened heart may produce hypotension, whereas damaged blood vessels that are narrowing or becoming less elastic will give hypertension, as you will see later in this chapter.

fig. 1.2.7 Measuring blood pressure with a manual sphygmomanometer that uses a column of mercury to indicate pressure.

Questions

1 Describe the changes in blood pressure as the blood passes through each main group of blood vessels, and explain why these changes occur.

2 Suggest why the cuff of a manual sphygmomanometer is placed as close to the heart as possible.

3 Explain why blood pressure is one of the most common health measurements taken.

Atherosclerosis

Cardiovascular diseases in the UK

Problems with the cardiovascular system have serious consequences. Data from the Office for National Statistics in England and Wales for 2005 showed that heart diseases were responsible for more deaths (20.2% of deaths) than any other cause. Second came cerebrovascular diseases, more commonly known as **strokes** – another disease of the cardiovascular system. Together these **cardiovascular diseases** (**CVDs**) caused more than a quarter of all deaths in England and Wales in 2005 (see **fig. 1.2.8**). What is more, almost a third of these deaths were in people younger than 75. These are known as premature deaths – people dying younger than expected. Many of these cardiovascular diseases are linked to a condition called **atherosclerosis**.

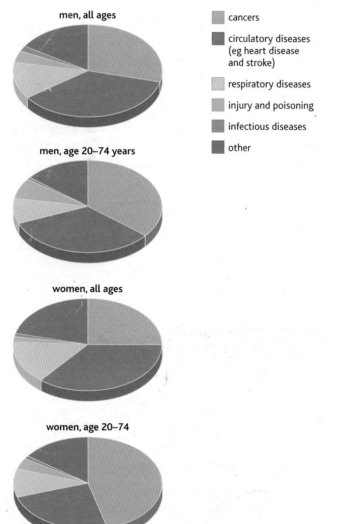

Legend:
- cancers
- circulatory diseases (eg heart disease and stroke)
- respiratory diseases
- injury and poisoning
- infectious diseases
- other

men, all ages

men, age 20–74 years

women, all ages

women, age 20–74

fig. 1.2.8 The most common causes of death in England and Wales in 2005.

How atherosclerosis forms

Atherosclerosis literally means hardening of the arteries, and is a build-up of yellowish fatty deposits (plaques) on the inside of arteries. It can begin in late childhood and continues throughout life. A **plaque** can build up until it restricts the flow of blood through the artery or even blocks it completely. Plaques are particularly likely to form in the arteries of the heart (coronary arteries) and of the neck (carotid arteries). The development of a plaque follows a pattern as shown in **fig. 1.2.9**.

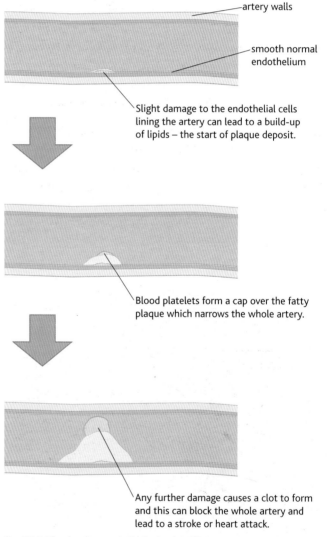

artery walls

smooth normal endothelium

Slight damage to the endothelial cells lining the artery can lead to a build-up of lipids – the start of plaque deposit.

Blood platelets form a cap over the fatty plaque which narrows the whole artery.

Any further damage causes a clot to form and this can block the whole artery and lead to a stroke or heart attack.

fig. 1.2.9 The development of atherosclerosis.

Damage to the endothelial lining of blood vessels leading to plaque formation can be caused by several factors, such as high blood pressure and chemicals in tobacco smoke. Atherosclerosis usually occurs in arteries rather than veins. This is because the blood in the arteries flows fast under relatively high pressure, which puts more strain on the endothelium lining the vessels and can cause small areas of damage. In the veins the pressure is lower so damage to the endothelium is much less likely.

Once the damage has occurred, the body's **inflammatory response** begins and white blood cells arrive at the site of the damage. These cells accumulate chemicals from the blood, in particular cholesterol. This leads to a fatty deposit known as an **atheroma** forming on the endothelial lining of the artery. Fibrous tissue and calcium salts also build up around the atheroma, turning it into a hardened plaque. This hardened area means that part of the artery wall hardens, so it is less elastic than it should be. This is atherosclerosis.

The lumen of the artery becomes much smaller as a result of the plaque. This increases the blood pressure, making it harder for the heart to pump blood around the body. The raised blood pressure makes damage more likely in the endothelial lining in other areas, and more plaques will form. More plaques will make the blood pressure even higher.

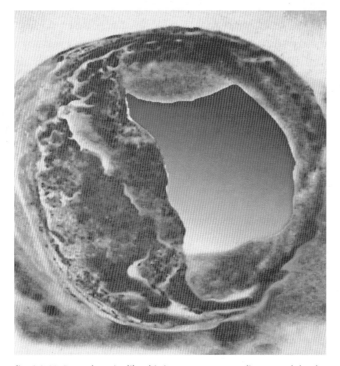

fig. 1.2.10 Fatty deposits like this in an artery cause disease and death to thousands of people every year.

Problems caused by atherosclerosis and high blood pressure

If an area of artery is narrowed by plaque, blood tends to build up behind the blockage. The artery bulges and the wall is under more pressure than usual, which can cause the wall of the artery to be severely weakened. This is known as an **aneurysm**. The weakened artery wall may split open, leading to massive internal bleeding. Aneurysms often happen in the brain or in the aorta, especially in the abdomen. The massive blood loss and drop in blood pressure are often fatal. However, sometimes aneurysms can be diagnosed and treated by surgery before they burst.

Raised blood pressure due to narrowed arteries can cause serious damage in a number of other organs, including the kidneys, the eye and the brain. The high pressure damages the tiny blood vessels where your kidney filters out urea and other substances from the blood. If the vessels feeding the kidney tubules become narrowed, the pressure inside them gets even higher and proteins may be forced out through their walls. Doctors test for protein in your urine as a sign of kidney damage if you have high blood pressure.

Similarly the tiny blood vessels supplying the retina of your eye are easily damaged and if they become blocked or leak the retinal cells are starved of oxygen and die. This can cause blindness.

Bleeding from the capillaries into the brain results in a stroke (see next page).

There are many factors that are linked to the development of atherosclerosis. You will look at these in more detail in chapter 1.4.

Questions

1 Plaque formation is one of the few examples in the human body of a positive feedback (where a change encourages even greater change, rather than returning to the normal state). Explain why this happens, and why it is so dangerous.

2 The build-up of a fatty plaque in the artery leads to changes in the blood flow and an increase in the blood pressure. Plan a way of modelling this which could be used on a television programme to explain high blood pressure to young people.

Cardiovascular disease

Heart disease

There are many kinds of heart disease, but the two most common ones are **angina** and **myocardial infarction** (heart attack). Both of these diseases are closely linked to the formation of atherosclerosis.

Plaques can build up slowly in the coronary arteries, reducing blood flow to the parts of the heart muscle beyond the plaques. For much of the time the person will be unaware that a problem is developing. Often symptoms are first noticed during exercise, when the cardiac muscle is working harder and needs more oxygen. The narrowed coronary arteries cannot supply enough oxygenated blood and the heart muscle has to resort to **anaerobic respiration**. This causes a gripping pain in the chest which can extend into the arms, particularly the left one, and jaw. It can often make you feel breathless as well. This is angina. The symptoms will subside once exercise stops, but the experience is painful and frightening.

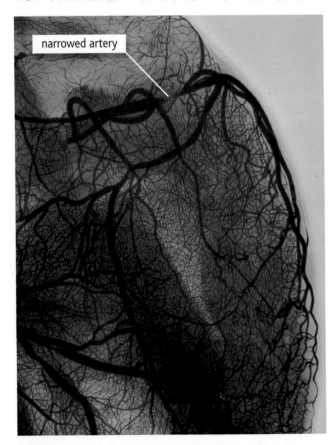

narrowed artery

fig. 1.2.11 Injecting the blood vessels with special dye allows doctors to see where the coronary arteries are narrowing and treat the problem before it becomes too severe.

Fortunately, most angina is relatively mild. It can be helped by eating a low-fat diet, taking regular exercise, losing weight and not smoking. The symptoms can be treated by drugs that cause rapid dilation of the coronary blood vessels so that they supply the cardiac muscle with the oxygen it needs. However, if the blockage of the coronary arteries continues to get worse, so will the symptoms of the angina. Other drugs are then used to dilate the blood vessels and reduce the heart rate (see chapter 1.4). Unfortunately drugs cannot solve a severe problem permanently, so heart bypass surgery may be carried out.

If one of the branches of the coronary artery becomes completely blocked, part of the heart muscle will be starved of oxygen and the person will suffer a myocardial infarction or heart attack.

Many heart attacks are caused by a blood clot as a result of atherosclerosis. The wall of an artery affected by a plaque is stiffened, making it much more likely to suffer cracks or damage. Platelets come into contact with the damaged surface of the plaque and the clotting cascade is triggered (see chapter 1.1). The plaque itself may break open, and the cholesterol that is released also causes the platelets to trigger the blood-clotting process. A clot may even form simply because the endothelial lining is damaged, for example by high blood pressure or smoking. A clot that forms in a blood vessel is known as a **thrombosis**. The clot can rapidly block the whole blood vessel, particularly if it is already narrowed by a plaque. A clot that gets stuck in a coronary artery is known as a coronary thrombosis. The clot can block the artery, starving the heart muscle beyond that point of oxygen and nutrients, and this often leads to a heart attack (see **fig. 1.2.12**).

During a heart attack there is chest pain in the same areas as during an angina attack but it is much more severe. The pain may occur even when not exercising, although exercise may trigger it, and it often lasts for several hours. Death may occur very rapidly with no previous symptoms, or it may happen after several days of feeling 'tired' and suffering 'indigestion'.

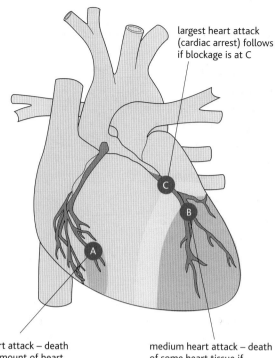

largest heart attack (cardiac arrest) follows if blockage is at C

small heart attack – death of small amount of heart tissue if blockage is at A

medium heart attack – death of some heart tissue if blockage is at B

fig. 1.2.12 The size and severity of the heart attack is closely related to the position of the blockage in the coronary artery.

It is very important to react quickly if you suspect someone is having a heart attack – call an ambulance and give them two full-strength aspirin tablets to help stop the blood clotting.

Strokes

A stroke is caused by an interruption to the normal blood supply in an area of the brain. This may be bleeding from damaged capillaries, or a blockage cutting off the blood supply to the brain, usually by a blood clot, an atheroma or a combination of the two. Sometimes the blood clot forms somewhere else in the body and is carried in the bloodstream until it gets stuck in an artery in the brain. The damage happens very quickly. A blockage in one of the main arteries leading to the brain causes a very serious stroke that may lead to death. In one of the smaller arterioles leading into the brain the effects may be less disastrous.

The symptoms of strokes vary, depending on how much of the brain is affected. Very often the blood is cut off from one part or one side of the brain only. Symptoms include dizziness, confusion, slurred speech, blurred vision or loss of part of the vision (usually just in one eye) and numbness. In more severe strokes there can be paralysis, usually just down one side of the body.

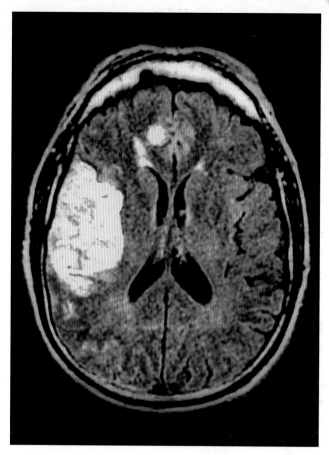

fig. 1.2.13 The damage caused in the brain by a major stroke can be seen on the left of this MRI brain scan. The healthy part of the brain is shown in blue.

The outcome of either a heart attack or a stroke usually depends on how soon the person is treated. The sooner patients are given treatment, including clot-busting drugs, the more likely they are to survive. For example, if treatment is given rapidly, 75% of patients who survive the first week after a heart attack can expect to be alive five years later.

Questions

1 Describe in detail the role of the blood-clotting process in cardiovascular disease.

2 Summarise the similarities and differences between a heart attack and a stroke.

1.3 Food and health

The nutrients you need

To grow and keep healthy, human beings, like other animals, have to take in all the nutrients they need from their food. This includes enough foods of the right types to make new tissue and to provide us with energy. We need **macronutrients** (**carbohydrates**, **proteins** and fats or **lipids**) and **micronutrients** (such as mineral salts and vitamins) along with water and **fibre**. Your health is dependent on the right intake of food. In this chapter you are going to look at the functions of the main food groups and their role in keeping you healthy.

fig. 1.3.1 **A wide variety of food is ideal to keep your body healthy, supplying all the different nutrients you need.**

Macronutrients

Macronutrients form the major part of your diet. Carbohydrates provide energy – they are broken down in digestion into glucose, most of which is used in cellular respiration to release energy. Some glucose is converted into glycogen, which is stored in the liver, muscles and brain. Any excess carbohydrate is converted to fat, which is stored readily in the body.

Lipids also provide energy, and any excess is also stored as body fat. Proteins are used for growth and repair of cells. They are broken down in digestion to their constituent **amino acids**. These are then rebuilt during protein synthesis to form the proteins the body needs. **Essential amino acids** are vital in a diet because the body cannot make them. They are mainly found in animal protein. (You will be looking at proteins in more detail in chapter 2.1.)

Micronutrients

Micronutrients are needed in much smaller quantities than macronutrients. Mineral salts are generally needed in minute amounts, but lack of them in the diet can lead to a variety of serious conditions. Calcium, found in milk and other dairy products and also in fish and hard water, is needed for the formation of your skeleton and teeth, for your muscles to contract properly and for blood clotting to take place. Sodium is needed for your nerves to work properly, for muscular contraction and to maintain your heartbeat. Sodium is crucial to the salt balance in your body. If sodium levels fall it can make you feel very unwell. Equally, too much sodium in your diet may lead to raised blood pressure, which in turn can cause health problems.

Vitamins are similarly required in very small amounts. They are usually complex organic substances, but they are absorbed directly into the bloodstream from the gut. The fat-soluble vitamins are absorbed along with the fat you eat. The lack of a particular vitamin in the diet for a long time will result in a deficiency disease such as scurvy if you lack vitamin C.

How much vitamin C?

You can investigate the levels of some vitamins in your food experimentally. For example, many foods and drinks are claimed to contain vitamin C. You can compare the levels of vitamin C in foods using a number of different practical methods, including titrations and colorimetry.

More recent research suggests that a lack of vitamins in our modern, overprocessed diets may have an important impact in many diseases, including CVDs. For example, vitamin C is important in the formation of connective tissue in the body, such as in the bones, teeth, skin and many internal body surfaces including the endothelial lining of blood vessels. A severe lack of vitamin C in the diet causes scurvy which can result in bleeding gums, easy bruising and painful joints. As you have seen, if the lining of an artery is damaged, atherosclerosis develops. Evidence is building that if your diet is low in vitamin C, your arteries are more likely to be damaged and you are more likely to be affected by CVD.

A study published in the *British Medical Journal* in 1997 looked at the association between blood vitamin C concentration and risk of heart attack in 1605 men from eastern Finland. The men had no sign of coronary artery disease when they were tested between 1984 and 1989. Their vitamin C levels were also tested. Between 1984 and 1992 a total 70 of the men had a heart attack (some fatal, some not). 13.2% of the men who showed low vitamin C levels had heart attacks, compared with 3.8% of the men who showed no sign of vitamin C deficiency.

Other vitamins are also linked to heart health – for example the vitamin B group seem to be important in preventing heart disease (see page 63). Fresh fruit and vegetables provide many of the vitamins our bodies need, and these studies are increasingly showing why it is important to make sure we get plenty of fresh fruit and vegetables in our diet.

Water and fibre

Water is also vital in the diet – while the average person can survive with little or no food for weeks, a complete lack of water is likely to kill you in a few days.

Roughage or fibre cannot be digested in the human gut, but it is essential because it holds water and provides bulk for the intestinal muscles to work on. Without enough roughage, food moves through the gut relatively slowly. This can lead to minor problems such as constipation and haemorrhoids. In the long term it may be linked with more serious conditions such as heart disease and bowel cancer.

HSW The need for vitamins

In the early years of the twentieth century Sir Frederick Gowland Hopkins took two sets of eight young rats and fed both sets a diet consisting of purified casein (milk protein), starch, sucrose, mineral salts, lard and water. In addition, one set was given 3 cm³ of milk every day for the first 18 days of the experiment. Milk contains vitamins as shown in **table 1.3.1**. At this point the milk supplement was withdrawn from the first set and given to the second set of young rats until the end of the experiment. The results, shown in **fig 1.3.2**, demonstrate the importance of vitamins in the growth of young rats.

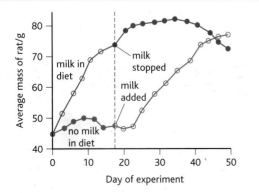

fig. 1.3.2 Gowland Hopkins's results on feeding young rats.

Questions

1 Gowland Hopkins concluded that there was something in fresh milk besides the basic macronutrients which was needed for growth. Why did he conclude this from his results?

2 Some evidence suggests that vitamins may play a protective role in heart disease. Explain how the level of vitamin C in your diet links with your model of CVD. Find another study which suggests this link and evaluate the research done.

Organic molecules in living things

What are organic compounds?

Put very simply, organic compounds contain carbon atoms. They also contain atoms of hydrogen, oxygen and, less frequently, nitrogen, sulfur and phosphorus. Almost all of your food – and your body – that is not water is made up of organic molecules. An understanding of why organic molecules are special will help you to understand the chemistry of carbohydrates, lipids and proteins.

Each carbon atom can make four bonds and so it can join up with four other atoms. Carbon atoms bond particularly strongly to other carbon atoms to make long chains, with other atoms such as hydrogen and oxygen along each side. The carbon atoms may also bond to form branched chains, rings or even three-dimensional shapes. In some carbon compounds small molecules (**monomers**) bond with many other similar units to make a very large molecule called a **polymer**.

The ability of carbon to combine and make large molecules provides the great variety and complexity found in living things. The four bonds of a carbon atom are usually arranged to give a tetrahedral shape which leads to the three dimensional shapes of organic molecules. However, in most drawings of organic molecules we ignore their three-dimensional shape and draw them flat (**fig. 1.3.3**).

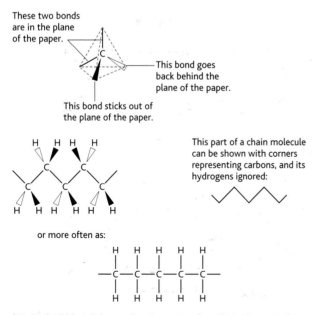

These two bonds are in the plane of the paper.

This bond goes back behind the plane of the paper.

This bond sticks out of the plane of the paper.

This part of a chain molecule can be shown with corners representing carbons, and its hydrogens ignored:

or more often as:

fig. 1.3.3 The bonds in a carbon atom, a carbon chain shown in three dimensions and 'flat'.

Carbohydrates

Carbohydrates are important in organisms as an energy source. Some are broken down to provide usable energy in plant and animal cells, while others are important for storing energy in the cells. The best-known carbohydrates are sugars, such as sucrose, glucose and starches. Sucrose is the white crystalline 'sugar' familiar to us all; glucose is the energy supplier in sports and health drinks; and starch is found in flour and potatoes. But the group of chemicals known as carbohydrates also includes many other compounds.

The basic structure of all carbohydrates is the same. They are made up of carbon, hydrogen and oxygen. There are three main groups of carbohydrates, depending on the complexity of the molecules: **monosaccharides**, **disaccharides** and **polysaccharides**.

Monosaccharides – the simple sugars

The monosaccharides are known as the simple sugars. Their molecules contain one oxygen atom and two hydrogen atoms for each carbon atom present. A **general formula** for a simple sugar is:

$(CH_2O)_n$

Here n can be any number, but it is usually low. **Triose** ($n = 3$) **sugars** have three carbon atoms and the formula $C_3H_6O_3$. They are important in the mitochondria, where glucose is broken down to triose sugars in cellular respiration. The **pentose** ($n = 5$) sugars ribose and deoxyribose are important in the nucleic acids DNA and RNA which make up genetic material. They have the formula $C_5H_{10}O_5$. The best-known monosaccharides are the **hexose** sugars ($n = 6$), such as glucose, galactose and fructose, which have the general formula $C_6H_{12}O_6$. Most of the monosaccharides taste sweet.

General formulae show you how many atoms there are in the molecule, and what type they are, but they do not tell you what the molecule looks like and why it behaves as it does. To show some of this information you can use **displayed formulae**. Although these do not follow every wiggle and kink in the carbon chain they can give you a good idea of how the molecules are arranged in three dimensions. This can reveal all sorts of secrets about the way biological systems behave as they do (see **fig. 1.3.4**).

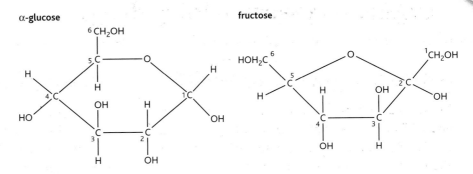

fig. 1.3.4 Hexose sugars have a ring structure. The arrangement of the atoms on the side chains make a significant difference to the way in which the molecule can be used by the body. We number the carbon atoms so that we can identify the different arrangements.

Disaccharides – the double sugars

Disaccharides are made up of two monosaccharides joined together. For example, sucrose is formed by a molecule of glucose bonding with a molecule of fructose. The two monosaccharides join in a **condensation reaction**, and a molecule of water (H_2O) is removed. The link between the two monosaccharides that results is a covalent bond known as a **glycosidic bond** (see **fig. 1.3.5**). This joining of monosaccharides gives a different general formula – disaccharides, and indeed chains of monosaccharides of any length, have the general formula $(C_6H_{10}O_5)_n$.

When different monosaccharides join together, different disaccharides result. Many disaccharides taste sweet. **Table 1.3.2** shows some of the more common ones.

Disaccharide	Source	Monosaccharide units
Sucrose	Stored in plants such as sugar beet and sugar cane	Glucose + fructose
Lactose (milk sugar)	The main carbohydrate in milk	Glucose + galactose
Maltose (malt sugar)	Found in germinating seed such as barley	Glucose + glucose

table 1.3.2 Some common disaccharides.

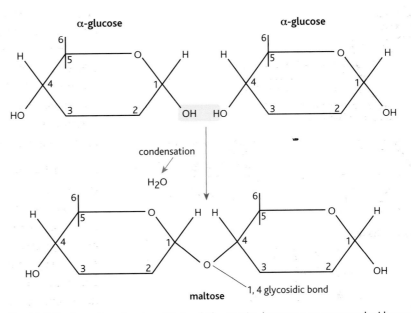

fig. 1.3.5 The formation of a glycosidic bond. The reaction between two monosaccharides results in a disaccharide and a molecule of water.

Polysaccharides

The most complex carbohydrates are the polysaccharides. They are made of many monosaccharide units joined by glycosidic bonds as a result of many condensation reactions (**fig. 1.3.6**). Molecules with 3–10 monosaccharides are known as oligosaccharides, while those containing 11 or more monosaccharides are known as polysaccharides. The sweet taste characteristic of both mono- and disaccharides is lost in a polysaccharide. But this linking of many sugar monomers to form a more complex polymer produces some very important biological molecules.

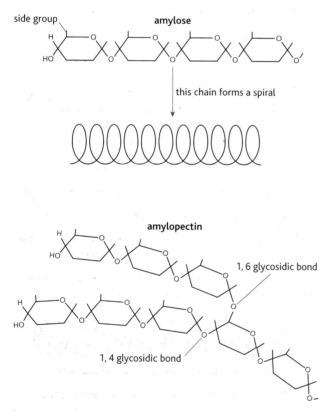

fig. 1.3.6 Amylose and amylopectin – a small difference in the position of the glycosidic bonds in the molecule makes a big difference to the properties of the compounds.

Polysaccharides can form very compact molecules, ideal for storing carbohydrate in cells. When simple sugars are needed for cellular respiration, the glycosidic bonds in these polysaccharides can be broken to release monosaccharide units in a process called **hydrolysis** (see below). Because polysaccharides are physically and chemically very inactive, storing them does not interfere with the other functions of the cell. They are not very soluble in water, which means they have very little impact on water relations and osmosis in a cell (see chapter 2.3.)

Starch is particularly important as an energy store in plants. The sugars produced by photosynthesis are rapidly converted into starch, which is insoluble and compact, but can be broken down rapidly to release glucose when it is needed. Plant storage organs such as potatoes are particularly rich in starch.

Starch is made up of long chains of glucose, but if you look at it more closely you will see that it is actually made up of a mixture of two compounds, amylose and amylopectin (see **fig. 1.3.6**).

- **Amylose** is an unbranched polymer. As the chain lengthens the molecule spirals, which makes it more compact. An amylose molecule will be made up of between 200 and 5 000 glucose molecules. The glucose molecules can only be released by enzymes working from each end of the amylose molecule.

- **Amylopectin** is also a polymer of glucose molecules but the amylopectin chains branch. These branching chains have lots of terminal (end) glucose molecules that can be broken off rapidly when energy is needed.

The combination of amylose and amylopectin in starch explains why starchy, carbohydrate-rich foods like pasta are so good for you when doing sport. The amylopectin releases glucose for cellular respiration rapidly when needed, and the amylose releases it more slowly over a longer period, keeping you going longer.

Starch is the main energy storage material in plants. A typical starch grain in a plant cell contains 70–80% amylopectin, and the rest is amylose.

Glycogen is sometimes referred to as 'animal starch' and is the only carbohydrate energy store found in animals (see **fig. 1.3.7**). Chemically it is very similar to starch, being made up of many glucose units. Like starch it is a compact molecule. However, the glycogen molecule has many side branches, which means it can be broken down very rapidly indeed. This makes it an ideal energy store for very active tissues such as your muscle and liver tissue, which need a readily available energy supply at all times.

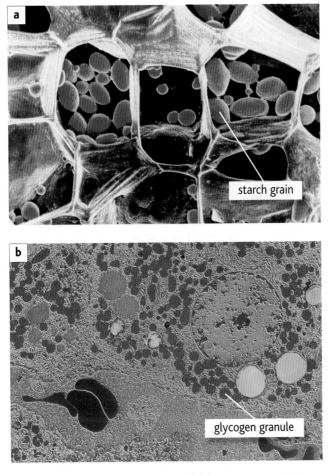

fig. 1.3.7 (a) Starch grains in a potato cell. (b) Glycogen granules in a liver cell. Both polysaccharides are important energy stores.

Amylose, amylopectin and glycogen – how do they differ?

Amylose, amylopectin and glycogen are all made of glucose molecules joined in long chains – so why are the molecules so different? It all depends on the position of the carbon atoms involved in the glycosidic bonds. Amylose is made up of glucose molecules joined purely by glycosidic bonds between carbon 1 on one glucose molecule and carbon 4 on the other (**1,4-glycosidic bonds**). This is why the molecules are long straight chains.

In amylopectin many of the glucose molecules are joined by 1,4-glycosidic bonds, but there are a few bonds between carbon 1 and carbon 6 of the glucose molecules (1,6-glycosidic bonds). This is what forms the branching chains in amylopection.

So starch has a combination of straight-chain amylose and branched-chain amylopectin molecules. In glycogen there are more 1,6-glycosidic bonds as well as 1,4-glycosidic bonds between the glucose molecules.

Breaking the bonds

To be useful as an energy store the bonds in carbohydrates need to be broken to release single sugars for cells to use. The glycosidic bond between two monosaccharide units is split by a process known as hydrolysis (**fig. 1.3.8**). The hydrolysis reaction is the opposite of the condensation reaction that formed the molecule, so water is added to the bond. Disaccharides can be split to form two monosaccharides, and the same reaction is used to gradually break down starch and glycogen into shorter and shorter chains. Hydrolysis takes place during digestion in the gut, and also in the muscle and liver cells. The single sugars that result can then be used in cellular respiration.

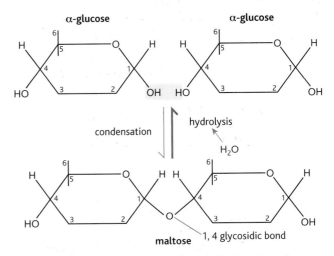

fig. 1.3.8 Glycosidic bonds are broken down by hydrolysis to release monosaccharide units for use in cellular respiration.

Questions

1 How do monosaccharides form disaccharides?

2 What is the relationship between glycosidic bond formation and hydrolysis?

3 What are the main differences between starch and glycogen? What are the implications of these differences in the way starch and glycogen molecules are used to provide energy in your body?

4 Explain why carbohydrates make good energy-storage molecules in plants and animals.

Lipids

The lipids are another group of macronutrients, organic chemicals that make up a healthy diet. Lipids include some of the highest-profile chemicals in health issues, including cholesterol, **oils** such as olive oil and **fats** such as butter. The media constantly remind us of the importance of a low-fat diet and the dangers of high cholesterol levels (see chapter 1.4).

Lipids are an important source of energy in the diet of many animals, and they form the most effective energy store because they contain more energy per gram than carbohydrates or proteins. Many plants and animals convert spare food into oils or fats for use later. For example, the seeds of plants contain oils to provide energy for the seedling when it starts to grow.

Lipids play a vital role in cell membranes (see chapter 2.3) and they can fulfil a protective function – organs such as the kidneys are surrounded by fat. Oils are important in waterproofing the fur and feathers of mammals and birds (**fig. 1.3.9**). Insects and plants use waxes for waterproofing their outer surfaces. Lipids are good insulators – a fatty sheath insulates your nerves so the electrical impulses travel faster. They also insulate animals against heat loss (eg blubber in whales). Lipids have a very low density, so the body fat of water mammals helps them to float easily.

fig. 1.3.9 Water birds rely on oils to stop their feathers becoming waterlogged.

All lipids dissolve in organic solvents, but are insoluble in water. So lipids don't interfere with the many water-based reactions that go on in the cytoplasm of a cell.

Fats and oils

Fats and oils are important groups of lipids. Chemically they are extremely similar, but fats (such as butter) are solids at room temperature and oils (such as olive oil) are liquids. Like carbohydrates, the chemical elements that make up all lipid molecules are carbon, hydrogen and oxygen. Lipids, however, contain a considerably lower proportion of oxygen than carbohydrates. Fats and oils are made up of two types of organic chemicals, **fatty acids** and **glycerol** (propane-1,2,3-triol).

Glycerol has the chemical formula $C_3H_8O_3$. **Fig. 1.3.10** shows how the atoms are joined together.

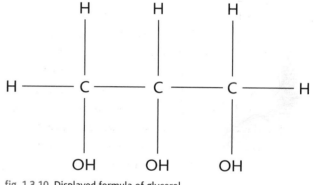

fig. 1.3.10 Displayed formula of glycerol.

Over 70 different kinds of fatty acids have been extracted from living tissues. All fatty acids have a long hydrocarbon chain – a pleated backbone of carbon atoms with hydrogen atoms attached, and a carboxyl group (-COOH) at one end. There are two main ways in which fatty acids vary:

- The length of the carbon chain can differ, although in living organisms it is frequently between 15 and 17 carbon atoms long.

- Perhaps more importantly, the fatty acid may be **saturated** or **unsaturated**.

In a saturated fatty acid each carbon atom is joined to the one next to it by a *single* covalent bond (see **fig. 1.3.11**). In an unsaturated fatty acid the carbon chains have one or more *double* bonds in them. A **monounsaturated** fatty acid has one double bond and a **polyunsaturated** fatty acid has more than one double bond (see **fig. 1.3.12**). An example of a polyunsaturated fatty acid is linoleic acid, an essential fatty acid in our diet that we cannot make from other chemicals.

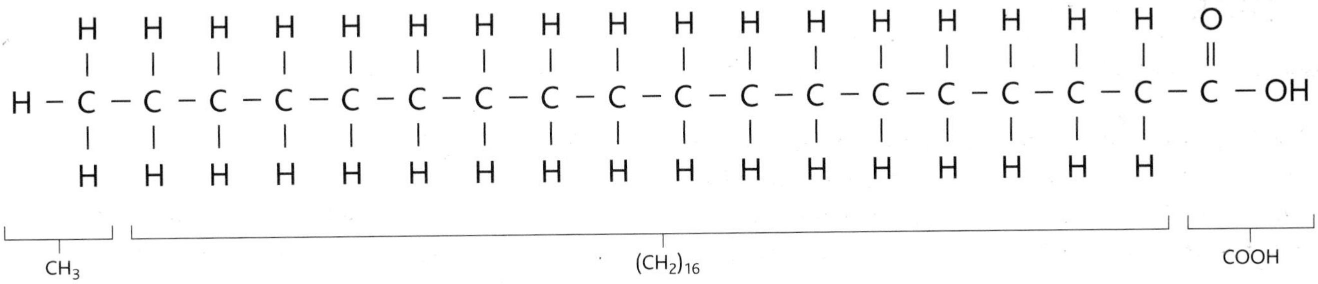

H H H H H H H H H H H H H H H H H O
| | | | | | | | | | | | | | | | | ||
H – C – C – C – C – C – C – C – C – C – C – C – C – C – C – C – C – C – C – OH
| | | | | | | | | | | | | | | | |
H H H H H H H H H H H H H H H H H

CH_3 $(CH_2)_{16}$ COOH

fig. 1.3.11 Displayed formula of stearic acid, a saturated fatty acid.

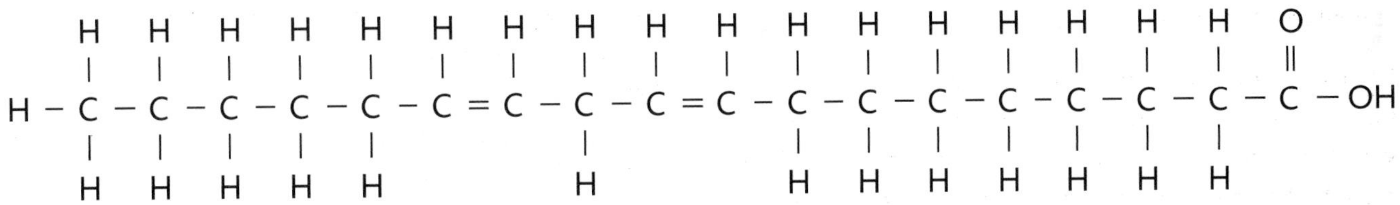

H H H H H H H H H H H H H H H H H O
| | | | | | | | | | | | | | | | | ||
H – C – C – C – C – C – C = C – C – C = C – C – C – C – C – C – C – C – C – OH
| | | | | | | | | | | | | |
H H H H H H H H H H H H H

fig. 1.3.12 Displayed formula of linoleic acid, a polyunsaturated fatty acid.

A fat or oil results when glycerol combines with one, two or three fatty acids to form a **mono-, di-** or **triglyceride**. A bond is formed in a condensation reaction between the carboxyl group (-COOH) of a fatty acid and one of the hydroxyl groups (-OH) of the glycerol. A molecule of water is removed and the resulting bond is known as an **ester bond**. This type of condensation reaction is called **esterification** (**fig. 1.3.13**). The nature of the lipid that is formed depends on which fatty acids are present. So, for example, lipids containing saturated fatty acids are more likely to be solid at room temperature than those containing unsaturated fatty acids.

For simplicity, fatty acids are represented by the general formula where 'R' represents the hydrocarbon chain. These fatty acids are drawn below in reversed form.

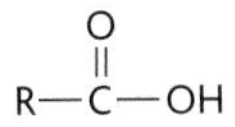

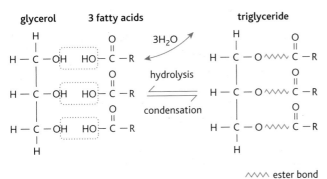

∿∿∿ ester bond

Note: there are only 6 atoms of oxygen in a triglyceride molecule.

fig. 1.3.13 Esterification and the lipid molecules that result from the reactions between glycerol and fatty acids.

Lipids and the heart

Fatty foods are very high in energy, and so a diet rich in lipids is likely to result in obesity. Coupled with this is the implication from medical research that saturated fats – found particularly in dairy produce and meat – can lead to fatty plaques in the arteries (see chapter 1.2), and possibly heart disease and death. Unsaturated fats, found mainly in plants, do not seem to have this effect and so health advice encourages us to replace much of the saturated fat in our diets with unsaturated fats. However, further research has shown that not all unsaturated fats are equally healthy. The lone double bond in the carbon chain of fatty acids in monounsaturated fats seems to have a positively beneficial effect, helping the body to cope better with saturated fats. Scientists now think that polyunsaturated fats, with two or more double bonds, have an even stronger positive effect on blood cholesterol levels and heart health. The links between diet and health are complex, and you will be looking at these in more detail in chapter 1.4.

Questions

1 Describe the main difference between a saturated and an unsaturated lipid, and the effect this difference has on the properties of a lipid.

2 Explain how triglycerides are formed.

Proteins

About 18% of your body is made up of protein. Proteins form your hair, your skin, your nails, the enzymes that control all the reactions in your cells, the enzymes that digest your food and many of the hormones that control the working of your organs. Proteins have a huge variety of functions in living organisms. They enable muscle fibres to contract so that you can move, they protect you from disease in the form of antibodies and the chemicals involved in blood clotting, they transport oxygen in your blood in the form of haemoglobin, and much more. If you understand how protein molecules are formed, and what affects their shapes and functions, you can begin to develop an insight into the detailed biology of cells and living things.

Like carbohydrates and lipids, proteins contain carbon, hydrogen and oxygen. In addition they all contain nitrogen. Many proteins also contain sulfur, and some contain phosphorus and various other elements. Proteins are very large molecules (**macromolecules**) made up of many small monomer units called amino acids joined together. **Amino acids** combine in long chains to produce proteins in condensation reactions, in a similar way to monosaccharides joining to form polysaccharides. However, there the similarity ends. Whilst each polysaccharide is made up of one or two different types of monosaccharide, there are about 20 different naturally occurring amino acids that can combine in different ways to form a vast range of different proteins.

Amino acids

All amino acids have the same basic structure. There is always an amino group (-NH$_2$) and a carboxyl group (-COOH) attached to a carbon atom (see **fig. 1.3.14**). The group known as the R group varies between amino acids. It affects the way the amino acid bonds with others in the protein, depending largely on whether the R group is polar or not. When you look at how amino acids link to form long chains, you can ignore the R group and concentrate entirely on the amino and carboxyl groups.

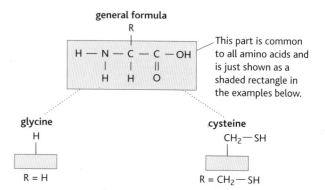

fig. 1.3.14 Some different amino acids. In the simplest amino acid, glycine, R is a single hydrogen atom. In a larger amino acid such as cysteine, R is much more complex.

Forming proteins from amino acids

Amino acids join together by a reaction between the amino group of one amino acid and the carboxyl group of another amino acid. They join in a condensation reaction and a molecule of water is lost. The bond formed is known as a **peptide link** and when two amino acids join, a **dipeptide** is the result (see **fig. 1.3.15**). More and more amino acids join to form **polypeptide** chains, which may contain around a hundred to many thousands of amino acids. When the polypeptide is folded or coiled or associated with other polypeptide chains, it forms a protein.

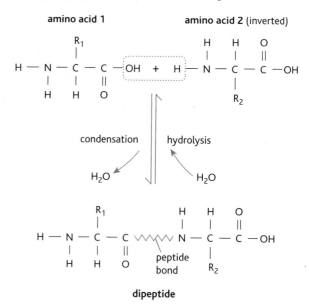

fig. 1.3.15 Amino acids are the building blocks of proteins.

Other bonds in proteins

The peptide link between amino acids is a strong bond, but other bonds also form between the amino acids in a chain. These bonds are often weaker than peptide bonds but there are a lot of them. They depend on the atoms in the R group, and include **hydrogen bonds**, **sulfur bridges** and **ionic bonds**. Together they are important in forming the three-dimensional structure of a protein, as you will see on the following pages.

Hydrogen bonds

In amino acids, tiny negative charges are present on the oxygen of the carboxyl groups and tiny positive charges are found on the hydrogen atoms of the amino groups. When these charged groups are close to each other the opposite charges attract, forming a hydrogen bond. Hydrogen bonds are weak but they can potentially form between any two amino acids positioned correctly, so there are lots of them holding the protein together very firmly. Hydrogen bonds are easily broken and reformed if pH or temperature conditions change. They are very important in the folding and coiling of the polypeptide chains.

Sulfur bridges

Sulfur bridges are formed when two cysteine or methionine molecules are close together in the structure of a polypeptide (see **fig. 1.3.17**). An oxidation reaction takes place between the two sulfur-containing groups, resulting in a strong covalent bond known as a sulfur bridge or disulfide link. These sulfur bridges are much stronger than hydrogen bonds but they occur much less often. They are important for holding the folded polypeptide chains in place.

fig. 1.3.16 If you blow-dry your hair, you break the hydrogen bonds in the protein and reform them with your hair curled a different way – but it doesn't last. If you have a perm, the chemicals break the sulfur bridges between the polypeptide chains and reform them in a different place – and this effect is permanent.

Ionic bonds

Normal ionic bonds can form between some of the strongly positive and negative amino acid side chains found buried deep in the protein molecules. These links are known as salt bridges. They are strong bonds but they are not as common as the other structural bonds.

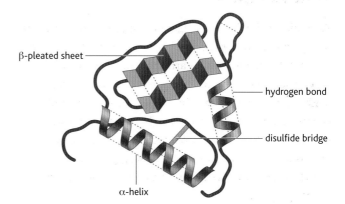

fig. 1.3.17 Some very complex molecule shapes result in our hair, the haemoglobin in our blood and our enzymes. The shapes of the molecules, maintained by hydrogen bonds and sulfur bridges, determine their functions.

Protein structure

Proteins can be described by their **primary**, **secondary**, **tertiary** and **quaternary structure** (see fig. 1.3.18).

The primary structure of a protein is the sequence of amino acids that make up the polypeptide chain. Hydrogen bonding between amino acids can produce a regular, repeating three-dimensional structure known as the secondary structure. One example is the right-handed helix (α-helix) or spiral coil with the peptide links forming the backbone and the R groups sticking out in all directions. In other proteins the polypeptide chains fold up into pleated sheets, with the pleats held together by hydrogen bonds between the amino and carboxyl ends of the amino acids. Most **fibrous proteins** have this sort of structure (see below). Sometimes there is no regular secondary structure and the polypeptide just forms a random coil.

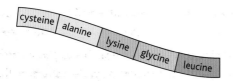

Primary structure – the linear sequence of amino acids in a peptide.

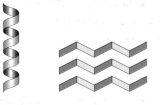

Secondary structure – the repeating pattern in the structure of the peptide chains, such as an α-helix or pleated sheets.

Tertiary structure – the three-dimensional folding of the secondary structure.

Quaternary structure – the three-dimensional arrangement of more than one tertiary polypeptide.

fig. 1.3.18 The three-dimensional structure of proteins.

Many proteins, including the **globular proteins** (see below), have a further level of three-dimensional organisation. The amino acid chain, including any α-helices and pleated sheets, are folded further into complicated shapes. These three-dimensional shapes are held in place by hydrogen bonds, sulfur bridges and ionic bonds between the amino acids. This organisation, the folding of the polypeptide coils, is the tertiary structure of the protein.

And finally, some enzymes in your body and your blood pigment haemoglobin are made up of not one but several polypeptide chains. The quaternary structure of a protein depends on the way these separate polypeptide chains fit together.

The different kinds of weak bonds that hold the three-dimensional shapes of proteins are easily affected by changes in conditions such as temperature or pH. Even small changes in these conditions can cause the bonds to break, resulting in the loss of the protein's three-dimensional shape. We say that the protein is **denatured**. Because the three-dimensional structure of a protein is important to the way it works, changing conditions inside the body can cause proteins to stop working properly.

Fibrous and globular proteins

Fibrous proteins have little or no tertiary structure. They are long, parallel polypeptide chains with occasional cross-linkages that form into fibres. They are insoluble in water and are very tough, which makes them ideally suited to their usual structural functions within organisms. They are found in connective tissue, in tendons and the matrix of bones (collagen), in the structure of muscles, in the silk of spiders' webs and silkworm cocoons, and as the keratin that makes up hair, nails, horns and feathers.

Globular proteins have complex tertiary and sometimes quaternary structures. They are folded into spherical (globular) shapes. The large size of these globular protein molecules affects their behaviour in water. Because their carboxyl and amino ends give them ionic properties you might expect them to dissolve in water and form a solution. In fact, the molecules are so big that instead they form a **colloid** (see chapter 1.1) Globular proteins play an important role in holding molecules in position in the cytoplasm.

Globular proteins are also important in your immune system – for example, antibodies are globular proteins. They form enzymes and some hormones and are also important for maintaining the structure of the cytoplasm. You can read about the essential relationship between the shape of an enzyme and how it carries out its role in catalysing a reaction in chapter 2.1.

Conjugated proteins

The shape of a protein molecule is usually very important in its function. Some protein molecules are joined with or **conjugated** to another molecule called a **prosthetic group** (see **fig. 1.3.19**). This structural change usually affects the performance and functions of the molecules. For example, **glycoproteins** are proteins with a carbohydrate prosthetic group. The carbohydrate part of the molecule helps them to hold on to a lot of water, and also makes them less likely to be broken down by protein-digesting enzymes. Lots of lubricants used by the human body – such as mucus and the synovial fluid in the joints – are glycoproteins whose water-holding properties make them slippery and viscous, which reduces friction. This also helps to explain why the mucus produced in the stomach protects the protein walls from digestion.

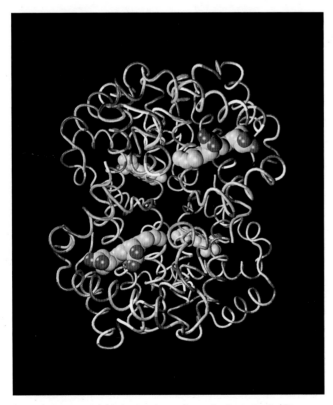

fig. 1.3.19 Haemoglobin is a conjugated protein that contains four prosthetic groups (shown here by the groups of balls). Each prosthetic group contains iron which can bond with oxygen.

Haemoglobin, the complex oxygen-carrying molecule in the red blood cells, is a conjugated protein with an inorganic iron-containing prosthetic group. It is the iron that enables the haemoglobin to bind and release oxygen molecules.

Lipoproteins are proteins conjugated with lipids and they are very important in the transport of cholesterol in the blood (see chapter 1.4). The lipid part of the molecule enables it to combine with the lipid cholesterol. There are two main forms of lipoproteins in your blood – low-density lipoproteins (LDLs) (around 22 nm in diameter) and high-density lipoproteins (HDLs) (around 8–11 nm in diameter). The HDLs contain more protein than LDLs, which is partly why they are denser – proteins are more compact molecules than lipids.

Questions

1 Explain how the order of amino acids in a protein chain affects the structure of the whole protein.

2 Explain why weak hydrogen bonds play a much bigger part in maintaining protein structure than sulfur and salt bridges which are much stronger.

3 Using examples, explain the relationship between structure and function for:
   a fibrous proteins
   b globular proteins
   c glycoproteins.

4 There are many systems in your body that keep internal conditions as constant as possible. Explain, with reference to proteins, why constant internal conditions are important.

Proteins are very complicated molecules, and yet we know the structure of many of them in great detail. How do scientists get this information? There are many different tools used to discover the details of molecular structure. Two techniques have played a large part in developing an understanding of proteins – paper **chromatography** and **electrophoresis**. Before either of these can be used, the protein must be broken down into the amino acids it contains. This is done using enzymes which break the peptide links of the protein.

Paper chromatography

Simple paper chromatography can be quite effective at separating amino acids. A piece of filter paper is soaked in water and then dipped into a solution of the amino acid mixture in a different solvent, or drops of the amino acid mixture are put on the paper. As the solvent moves up the paper the amino acids are distributed between the water in the filter paper and the solvent moving up the papers depending on their solubility in water. Different amino acids will move different distances. At the end of the process the paper is dried and sprayed with ninhydrin, a dye that reacts with the amino acids to make them purple and so reveal their position as dots on the paper. We can work out which amino acids are present by looking at how far they have moved and comparing them with the results from known amino acids.

Two-way chromatography takes this one step further. After the paper has been dried it is turned at right angles and placed in a different solvent before drying and spraying it with ninhydrin. This creates a two-dimensional map of the molecules, giving a much better separation of the different amino acids.

Electrophoresis

A more sophisticated form of chromatography, called electrophoresis, gives much better results. Known amino acids are placed on a special solid or gel which they don't react with. This is placed in a solution that keeps pH constant, and an electric current is passed through it. The amino acids move at different rates according to the charge on their R group. Once the medium has dried, the amino acids can be revealed using a ninhydrin spray. The distance each amino acid has travelled under these known conditions can then be measured (see **fig. 1.3.20**). This technique is widely used in research and a modified version is used in DNA fingerprinting.

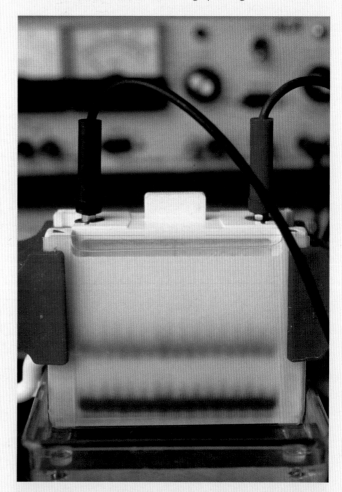

fig. 1.3.20 The amino acids in a protein revealed after electrophoresis.

To work out the precise order of the amino acids in a particular protein scientists use special enzymes that break peptide links one at a time, starting from either the amino or the carboxyl end of the molecule. One terminal amino acid at a time is labelled (tagged), removed and identified by chromatography or electrophoresis.

Interpreting chromatograms and electrophoresis plates

To identify the amino acids after chromatography or electrophoresis, you need to work out their R_f value and compare this with the R_f values of known amino acids using the same solvent. The R_f value is the ratio of the distance moved by the amino acid to the distance moved by the solvent alone. The R_f value lies between 0 and 1 and it is calculated as follows:

$$R_f = \frac{\text{distance travelled by solute (amino acid)}}{\text{distance travelled by solvent}}$$

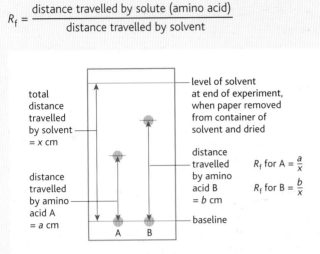

fig. 1.3.21 In this example the R_f values for amino acids A and B have been worked out.

X-ray crystallography

X-ray crystallography is another technique that has been vital in helping us to build up a picture of what protein molecules are like. X-rays are fired into a pure protein crystal. The X-rays are reflected by the atoms of the protein molecules and this scattered pattern is recorded on a photographic plate. The crystal is then turned and the whole process is repeated many times around the structure to build up a three-dimensional image. The difficult part is looking at all the pictures and deciding what they mean. Interpreting the images is highly skilled work, once carried out by specialised scientists and now mainly done by computers. X-ray crystallography has revealed that proteins usually have regular and predictable molecule shapes.

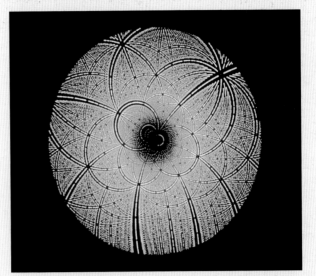

fig. 1.3.22 A skilled interpreter can look at the size and density of these dots and use them to build up a picture of a molecule.

Uncovering the structure of insulin

Frederick Sanger, a British biochemist, was the first person to work out the amino acid sequence of a complete protein. Sanger worked on insulin, a relatively small protein. Initially he thought the insulin molecule contained four long chains. He split the bridges joining the chains and investigated each chain individually, discovering in the process that there were only two chains, not four. He separated each one into smaller and smaller links and each time he identified the final amino acid in the chain. In this way he worked out the exact sequence of amino acids in each chain and found that one contained 21 amino acids and the other 30 amino acids.

Sanger then worked out the exact positions of the sulfur bridges which joined the chains. After over 12 years of research, he produced a diagram showing the structure of insulin. In 1958 he was awarded his first Nobel prize for his ground-breaking work. The methods Sanger developed have been used to build up an understanding of the amino acid sequences and the structures of many other proteins since. Modern techniques with automated electrophoresis have speeded up the process considerably.

Questions

1 Why are proteins difficult to analyse?

2 Describe the difference in the type of information that can be obtained by electrophoresis and X-ray crystallography.

Why is a balanced diet important?

The right balance of food in your diet is central to your health. If you eat too little food (undernutrition) or too much (overnutrition) over a period of time then you may suffer from **malnutrition**.

A balanced diet should supply the energy you need for everyday activity. In the western world this isn't usually a problem. However, millions of people, particularly those who live in the developing world, cannot get enough food to provide the energy they need. As a result they are seriously underweight, which can reduce resistance to disease and shorten lifespan. These people also often cannot get sufficient essential amino acids, minerals and vitamins and so suffer deficiency diseases.

Eating too much food can also create many health problems. If more energy is taken in than is required, the excess is stored as fat and **obesity** may result. Obesity is a rapidly increasing problem, mainly in the developed world. Obesity is linked to an increased risk of coronary heart disease, high blood pressure, diabetes and other disorders that can reduce life expectancy.

A balanced diet needs to be balanced for more than just energy. Strangely, the average diet in the developed world, while supplying more than enough energy from fat and refined carbohydrate, may be low in vitamins because relatively little fresh fruit and vegetables are eaten. These foods also provide the fibre we need, so a diet like this can lead not only to constipation but also to many diseases of the gut and bowel.

How much is enough?

In the UK there are recommended national guidelines for nutrition that try to clarify which nutrients people need and in what quantities. Actual values vary between individuals, depending on factors such as age and lifestyle, but estimates for groups of the population are based on advice given by the Committee on Medical Aspects of Food Policy (COMA). This committee has been replaced by the Scientific Advisory Committee on Nutrition (SACN), which reviews scientific evidence from research into diet and disease, and then uses this to advise the Government on policy.

In 1991 COMA published a detailed report of the estimated nutritional requirements of different groups in the population. From these they developed **dietary reference values** (**DRVs**) that show the range of requirements and appropriate intakes for the population. There are several different kinds of DRVs (see **fig. 1.3.23**).

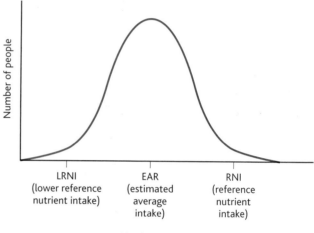

fig. 1.3.23 This graph shows the distribution of nutritional requirements within a population.

Occasionally a safe intake value is quoted, when there is either lack of agreement about the data or not enough scientific evidence to provide an RNI. The safe intake is judged to be probably sufficient for everyone, with no evidence that levels above this are more beneficial.

The COMA report includes RNIs for protein, nine vitamins and eleven minerals, EARs for energy, safe intakes for four further vitamins and minerals, and desired population intakes for fat and carbohydrates. The RNIs for protein in **table 1.3.3** can be used to assess whether the diet of a group within the population is adequate or not. It also shows how requirements change through life.

Age	RNI (grams/day)
0–3 months	12.5
4–6 months	12.7
7–9 months	13.7
10–12 months	14.9
1–3 years	14.5
4–6 years	19.7
7–10 years	28.3
Men 11–14 years	42.1
Men 15–18 years	55.2
Men 19–49 years	55.5
Men 50+ years	53.3
Women 11–14 years	41.2
Women 15–18 years	45.4
Women 19–49 years	45.0
Women 50+ years	46.5
Pregnant women	51.0
Breastfeeding women	53–56

table 1.3.3 Reference nutrient intakes (RNIs) for protein.

The requirements for different nutrients change throughout life, and this is clearly reflected in the DRVs. In addition individuals vary; for example, girls and women who have particularly heavy or frequent menstrual periods lose more iron than normal, and so they may need iron supplements to prevent themselves suffering from anaemia even if their dietary intake of iron is around the RNI.

Carbohydrates, proteins and lipids, along with any alcohol in the diet, provide us with energy. There is evidence that the proportion of energy-giving foods in the diet can increase or decrease the likelihood of problems such as heart disease developing. So a balance of the energy-giving nutrients is also recommended as part of the DRVs. If you look at the average British diet (see **fig. 1.3.24**) and compare the reality with the DRV you can see that the population as a whole is eating food that contains too much energy, not enough vitamins and minerals and too much fat. There is some fairly convincing scientific evidence linking some of these dietary habits to disease (see chapter 1.4). So Government policies are developed to encourage people to make changes to their diet and eat more healthily.

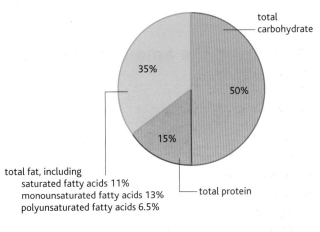

suggested energy mix for the UK diet (excluding alcohol)

total carbohydrate 50%

35%

15%

total fat, including
saturated fatty acids 11%
monounsaturated fatty acids 13%
polyunsaturated fatty acids 6.5%

total protein

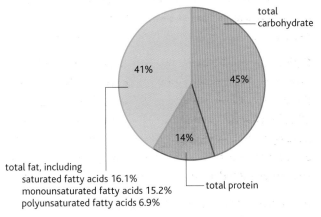

actual energy mix for the UK diet (excluding alcohol)

total carbohydrate 45%

41%

14%

total fat, including
saturated fatty acids 16.1%
monounsaturated fatty acids 15.2%
polyunsaturated fatty acids 6.9%

total protein

fig. 1.3.24 The recommended dietary make-up for the UK population and the actual percentage of energy gained from various nutrients in British households in 1993.

Questions

1 Look at **table 1.3.3**.
   a Why would an average protein requirement for an adult be a fairly meaningless measure?
   b Why do you think the RNI for protein is so different for men and women in the 15–18-year age group?
   c In what circumstances would the RNI for women aged 19–49 be inaccurate?

2 Nutrition guidelines are provided as averages for different groups of people. Discuss the advantages and disadvantages of this.

3 Suggest why governments publish nutritional guidelines.

4 Suggest why health advice usually quotes the EAR for energy but the RNI for particular nutrients such as a vitamin.

The energy budget

Measuring the amount of energy stored in foods

Different types of foods contain different amounts of energy stored within the chemical bonds. As these bonds are broken, this energy is released. The amount of energy contained in a food can be measured using a process known as **calorimetry**.

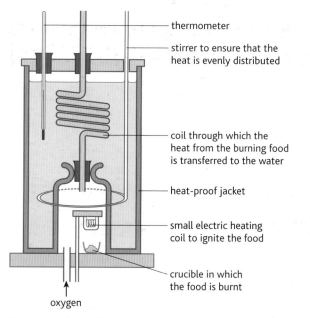

thermometer

stirrer to ensure that the heat is evenly distributed

coil through which the heat from the burning food is transferred to the water

heat-proof jacket

small electric heating coil to ignite the food

crucible in which the food is burnt

oxygen

fig. 1.3.25 A food calorimeter for measuring the energy in food.

Calorimetry measures the amount of energy released when a known quantity of food is completely oxidised by burning it in pure oxygen. The energy released is transferred to the surrounding water as heat. The resulting rise in the temperature of the water can be measured and used to calculate the energy value of the food, based on the fact that 4.2 kJ of heat energy raise the temperature of 1 kg of water by 1 °C. Carbohydrates yield 17.2 kJ/g, lipids have an energy value of 38.5 kJ/g and proteins 17.2 kJ/g. The data from calorimetry experiments are used to produce tables showing the energy values of an enormous range of foods.

The energy units measured in calorimetry are kilojoules, kJ. The pre-SI units of energy were called calories and kilocalories – most confusingly what most people referred to as a Calorie was in fact a kilocalorie! The 'Calorie' is still a common term in everyday language, particularly when people are talking about weight-reducing diets. Biologists no longer refer to

calories at all, but most food packaging indicates the energy value of food in both kilocalories and kilojoules. The relationship between the two is very simple: 4.2 kilojoules = 1 kilocalorie.

fig. 1.3.26 The energy content of pre-wrapped food is usually shown clearly on the label.

How much energy do you need in your diet?

Your body needs a certain amount of energy every day to keep your metabolism 'ticking over', so that your organs function correctly and the reactions of life in your cells can take place. This energy is known as your **basal metabolic rate** (**BMR**) and is measured when you are at complete rest. It is calculated from the temperature changes that result as the human body gives out heat over a period of hours or days in a heatproof room – another type of calorimeter. In babies and young children the BMR is proportionately higher than in adults because they use more energy in growth.

The BMR is related to the total body mass and the lean body mass. People with a high proportion of muscle will have a higher BMR because muscle tissue requires more energy for maintenance than fat. This is one reason why men usually have higher BMRs than women, because they tend to have a higher proportion of muscle to fat. As people age, not only is their tissue replaced less often but they also tend to lose muscle and so the BMR tends to fall with age. The BMR makes up, on average, about 75% of the metabolic needs of the body. But if you are very active then your BMR may make up 50% or less of your daily energy needs, because you are doing so much exercise as well.

Measurements of BMR show that an 'average' man needs to take in about 7500 kJ/day, and an 'average' woman needs about 5850 kJ/day. This reflects the energy needed if the person concerned lies on a bed all day and night and expends no extra energy above that needed to breathe and excrete – not even to feed! You can see from this that the BMR is of relatively little use on its own in assessing the energy intake needed in a healthy diet. To make the measure more useful, the **physical activity level** (**PAL**) must also be taken into account. Multiplying the BMR by a factor that reflects the physical activity level gives the estimated average requirement (EAR) for energy. A PAL of 1.4 is used as an average for adults in the UK, reflecting the rather sedentary way of life most people have. The EARs for energy of people of different ages are shown in **table 1.3.4**.

Age	EAR (kJ/day) (males)	EAR (kJ/day) (females)	Age	EAR (kJ/day) (males)	EAR (kJ/day) (females)
0–3 months	2 280	2 160	11–14 years	9 270	7 720
4–6 months	2 890	2 690	15–18 years	11 510	8 830
7–9 months	3 440	3 200	19–50 years	10 600	8 100
10–12 months	3 850	3 610	51–59 years	10 600	8 000
1–3 years	5 150	4 860	60–64 years	9 930	7 990
4–6 years	7 160	6 460	65–74 years	9 710	7 960
7–10 years	8 240	7 280	75+ years	8 770	7 610

table 1.3.4 The estimated average energy requirements (EARs) for people of different ages in the UK.

fig. 1.3.27 Pregnancy and breastfeeding increase the energy needs of these women.

If your energy intake and energy requirements are the same, your energy balance will mean that you neither gain or lose weight. If you do not match your energy intake to the requirements of your body then you will either gain weight (if you eat too much) or lose weight (if you eat too little).

Questions

1 Dietary advice usually suggests cutting back on foods that are high in fat content as a good first step if you want to lose weight. Why?

2 Why are the figures given in **table 1.3.4** only useful as guidelines for the energy intake required each day?

Weight issues

In the developed world, where there is plenty of food and people can easily eat more than they need to supply the metabolic needs of the body, many people have a positive energy balance. The excess food energy is converted into a store of fat and these people become overweight.

Measuring a healthy weight – the body mass index

What do we mean by 'overweight'? It isn't just how much you weigh – doctors and scientists look at your **body mass index** (**BMI**). This compares your weight to your height in a simple formula:

$$BMI = \frac{\text{weight in kilograms}}{(\text{height in metres})^2}$$

For an adult, the following definitions apply:

- a BMI of less than 18.5 kg/m² means you are underweight
- a BMI of 18.5–25 kg/m² is the ideal range
- a BMI over 25 and up to 30 kg/m² means you are overweight
- a BMI of 30–40 kg/m² is considered obese
- a BMI over 40 kg/m² defines you as morbidly obese.

The BMI was developed in the mid-1800s by a Belgian called Adolphe Quetelet. It was originally used to classify normal, relatively inactive people of average body composition. The normal charts apply to adults only – there are special charts for children and

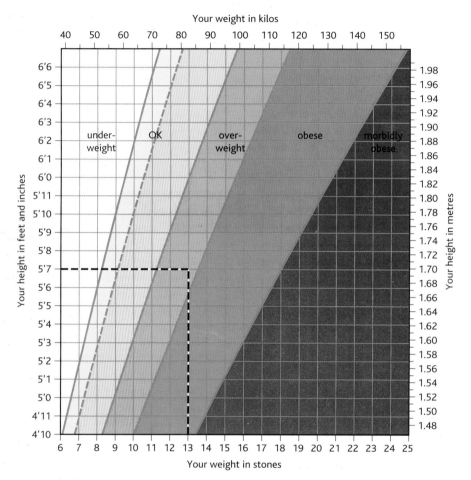

fig. 1.3.28 Using a graph like this gives an adult a good idea of whether their BMI is in the healthy range.

teenagers. Young people grow and their body composition changes as they mature, so both age and gender are important in calculating what is normal until they become adults. The BMI became widely used for deciding whether people are a healthy weight for their height and even for predicting the likelihood of CVDs – but doctors increasingly feel it is a very limited tool. Most top athletes would have BMIs in the obese range, because BMI makes no allowance for the difference between fat and muscle. BMI values also underestimate body fat in older people who have lost a lot of their muscle mass. New evidence suggests it is not a good predictor of CVDs either (see chapter 1.4).

Being seriously underweight is not good for you and can lead to muscle wasting, heart damage and other health problems. However, it is the other end of the scale that is causing the most concern. The available data show that around 61% of all adults in England (that's almost 24 million people) are either overweight or obese, and that the proportion of the population affected is continuing to rise. The trend towards obesity is being seen across the developed world, with the US leading the way but the UK and the rest of Europe following closely behind.

HSW Tackling obesity

In 2007 a major report into obesity in the UK was published. It was commissioned by the British Government and put together by 250 experts, including Sir David King, the Government's chief scientist. This suggests that if current trends continue the majority of the population will be obese by 2050.

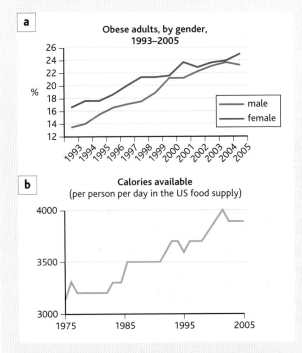

a Obese adults, by gender, 1993–2005

b Calories available (per person per day in the US food supply)

fig. 1.3.29 (a) This graph shows trends in adult obesity levels in the UK 1993–2005, along with (b) evidence of changing dietary trends in the US.

The report suggests that the change in energy balance is linked to modern lifestyles rather than to simply poor individual choices. Energy-rich food is widely available and cheap. The twenty-first century way of life in the UK involves little exercise. Few jobs now require manual labour, household tasks are often automated and people often drive instead of walking. So as the average energy input has increased (or stayed the same) the energy output has fallen and people are gaining weight. The solutions suggested in the report include a tax on fatty foods, town planning to make walking and cycling easier and educating children to prevent childhood obesity – but there is no clear scientific evidence that any of these solutions work. There is another question – is it ethical for scientists or the Government to try and influence people about the food they eat, the way they travel to work and in general how they lead their lives?

From obesity to underweight

In stark contrast, around two-thirds of the world's population do not get enough food to provide them with the recommended minimum daily intake, and for many their food barely yields sufficient energy to cover the BMR. They have a negative energy balance and will lose weight quickly if they attempt any physical activity. According to the Food and Agriculture Organization of the United Nations, 850 million people worldwide were undernourished in the years between 1999 and 2005. Yet from 2006 the number of overweight people in the world will be greater than the number who lack sufficient food.

Questions

1 What is meant by the term BMI?

2 Ali weighs 65 kg and is 1.68 m tall. Calculate Ali's BMI. What does this tell you about him?

3 Using the graph in **fig. 1.3.27**, what are the highest and lowest weights that would be healthy for an individual who is:
   a 6 feet tall
   b 1. 58 metres tall?

   At what weight would these individuals be defined medically as obese?

4 Which types of data that you know affect body mass and energy requirements are missing from the simple calculation of BMI? How does this limit the usefulness of the BMI?

5 Use **fig. 1.3.29** (a) to calculate the percentage increase in male and female obesity between 1995 and 2005.

6 US trends are often followed soon after in the UK.
   a What does the evidence in **fig. 1.3.29** (b) tell you about recent trends in energy (food) availability in the US?
   b Use the information to suggest a hypothesis that explains increasing obesity in the UK in **fig. 1.3.28** (a)
   c What more evidence would you want to see before drawing any firm conclusions?

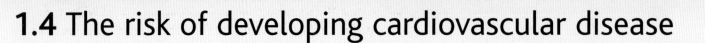

1.4 The risk of developing cardiovascular disease

What is risk?

Cardiovascular diseases or CVDs (which as you saw in chapter 1.2 include heart disease, atherosclerosis and strokes) kill many thousands of people in the UK every year. Some people are known to be at higher risk of developing CVDs than others. Before looking at what affects this risk, it is useful to think about exactly what is meant by the term 'risk'.

The word risk is used regularly in everyday conversation – 'I'm not risking that!' or 'What's the risk?' are just two examples. But in science risk has a very specific meaning. Risk describes the **probability** that a particular event will happen. Probability means the chance or likelihood of the event, calculated mathematically. The most common example used to explain how to calculate probability in maths is a die. The probability of throwing a 1, or of throwing any other number up to 6, can be expressed in one of three different ways, as:

1 in 6 (1:6)

0.166 66 recurring (0.17), or

17%.

fig. 1.4.1 The chance or probability of you throwing a 1 with a normal six-sided die is 1 in 6. If you throw a 1, this doesn't affect the probability of throwing 1 again next time. The probability is always 1 in 6.

Just as you can calculate your 'risk' of throwing a 1 with a die, it is possible to work out your risk of developing certain diseases, or of dying from a particular cause. For example, the National Safety Council in the US has shown that, based on figures from 2003, an American citizen had a 1 in 347 076 chance of dying as the result of a fall involving their bed or another piece of furniture. This was worked out by dividing the American population as a whole by the number who died like this.

How do we perceive risk?

The actual risk of doing something is not always the same as the perceived risk. Most people don't think twice before getting into their car – yet in the UK you have a lifetime risk of 1 in 237 of being killed in a road traffic accident. On the other hand many people regard motorcycles as very dangerous – yet your lifetime risk of dying in a motorcycle accident is 1 in 1020. Perception of risk is based on a variety of factors which include familiarity with an activity, how much you enjoy the activity and whether you approve of it. The mathematical risk may play very little part in building up your personal perception of risk.

The chance of winning the lottery is about 1 in 14 million – but this doesn't stop people buying tickets. Similarly, the mathematical risk of an early death if you smoke cigarettes doesn't stop people smoking.

Epidemiology

If you know the number of people in a population who are affected by a particular disease it is possible to calculate the average risk of a person within that population developing that disease. But the risk is higher for some people than others, depending on their lifestyle and the genes they have inherited.

By looking at people who have certain things in common, eg smoking, and comparing their risk of disease with the average risk for the whole population, it is possible to identify the factors that may be involved in causing that disease. This is known as probability and it plays a very important role in identifying **risk factors** for disease. Using these techniques, it appears that there are a number of factors that increase the likelihood that a person will develop CVD. It is a multifactorial disease as many things influence your chances of being affected.

The data in **fig. 1.4.2** and **table 1.4.1** show the risk of dying (**mortality**) from heart disease or strokes in different areas of the UK, and the level of smoking in the same areas.

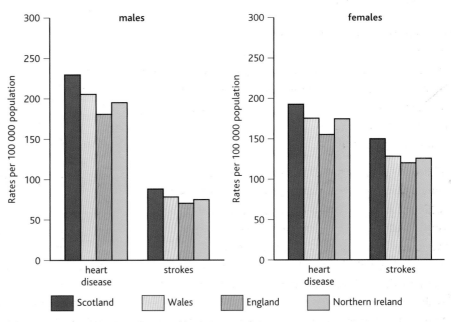

fig. 1.4.2 The numbers of deaths caused by two types of CVD.

Country	Men who are regular smokers (%)	Women who are regular smokers (%)
Scotland	30	31
Wales	27	24
England	29	25
Northern Ireland	26	27

table 1.4.1 The percentage of men and women who smoke regularly in different areas of the UK.

You can see that there is a similarity between the mortality from heart disease or stroke and smoking patterns, suggesting a link or relationship between the two. A link like this is called a **correlation**. This does not prove, however, that one is the **cause** of the other. They could both be caused by something else, which would explain why they change in the same way. Correlation is not the same as causation. Epidemiology is good at establishing risk and correlations, but further work needs to be done to find out whether there is a causal link.

Questions

1 In the UK the mortality from heart disease is about three times greater for smokers than for non-smokers. Explain why this does not mean that an individual person who smokes will die from heart disease.

2 The risks linked to smoking are well known, so why do you think people keep smoking?

3 Using the data from **table 1.4.1** draw a bar chart to show the percentages of men and women who smoke regularly in England, Scotland, Wales and Northern Ireland. Compare this with the data on deaths caused by heart disease and strokes in **fig. 1.4.2**. What correlation, if any, can you see between the data? Suggest other information you would need to consider before you could establish that smoking causes serious conditions such as heart disease and strokes.

Epidemiological studies and CVDs

Designing studies

Most epidemiological studies are based on a very big sample size – as a general rule, the bigger the study, the more meaningful the results.

The ideal is to investigate one factor or variable, keeping all other variables the same (controlled). However, controlling variables isn't always easy when you are working with human beings! The way people live is complex and varies enormously, so it is hard to detect what the effect of one factor is on people in general. The larger the number studied, the more likely it is that patterns may emerge, despite all the other differences between people. Evidence based on large amounts of data is more likely to be statistically significant.

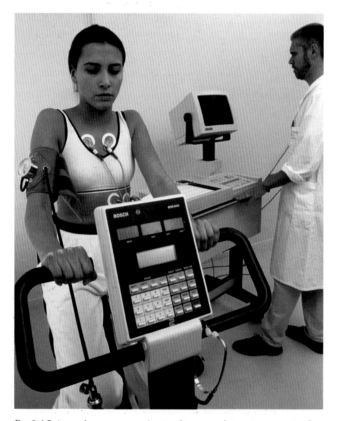

fig. 1.4.3 It can be very expensive to observe and measure aspects of health in many people over a long period of time. This is why many studies are retrospective – scientists look back at information on health and death rates collected years earlier.

Some epidemiological studies are also carried out over a long time. **Longitudinal studies** are very valuable because they follow the same group of individuals over many years (see **fig. 1.4.4**). This means the

impact of their known lifestyle on their health can be tracked over time. For example, the Münster Heart Study looked at cardiovascular disease in 10 856 men aged 36–65 in Europe, following them for time periods which ranged from 4 to 14 years. The results from this study are given a great deal of weight because so many people were involved over a long period of time. The Framlingham Study in the US also provided huge amounts of data – the main limitation being that they were all from similar American citizens.

An ambitious new study has been set up in the US to follow 100 000 children from birth until they are 21 years old. In 2008 potential children will be selected to be representative of the whole of the US population. One major objective of the study is to examine how environmental inputs and genetic factors interact to affect the health and development of children. This is believed to be the biggest longitudinal study ever set up.

Sometimes scientists look at all the available studies in a particular area and analyse all the available data in a massive literature study. This combines small and large studies and can give more reliable evidence than many of the studies do individually.

Evaluating scientific studies

When considering a study you need to examine the methodology to see if it is **valid**, that is, properly designed to answer the questions being asked. You also need to see if the measurements have been carried out with **precision**. It is important to find out if other scientists have been able to repeat the methodology and have had similar results – if so, the results are considered more **reliable**.

It is also important to know who carried out the research, who funded it and where was it published, and decide whether any of these might have affected or **biased** the study. You then need to **evaluate** the data and conclusions from the study in the light of all these factors.

In the next few pages you are going to look at some of the evidence that scientists have gathered suggesting factors that may – or may not – affect your risk of developing heart disease. In each case you need to look carefully at the type of evidence that is presented and think about what else you need to know to draw firm conclusions.

Risk factors for CVDs

The results from many epidemiological studies have identified a range of risk factors linked to CVDs. These factors break down into two main groups – those you can't help and those you can do something about.

There are three main risk factors for CVDs which cannot (at present) be altered.

- Genes: studies show there is a genetic tendency in some families and in some ethnic groups to develop CVDs. The genetic tendency varies – the arteries may be more easily damaged, there may be a tendency to develop hypertension which can cause arterial damage and make CVDs more likely, or the cholesterol metabolism may be faulty (see page 60).

- Age: as you get older, your blood vessels begin to lose their elasticity and to narrow slightly, making you more likely to suffer from CVDs, particularly heart disease.

- Gender: statistically, under the age of 50, men are more likely to suffer from heart disease (and other CVDs) than women. The female hormone oestrogen, which plays an important role in the woman's menstrual cycle, appears to reduce the build-up of plaque. This gives women some protection against CVDs until they go through the menopause and oestrogen levels fall.

Looking at the data

Identical twin studies are an excellent resource when investigating whether there is a genetic factor at work, because identical twins have exactly the same genes. Any differences should therefore be due to the environment in which they live. A major twin study in Sweden showed that if one twin died of heart disease between the ages of 36 and 55, then the risk of the other twin also dying of heart disease was eight times higher than if neither was affected by this CVD (see **fig. 1.4.4**). However, as the twins got older, one dying of heart disease had less effect on the risk of the other twin also dying of heart disease. In other words, there appears to be a clear genetic link to heart disease in younger men, but it gets less in much older men.

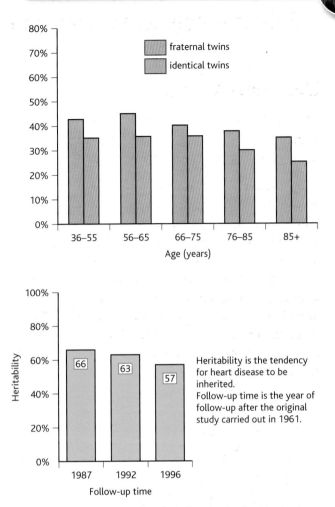

fig. 1.4.4 **Results from an epidemiological study of male twins in Sweden published in 1994.**

Epidemiological studies have also identified a number of lifestyle factors linked to CVDs, some of which you will look at in the following pages. These lifestyle factors are important for health because they are the factors that we can do something about.

Questions

1 When scientists design a major study, what can they do to try and make sure their results will be both valid and reliable?

2 Using the data in **fig. 1.4.4**, answer the following:

a How do the figures for identical and fraternal twins differ in the top graph? What does this suggest about a genetic link to heart disease?

b What does the bottom graph show you about the apparent heritability of heart disease in men? What might affect the fall in apparent heritability as the men got older?

HSW Lifestyle factors in CVDs

The way you live your life, or your **lifestyle**, can affect your risk of getting a CVD in the future. Epidemiological studies have shown links with smoking, stress, diet and weight, lack of activity and high blood pressure. Here we will look at the evidence for some links. On the following pages we will look at the links with diet in more detail.

Smoking and CVDs

Studies have shown that smokers are far more likely to develop CVDs than non-smokers with a similar lifestyle, and nine out of ten people needing heart bypass surgery are smokers. **Fig. 1.4.5** is based on a large study carried out in the US. Professor Patrick McBride and his team did a complete literature review of studies on the impact of smoking on the incidence of CVDs, and took into account data from several different US Surgeon General's reports. They also carried out their own surveys in general practices.

Their results showed that smoking increases your risk of developing all the different CVDs, but that the impact on some people is greater than on others when all other factors are taken into account.

The studies show a correlation between smoking and CVDs. Causation was established by further research. For example, studies found that some of the chemicals found in tobacco smoke can damage the lining of the arteries, making the build-up of plaque more likely. Smoking also causes the arteries to narrow, raising the blood pressure and increasing the risk of CVDs.

Weight and CVDs

Being overweight puts a greater strain on the heart because your heart has to work harder to move the blood through all the extra tissue. Losing weight should improve the situation. Being overweight also increases the likelihood that you will suffer from high blood pressure, which in turn increases the risk of CVDs.

The most common measure for obesity is the BMI (see chapter 1.3). However, research published in the scientific journal *The Lancet* by a team led by Francisco Lopez-Jimenez in the US has shown BMI to be a poor predictor of CVD risk. They looked at data from 40 studies including 250 000 people, and found that other measures of obesity such as the waist-to-hip ratio may be much better predictors of risk. A waist-to-hip ratio (**fig 1.4.6**) of over 0.8 in women or over 1.0 in men seems to indicate an increased risk of heart disease. This isn't a random correlation – it is linked to the pattern of fat laid down in the body and to the balance of HDLs and LDLs in your blood.

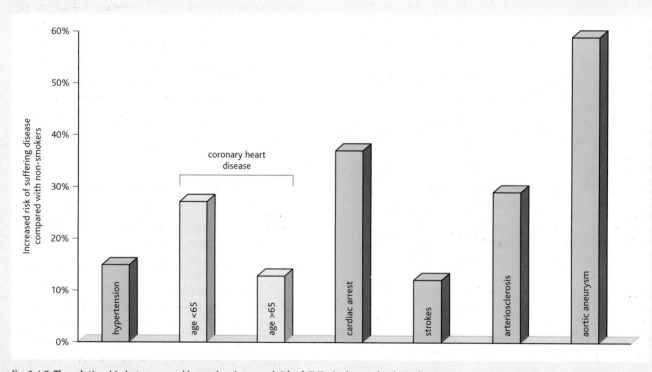

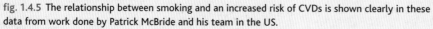

fig. 1.4.5 The relationship between smoking and an increased risk of CVDs is shown clearly in these data from work done by Patrick McBride and his team in the US.

fig. 1.4.6 The simple waist-to-hip ratio can give a good prediction of your risk of CVDs.

An increasing number of studies suggest that simply being overweight has little or no impact on your risk of cardiovascular disease. Many scientists now think it is where fat is stored on your body, your levels of exercise and the levels of different fats in your blood that seem to be the best predictors of future CVDs. At least, that is the picture at the moment.

Exercise and CVDs

Your heart is a muscle, and like any other muscle it gets stronger and works more efficiently if it is used regularly. Regular exercise can slow the heart rate, lower blood pressure, lower blood cholesterol levels and balance the lipoproteins in your blood. All of these lower your risk of CVDs. A study on 10 269 male Harvard graduates aged between 45 and 84 showed that the men who increased their levels of physical activity had a 23% lower mortality after 20 years than their peers who did not exercise, and the main cause of the deaths was CVDs. Another study on 9777 men aged 20–82 showed that men classed as unfit on a treadmill test at their first

fig. 1.4.7 Whatever your age or fitness level, increasing your exercise will lower your risk of CVDs.

examination who were fit on their second examination almost five years later had a 52% lower risk of dying of CVDs than men who remained unfit. And a study on 72 488 female nurses showed the same benefits for women – the more active women had a significantly lower risk of CVDs.

Stress and CVD

High stress levels increase the risk of cardiovascular disease by causing prolonged high blood pressure and a faster heart rate. The link between stress and heart disease was established by the Whitehall I and Whitehall II Studies. These were longitudinal studies carried out on thousands of civil servants working in Whitehall in the UK.

The Whitehall I Study was carried out on 18 000 civil servants and set up in 1967. The hypothesis was that more senior staff would be under greater stress because of their added responsibilities, and so more likely to be affected by heart disease. But the results showed that people were more likely to die relatively young of heart disease if they were in one of the lower levels of the civil service.

The Whitehall II Study was set up in 1985 to try and explain the results from the first study, which was based on an initial medical and follow-up recording of causes of death among employees.

This study is continuing. Early results suggest that it is not the stress of a high-profile job that is the risk factor, but rather the stress of having very little control over one's life and work. People in the lower grades have little choice or control over what they do at work. Although more senior staff are making important decisions, they have a lot more control – and it looks as if this is better for their cardiovascular health.

Diet and CVDs

There have been many studies on how diet is linked to CVDs, some looking at general diet and some looking at specific foods. Designing studies that can show clear links with just one type of food is difficult.

Many studies have looked at the general diet people eat and at the incidence of heart disease. For example, one study produced the graph in **fig. 1.4.8**. This shows that in countries where people eat a lot of fatty meat and dairy foods (mostly saturated fats), many people die of heart disease. This suggests that high levels of saturated fats in the diet may be a risk factor.

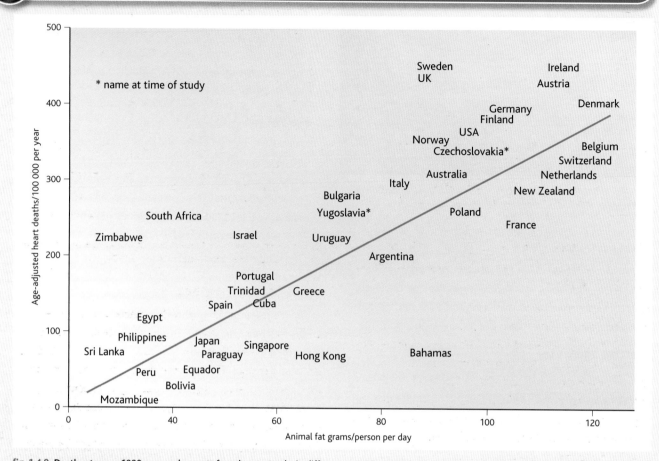

fig. 1.4.8 Death rates per 1000 men and women from heart attacks in different countries compared with the average intake of animal (saturated) fats.

The link between a diet high in saturated fats and a raised incidence of CVDs shows a correlation, but not a cause. In a number of studies scientists discovered that a high intake of saturated fats was often associated with high blood cholesterol levels. Cholesterol is involved in plaque formation in atherosclerosis, so this suggested a cause for the link between a high-fat diet and CVDs (see **fig. 1.4.9**).

Our picture of the relationship between fat in the diet and cholesterol in the blood has been further complicated in recent years as we have learned more about two particular types of lipoprotein in the body.

- Low-density lipoproteins (LDLs) are formed from *saturated* fats, cholesterol and protein. They carry more cholesterol than HDLs and bind to cell membranes before being taken into the cells. If your LDL levels are high, your cell membranes become saturated and so more LDL cholesterol is left in your blood.

- High-density lipoproteins (HDLs) are formed from *unsaturated* fats, cholesterol and protein. They carry cholesterol from your body tissues to your liver where it is broken down, lowering blood cholesterol levels. HDLs can even help to remove cholesterol from fatty plaques on the arteries, reducing the risk of heart disease from atherosclerosis.

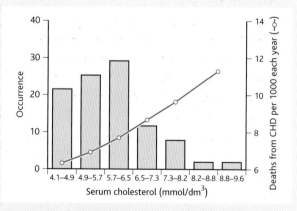
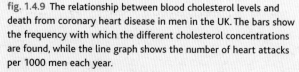

fig. 1.4.9 The relationship between blood cholesterol levels and death from coronary heart disease in men in the UK. The bars show the frequency with which the different cholesterol concentrations are found, while the line graph shows the number of heart attacks per 1000 men each year.

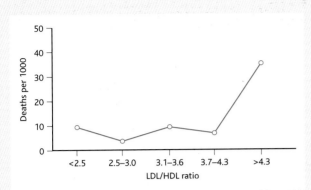

fig. 1.4.10 This evidence from the Münster Study appears to show a clear link between the LDL/HDL ratio and deaths from coronary heart disease.

The balance of these lipoproteins in your blood is now recognised as a clear indication of your risk of developing CVDs (see **fig. 1.4.10**).

Further studies have shown that some kinds of unsaturated fats are better than others. Monounsaturated fats (found in olive oil, olives, peanuts and some margarines) improve the balance of LDLs and HDLs in your blood and help to reduce the overall levels of cholesterol. Polyunsaturated fats (found in foods such as corn oil, sunflower oil, many margarines and oily fish) seem to be even better than monounsaturated fats at reducing overall cholesterol levels by removing it from your blood and balancing the HDLs and LDLs.

Whatever you eat, you cannot guarantee what will happen to your blood cholesterol and LDL/HDL levels. This is because the way your body deals with the fats you eat and with the levels of cholesterol and the balance of lipoproteins in your blood are all linked to your genetic make-up. Some people have livers that can deal with almost any amount of fat in the diet and maintain a good balance of LDLs and HDLs. Other livers simply cannot cope and almost any amount of fat in the diet will be reflected in raised blood cholesterol levels.

Links between factors

Although it is difficult to separate out factors and so identify which is having an effect, many epidemiological studies are starting to show increased risks due to a combination of factors. For example, evidence now suggests that smoking not only increases your risk of CVDs because of its effect on your blood vessels and on your blood pressure, it also increases the impact of raised LDL/HDL ratios on your cardiovascular system (see **fig. 1.4.11**).

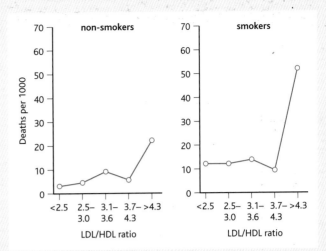

fig. 1.4.11 Data from the Münster Heart Study shows how smoking and diet can interact to increase risk of death from CVDs.

So how can we stay healthy?

As epidemiological studies of links between risk factors and CVDs become more sophisticated, and as scientific research discovers more reasons why some factors could cause CVDs, the advice changes about what is 'good' and what is 'bad' for us. Current evidence suggests that eating foods containing more unsaturated than saturated fats, and those that give a higher ratio of HDLs to LDLs, reduces the risk of CVDs later in life. Not smoking, reducing constant stress and getting plenty of exercise are also likely to help.

Questions

1 Look at **fig. 1.4.5**. Which CVDs appear to be most affected by smoking? Why do you think these are the most affected?

2 A lot of evidence is shown on these pages. Look carefully at each data set and find out as much as you can about the size of the study, who carried out the research, when the work was done, etc. Decide which studies carry the most weight scientifically.

3 Choose two of the studies highlighted here. Look for research that has produced conflicting evidence. How valid is this research? Which evidence would you pay most attention to and why?

A controversial link

Proteins in milk – another possible risk factor?

Research on the links between diet and CVDs continues, and some scientists have suggested new risk factors. For example, Professor Bob Elliot and Dr Corrie McLachlan from New Zealand developed a theory that it might be the proteins in milk rather than the fat that cause problems in the heart. About 80% of the protein in milk is casein. The main forms of casein are known as beta-casein A1 and A2, and the difference between them is just a single amino acid. A2 seems to cause no problems, but Elliot and McLachlan thought that the A1 form might be a factor linked to CVDs, and in particular coronary heart disease.

The amounts of A1 and A2 in the diet depend largely on the balance found in milk. The evidence for McLachlan's ideas comes mainly from observations of the levels of A1 consumption in different countries (see **fig. 1.4.12**). Finland has the highest consumption of A1 in the world and the highest levels of heart disease. In Iceland, where the people are of similar origin but cows produce milk with predominantly beta-casein A2, there is a low to moderate level of heart disease. The Masai people in Kenya drink large amounts of milk but have very low levels of heart disease, and all of their milk is A2.

Results like these show a remarkable correlation, but do they indicate a causal relationship? Many scientists are not convinced by the evidence so far, not least because it flies in the face of accepted theories of the cause of heart disease. The companies that produce drugs to control blood cholesterol will take some convincing – if McLachlan's ideas are right it will be possible to prevent many cases of heart disease simply by drinking the right kind of milk.

fig. 1.4.13 Some Holstein Friesian cows (the main dairy breed used in Europe) produce A1-free milk; others do not.

Several different teams have been working on these theories but so far the results have been inconclusive. Some groups have results suggesting strong links and others show little evidence of any correlation or causation.

There are scientists who are suspicious of these ideas because McLachlan (who died in 2003) was Chief Executive of a company marketing milk containing only 'healthy' forms of casein. They suspect he (and his successors) could be using this theory to generate business.

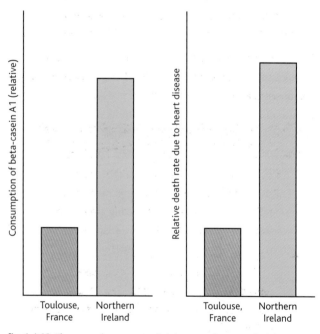

fig. 1.4.12 These graphs suggest a link between beta-casein A1 consumption and death rate due to heart disease.

However, Keith Woodford, Professor of Farm Management and Agribusiness at Lincoln University in New Zealand, published an in-depth consideration of the evidence on both sides. He is becoming increasingly convinced that the A1/A2 milk theory may help to explain a great deal about one of the causes of heart disease.

The right idea at the wrong time

In the 1960s, the American researcher Kilmer McCully was working on a rare, fatal genetic disease of young children. He noticed that their arteries had atherosclerosis, linked to problems in metabolising the amino acid homocysteine. The atherosclerosis in these children was identical to the damage seen in adult atherosclerosis. McCully wondered if there might be a link and came up with a new hypothesis – that a problem in homocysteine metabolism could be part of the cause of CVDs. If so, there would be a simple cure. Increasing the vitamin B6 and folic acid in the diet allows homocysteine to be metabolised instead of building up in the blood.

McCully published his findings in 1969. Not only were they ignored, but he was sidelined away from his area of research. He had had a great idea, but at the wrong time! Senior scientists at the time were convinced that cholesterol was the answer to heart disease and their reputations relied on this. Drug companies had spent millions of pounds developing cholesterol-lowering drugs, so the idea of preventing CVDs with cheap vitamin B6 and folic acid tablets or a change of diet was not good news. What's more, McCully was only a young researcher. He was not well known or influential in his field. So for years he struggled on with his work alone, convinced that his idea was right.

In the 1960s deaths from heart disease began to fall in the US, followed in the 1970s in the UK – yet average blood cholesterol levels had hardly changed. But in the mid-1960s US cereal manufacturers started to add B vitamins to breakfast cereals, followed by UK manufacturers in the 1970s. McCully thinks this is the answer!

Then new evidence suggested that high-fat diets and raised blood cholesterol are only part of the CVD story. People began to take notice of McCully's homocysteine theory. A major trial involving 19 European countries was set up in the 1990s. The evidence (see **fig. 1.4.14**) suggests that McCully's theory is right. Raised homocysteine levels indicate an increased risk of CVDs – and increasing your intake of B6 and folic acid reduces the risk again.

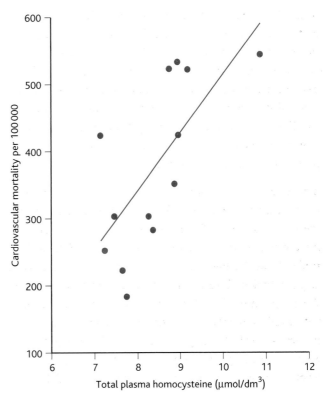

fig. 1.4.14 **The link between raised levels of the amino acid homocysteine in the blood and the risk of dying from CVDs.**

The story of the link between homocysteine and CVDs is an interesting example of how science works – and sometimes doesn't work very well!

Questions

1 Using **fig. 1.4.14**, what is the impact of a 3 µmol/dm³ increase in homocysteine levels on CVD mortality rates? Does this graph show correlation or causation between the two?

2 Using both of these examples, explain why scientists may be suspicious of new ideas which seem to contradict an accepted model. Explain what is needed to prove that a new theory is both valid and reliable.

The benefits and risks of treatment

Once a patient has signs of cardiovascular disease there are a number of different treatment options available. Changing lifestyle, such as improving diet, giving up smoking and taking more exercise can help but there are also various drugs that can be given. The drugs aim to reduce the risks associated with CVDs by helping to prevent problems developing. However, all medicines carry some risk.

Controlling blood pressure

Drugs that reduce blood pressure are known as **antihypertensives**. A number of commonly prescribed antihypertensive drugs are described below.

- Treatment often begins with **diuretics**, which increase the volume of urine produced. This gets rid of excess fluids and salts, so that the blood volume falls. With less blood, a smaller volume is pumped from the heart and the blood pressure falls.

- **Beta blockers** interfere with the normal system for controlling the heart. They block the response of the heart to hormones such as adrenaline, which normally act to speed up the heart and increase the blood pressure. So beta blockers make the heart rate slower and the contractions less strong, so the blood pressure is lower.

- **Sympathetic nerve inhibitors** affect the sympathetic nerves which go from your central nervous system to all parts of your body. Sympathetic nerves stimulate your arteries to constrict, which in turn raises your blood pressure. The inhibitors prevent these nerves signalling to the arteries, which helps to keep the arteries dilated and your blood pressure lower.

- Angiotensin is a hormone which stimulates the constriction of your blood vessels and so causes the blood pressure to rise. ACE inhibitors block the production of angiotensin, reducing the constriction of your blood vessels and so keeping your blood pressure lower.

The benefits of these drugs in reducing blood pressure are clear. Not only is the risk of CVDs substantially reduced, but so also is the risk of damage to organs such as the kidneys and eyes from the high blood pressure.

However, the risks of these treatments themselves are twofold. If the treatment is not monitored carefully, your blood pressure may become too low. That can lead to falls and injuries, which in elderly patients particularly can be serious and even life-threatening. The other major risk is of the side-effects that may result from the way your body reacts to the drugs. Each type of drug has its own possible side-effects. For a drug to be given a licence for use, the benefits of the treatment must be judged to outweigh any side-effects.

The side-effects from commonly used antihypertensives include coughs, swelling of the ankles, impotence, tiredness and fatigue and constipation. These are not serious compared with the health risks from high blood pressure – but to the patient they may feel very important. High blood pressure often doesn't make you feel ill, but the medication needed to control it can affect your quality of life. Doctors find many patients stop taking their medication – the side-effects make them ignore the much larger but invisible risk of CVDs.

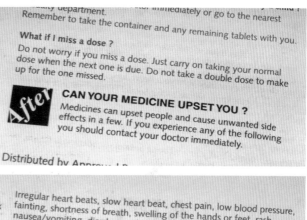

...y department. ...or immediately or go to the nearest
Remember to take the container and any remaining tablets with you.

What if I miss a dose ?
Do not worry if you miss a dose. Just carry on taking your normal dose when the next one is due. Do not take a double dose to make up for the one missed.

CAN YOUR MEDICINE UPSET YOU ?
Medicines can upset people and cause unwanted side effects in a few. If you experience any of the following you should contact your doctor immediately.

Distributed by Appro... Lt...

...ck Irregular heart beats, slow heart beat, chest pain, low blood pressure, fainting, shortness of breath, swelling of the hands or feet, rash, nausea/vomiting, diarrhoea, abdominal pain, flatulence, feeling of sickness, fatigue, dizziness, light-headedness, headache, sleep disturbances, mood changes including depression and anxiety, loss of
o energy, pins and needles, cramp, conjunctivitis or dry eyes, visual
or disturbances, taste abnormalities, hearing disturbances, fever and sexual dysfunction.

If the symptoms persist or become troublesome you should inform your doctor.
You should also tell your doctor or pharmacist if you notice any other troublesome side effects.

fig 1.4.15 All medically licensed drugs come with instructions which include the possible side-effects known to be caused by that drug.

Statins

Statins are a group of drugs that lower the level of cholesterol in your blood. They block the enzyme in the liver that is responsible for making cholesterol, and are particularly effective at blocking the production of LDLs. Statins also improve the balance of LDLs to HDLs and reduce inflammation in the lining of the arteries. Both of these functions reduce the risk of atherosclerosis developing.

Fig 1.4.16 shows the results from a trial using statins with a group of 6605 Asian Indians in the US. This shows the results for men and women, and also other groups that are high risk categories for cardiovascular disease. Although statins reduce the incidence of serious cardiovascular disease in all categories, they seem to have a greater effect for some groups than for others.

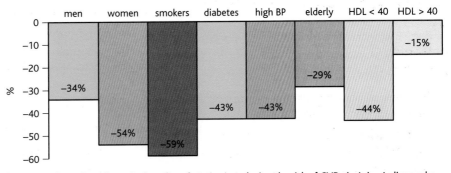

fig 1.4.16 These data show the benefits of statins in reducing the risk of CVDs in Asian Indians, who are a particularly high-risk group. Statins had a strong positive benefit to a range of patients.

Some statins have proved so effective that they are no longer prescription only, so people can buy them and take them as a precautionary measure. This suggests that the risks associated with the drugs are small. A team at Glasgow University led by Ian Ford has shown that men who took pravastatin (a particular statin) for 5 years had a lower risk of death or heart attack even 10 years after they stopped taking the drug. The study involved 6595 middle-aged men and it showed that for the first 5 years the overall risk of heart attack or death from any type of heart disease was 11.8% for the men who took pravastatin compared with 15.5% for men who took a **placebo** (an inactive substance that resembles the drug but has no action in the body). The risk was reduced the most while the men were taking the drug, but some level of protection lasted for up to 10 years afterwards.

Most people use statins with little or no ill effect. Side-effects of muscle and joint aches and nausea, constipation and diarrhoea are sometimes reported. However, there are two serious but very rare side-effects. In a tiny number of people statins trigger a form of muscle inflammation which can be fatal – for example, the US Food and Drug Administration (FDA) reported 3339 cases of these muscle reactions between January 1990 and March 2002, during which time millions of Americans took statins daily. Statins can also cause liver problems in a small group – for example, the risk of liver damage in people taking lovastatin is two in a million. Of 51 741 liver transplant patients in the US between 1990 and 2002, liver failure appeared to be due to statins in only three cases. The other risk is more subtle. There is a risk that if people take statins to lower their blood cholesterol, they will no longer try to eat a healthy diet, and statins give no protection against the other ill effects of a bad diet.

Plant stanols and sterols are now widely sold in spreads and yoghurts. These compounds are very similar in structure to cholesterol. They reduce the amount of cholesterol absorbed from your gut into your blood, which can make it easier for your body to deal with cholesterol and reduce the levels of LDLs in the blood. Products like these are sold as a food, not as a drug, so although there is scientific evidence that they are effective in many people, they have not been subjected to the levels of testing that drugs such as statins undergo. **Metadata analysis** (looking at a large number of studies and putting the data together) has shown that these products do work if they are eaten regularly in the recommended amounts (2 g a day of plant sterols and stanols). It has been estimated that used correctly these products can lower your risk of heart disease by about 25%.

Anticoagulants and platelet inhibitory drugs

Following heart surgery, or after suffering from a blood clot (thrombosis), drug treatments are used to help prevent the blood clotting too easily. Here are two examples.

Warfarin is an **anticoagulant** that interferes with the manufacture of prothrombin in the body. Low prothrombin levels make the blood clot less easily (see chapter 1.1). Warfarin has been used in rat poison – in high doses the blood will not clot at all and the rats bleed to death after the slightest knock. In humans the dose is carefully monitored to make sure that the clotting of the blood is reduced but not prevented completely.

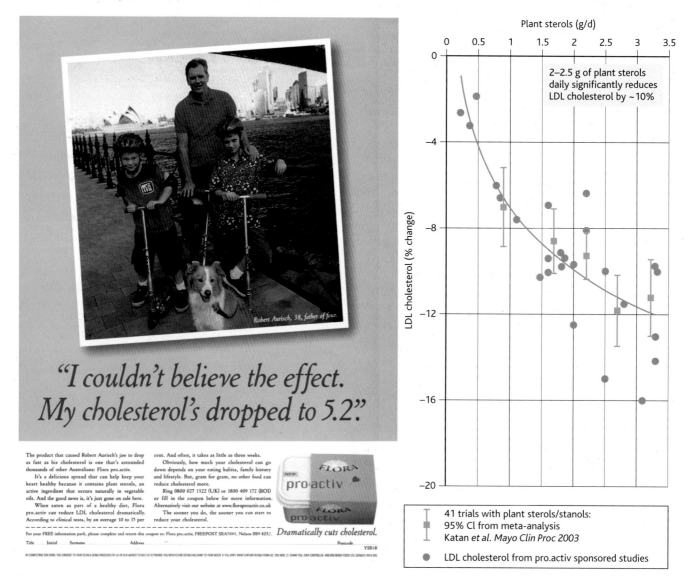

fig. 1.4.17 Some foods like this spread are promoted for their impact on health. Scientific evidence from a number of studies shows that they do work if they are used properly.

Platelet inhibitor drugs make the platelets less sticky, and so reduce the clotting ability of the blood. The cheapest and most common of these is aspirin (**fig. 1.4.18**) but clopidogrel is also commonly used.

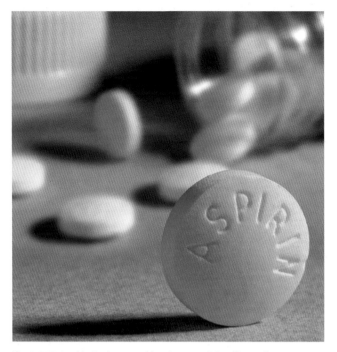

fig. 1.4.18 Aspirin is cheap, and has been traditionally used as a painkiller. It is also a very effective way of preventing many cardiovascular problems.

The risks of taking aspirin are well known – it irritates the stomach lining and causes bleeding in the stomach which can become serious. A combination of aspirin and clopidogrel can reduce the risk of developing a range of cardiovascular diseases by 20–25% in some low-risk patients. However, based on data from a number of studies it appears that for some patients the risk of side-effects are much higher when the two drugs are combined. For example, for every 1000 patients at high risk of CVDs treated for 28 months, 5 cardiovascular events would be avoided – but 3 major stomach bleeds would be caused. In lower-risk patients, 23 cardiovascular events would be avoided while 10 major bleeds would be caused.

The balance of preventing the blood from clotting too easily while allowing it to clot when necessary is a very fine one. For example, when people are treated with anticoagulant drugs such as warfarin, they have to be monitored very carefully to make sure that they do not bleed internally, particularly in the brain. The decision whether to give warfarin will depend on many factors, including the patient's age and condition as well as other medication they may be taking.

Questions

1 Explain why the side-effects of medication may result in a patient giving up on the treatment. Use the terms 'perceived risk' and 'actual risk' in your answer.

2 Suggest advantages and disadvantages of statins being made available without prescription.

3 Products containing plant stanols are freely available without prescription. How do the risks and benefits of the decision to use such products differ from the risks and benefits of using statins without a doctor's prescription?

4 The graph in **fig. 1.4.17** comes from the website of a company that makes products containing plant stanols. However, the data appear scientifically acceptable – why? What do they show you about the effect of plant stanols on blood cholesterol levels?

5 a Explain why placebos may be used in drug trials.

   b The study shown in **fig. 1.4.16** was stopped 2 years early because it was deemed unfair to the patients taking the placebo. Why do you think it was unfair? Is it ever unethical to use a placebo in a trial?

6 Look at **fig. 1.4.16** and answer these questions.

   a Explain why statins have a greater effect on reducing the risk of CVDs in people with a lower HDL level.

   b Considering that all medical drugs have associated side-effects, what does this graph suggest about which groups should be targeted with statins to reduce CVDs overall in the population?

Using the evidence

There is plenty of scientific evidence about the main factors that increase the risk of heart disease. A lot of that evidence is used by the Government and health organisations to produce advice on how to improve our health. Why do they do this?

Prevention is better than cure

Cardiovascular disease has a devastating effect on individuals, on families and on society. It costs a lot of money to treat people in hospital. While those people are too ill to work, they are not only losing money for their families, they are losing money for the companies they work for. Treating people with drugs to prevent them from needing surgery is cheaper for the health service, but it is even cheaper (and obviously better for the individual concerned) if we can stop ourselves reaching the point where we need the drugs. So prevention is better than a cure (treatment) for CVDs for many reasons, but persuading people to change their lifestyle habits is difficult.

For example, there is a large body of reliable evidence that shows that smoking is one of the highest risk factors for CVDs. However, if you stop smoking, your risk of developing heart disease is almost halved after just one year. What's more, research carried out by a team led by Azra Mahmud from Trinity College Dublin, published in 2007, suggests that after 10 years the arteries of smokers who quit are the same as if they had never smoked. There is a lot of support available for people who want to stop smoking. Yet, in spite of knowing the risks, around one in four Britons still smokes cigarettes and around 70 000 people die each year in the UK of CVDs linked to their smoking.

Health education programmes can help to make sure that everyone is aware of the risks associated with different lifestyle choices, but each individual has to make their own choices and take their own risks.

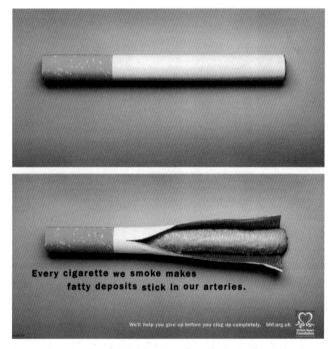

Every cigarette we smoke makes fatty deposits stick in our arteries.

We'll help you give up before you clog up completely. bhf.org.uk

fig 1.4.20 With adverts like this, why do so many people continue to smoke?

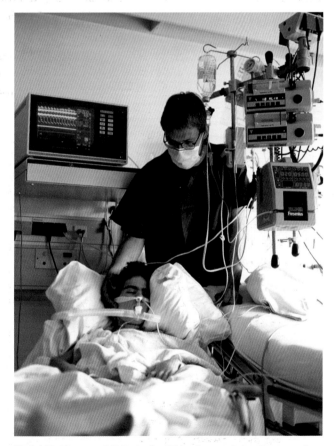

fig 1.4.19 The cost of CVDs in the UK has been estimated at around £1.7 billion a year.

Overweight or underfit?

Most people are aware that obesity is linked to CVDs, and many go on diets to try and lose weight. Most people also know that taking regular exercise helps protect against CVDs – but far fewer people take regular exercise than go on a slimming diet. The results of a study on 20 000 men aged from 30 to 83 years over an average of 8 years are given in **fig. 1.4.21**. Fitness was defined by how much oxygen they used during exercise. The results show that if you are obese and fit you have a lower risk of dying from CVDs than someone who is lean but unfit. Obviously being lean and fit is best of all!

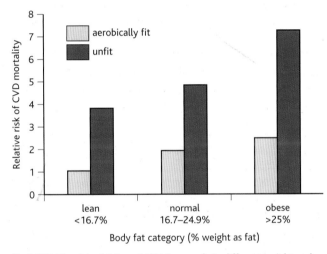

fig 1.4.21 **The risk of dying of CVD for people in different weight and fitness ranges.**

HSW Salt and CVDs

The Yanomami Indians in Brazil eat far less salt than people in the developed world and have much lower blood pressure levels. This evidence was used to draw a link between high salt levels, high blood pressure and CVDs.

After much more research, many scientists agree that high levels of dietary salt raise the blood pressure in around 25–30% of people in the UK. The rest of the population are not salt-sensitive.

Yet advertising campaigns warn of the dangers of salt and the UK food industry is under pressure to reduce salt in processed food. When research shows low salt levels will help only 30% or so of the population, is it ethical to manipulate the diet of the whole population to protect a minority?

So why don't people change their lifestyle?

Part of the problem is that people find it difficult to distinguish between perceived risk and actual risk. They can see a risk applies to an average of a group, but not to themselves as individuals. Familiarity plays a part in this. If you have seen people smoking, eating a high-fat, high-salt diet, never exercising and yet appearing perfectly well, the evidence of your own experience seems to contradict and therefore to override evidence from research reported in the media. People then underestimate the risk of CVDs associated with smoking, obesity, lack of exercise or a high-salt diet.

There are other reasons that result in the misjudging of risk. Sometimes people will continue smoking because they don't want to put on weight. Here the health risks of obesity are overestimated in comparison with those of smoking, or the risks of smoking are ignored in the light of not wanting to look fat! And the nicotine in tobacco smoke is addictive to many people, which makes it very difficult to give up smoking.

Peer pressure, personal experience and fatalism (the idea that if something is going to happen it will) all affect the desire of people to change their behaviour. When people calculate their personal risk/benefit situation, it is all too easy to consider the immediate benefit (pleasure in smoking, or not wanting to bother to exercise) more important than the seeming remote risk of heart disease.

Questions

1 Using data in **fig. 1.4.21** by what percentage does fitness reduce the risk of death from CVDs in each weight category? Why do you think people are more likely to try to lose weight than take more exercise?

2 What are the limitations of drawing conclusions about the effect of salt on blood pressure from a study comparison of the Yanomami in Brazil and the UK population? Do you think people in the UK under- or overestimate the risk of eating too much salt, and what influences these perceptions?

3 The Government spends millions of pounds on health advertising each year. Discuss whether or not this is a waste of money.

Examzone: Topic 1 practice questions

1 The diagram below shows a section of a human heart at a specific stage in the cardiac cycle.

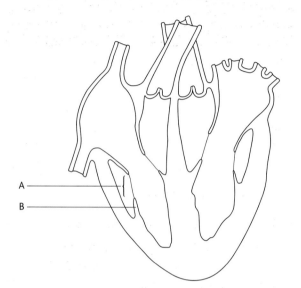

a Name the stage of the cardiac cycle shown in the diagram and give two reasons for your choice. **(3)**

b Give *one* function of each of the parts A and B.
(2)

(Total 5 marks)

2 Give an account of the structure and function of the polysaccharides starch and glycogen.

(Total 10 marks)

3 Read through the following account of the properties of water, then write out the full account completing the spaces with the most appropriate word or words.

Water has the chemical formula .... Water molecules are described as .... because they have a slight positive charge at one end of the molecule and a slight negative charge at the other end. As a result, individual molecules form .... bonds with each other.

Water is an important .... in living organisms because most biochemical reactions take place in aqueous solution. Water also has a high ....
which means that its temperature remains relatively stable despite large changes in the temperature of the surrounding environment.

(Total 5 marks)

4 The diagram below shows the structure of a triglyceride molecule.

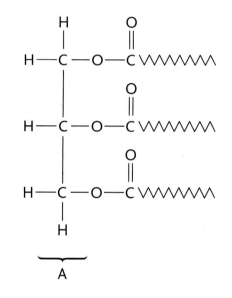

a i Name the part of the molecule labelled A. **(1)**

ii Name the type of bond formed between part A and a fatty acid. **(1)**

b Stearic acid and oleic acid are both examples of fatty acids. Each has a hydrocarbon chain containing 17 carbon atoms. Stearic acid is a saturated fatty acid but oleic acid is an unsaturated fatty acid.

Give *two* ways in which the structure of a stearic acid molecule differs from the structure of an oleic acid molecule. **(2)**

c Describe *two* functions of lipids in animals. **(4)**

(Total 8 marks)

5 A heart attack is caused when a blockage in the blood vessel supplying the heart muscle prevents the heart muscle receiving sufficient oxygenated blood.

a Name the blood vessel which supplies oxygenated blood to the heart muscle. **(1)**

b A study was carried out to assess the risk of heart attacks in American men. A large number of men was selected who had no history of heart disease and their blood cholesterol level was recorded. At each cholesterol level, the men were divided into smokers and non-smokers.

At each cholesterol level, the percentage of the men who had a heart attack within eight years from the start of the study was then monitored.

The results of this study are shown in the graph below.

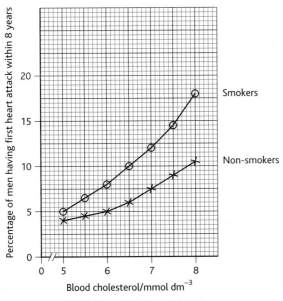

Adapted from MRC Research Update 9 (1997)

i Describe the effect that an increase in blood cholesterol level seems to have upon the risk of heart attack in non-smokers. **(2)**

ii Compare the relationship between blood cholesterol levels and risk of heart attack in smokers and non-smokers. **(4)**

iii Suggest *two* factors that were not monitored in this study, other than diet, which might be significant when assessing the risk of heart attacks in men. **(2)**

c The table below shows the number of men contracting serious heart disease per 100 000 men per year in four different countries. The mean ratio of polyunsaturated fats to saturated fats in the diet in these countries is also shown.

Country	Number of men contracting heart disease per 100 000 per year	Ratio of polyunsaturated to saturated fats in typical diet
Finland	210	0.18
Netherlands	140	0.30
Italy	100	0.38
Japan	30	0.95

Adapted from MRC Research Update 9 (1997)

i Explain what is meant by a saturated fat. **(1)**

ii State which of the four countries has the lowest intake of saturated fat compared with polyunsaturated fat in the diet. **(1)**

iii Use the data to suggest a possible relationship between the risk of heart disease and the types of fat included in the diet. **(1)**

(Total 12 marks)

Topic 2 Genes and health

This topic deals with how genes control the formation of proteins, including enzymes and the proteins in cell membranes. It shows how a fault in a gene can affect the protein that is made so that it results in a genetic disease. The example used here is cystic fibrosis.

What are the theories?

The nucleic acids contain the code that translates into chains of amino acids that form proteins. The sequence of amino acids controls the 3D structure of the protein and therefore how well it carries out its role. This topic explains how changes (mutations) in the nucleic acid code can result in changes in the amino acid sequence, which affects the shape of the protein and therefore its effectiveness.

Membranes control what crosses them as a result of their structure, which includes proteins that form channels through the membranes. Inheriting a faulty allele for the CFTR channel protein from both parents results in the symptoms of cystic fibrosis. You will learn how gene therapy raises the hopes of a cure for this disease, and how genetic counselling can be used to help people with faulty alleles make decisions about their future.

What is the evidence?

This topic introduces the 'detective' stories of how the genetic code was cracked and how membrane structure was discovered through the work of many scientists. You will also learn how Mendel began the interpretation of inheritance that we now know as the study of genetics. And you will have the opportunity to carry out practical work on membranes and on enzymes to provide evidence for the importance of their structure and how they work.

What are the implications?

The understanding of the genetic code, and the potential for using gene therapy to cure genetic diseases raises many questions. For example, should we manipulate other organisms to help us study the effectiveness of potential treatments? Although gene therapy is only considered for a few debilitating diseases at the moment, it could be used to change other inherited characteristics (such as height or intelligence) within an individual, or in their offspring. Would this be ethical?

Genetic screening produces information that could affect a person's future. Who should have access to that information and what kind of decisions does someone with that information need to take?

The map opposite shows you all the knowledge and skills you need to have by the end of this topic. The colour in each box shows which chapter they are covered in and the numbers refer to the Edexcel specification.

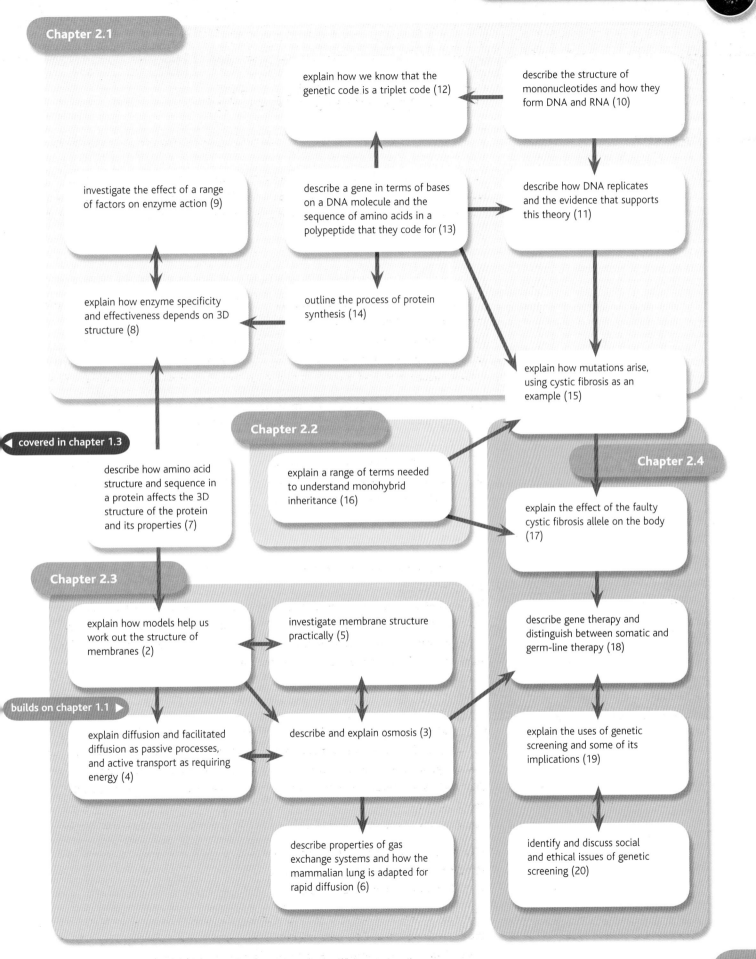

Chapter 2.1

explain how we know that the genetic code is a triplet code (12)

describe the structure of mononucleotides and how they form DNA and RNA (10)

investigate the effect of a range of factors on enzyme action (9)

describe a gene in terms of bases on a DNA molecule and the sequence of amino acids in a polypeptide that they code for (13)

describe how DNA replicates and the evidence that supports this theory (11)

explain how enzyme specificity and effectiveness depends on 3D structure (8)

outline the process of protein synthesis (14)

explain how mutations arise, using cystic fibrosis as an example (15)

◀ covered in chapter 1.3

Chapter 2.2

Chapter 2.4

describe how amino acid structure and sequence in a protein affects the 3D structure of the protein and its properties (7)

explain a range of terms needed to understand monohybrid inheritance (16)

explain the effect of the faulty cystic fibrosis allele on the body (17)

Chapter 2.3

explain how models help us work out the structure of membranes (2)

investigate membrane structure practically (5)

describe gene therapy and distinguish between somatic and germ-line therapy (18)

builds on chapter 1.1 ▶

explain diffusion and facilitated diffusion as passive processes, and active transport as requiring energy (4)

describe and explain osmosis (3)

explain the uses of genetic screening and some of its implications (19)

describe properties of gas exchange systems and how the mammalian lung is adapted for rapid diffusion (6)

identify and discuss social and ethical issues of genetic screening (20)

2.1 Nucleic acids – the molecules of life

Nucleic acids

Reproduction is one of seven key processes in living organisms. If the individuals in a species don't reproduce, then the species will die out. Multicellular organisms also need to grow and replace worn-out cells. Within every cell is a set of instructions for the assembling of new cells, both to form offspring and to produce identical cells for growth. Over the last 50 years or so scientists have made enormous strides towards understanding the form of these instructions, the genetic code. In the unravelling of the secrets of the genetic code, people have come closer than ever before to understanding the mystery of life itself.

Nucleic acids are the information molecules of the cell. They carry all the information needed to form new cells. The information is stored in the chromosomes in the nucleus of the cell. It takes the form of a code in the molecules of **DNA – deoxyribonucleic acid** (fig. 2.1.1). Parts of the code are copied into one form of **RNA – ribonucleic acid**, then to another form, and finally used to make proteins that build the cell and control its actions.

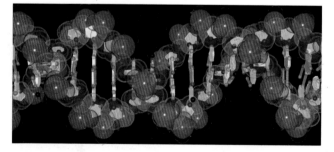

fig. 2.1.1 The DNA double helix is one of the most striking images of science from the last 100 years.

Nucleotides – the building blocks of nucleic acids

Both DNA and RNA are polymers. The chemical structure of the simple monomer units making up these two molecules is very similar. The single units are called **nucleotides** or **mononucleotides**. Each mononucleotide has three parts – a 5-carbon or **pentose sugar**, a nitrogen-containing base and

phosphoric acid. The pentose sugar in RNA is ribose, and in DNA it is deoxyribose. Deoxyribose, as its name suggests, contains one oxygen atom fewer than ribose (see fig. 2.1.2).

There are two types of nitrogen-containing bases found in nucleic acids. The **purine** bases have two nitrogen-containing rings, while the **pyrimidines** have only one. These rings have the chemical property of being bases because of the nitrogen atoms they contain (see fig. 2.1.4). DNA contains combinations of four different bases with equal numbers of pyrimidines and purines. The purines are **adenine** (A) and **guanine** (G) and the pyrimidines are **cytosine** (C) and **thymine** (T). In RNA the purine bases are the same but the pyrimidines are cytosine and **uracil** (U).

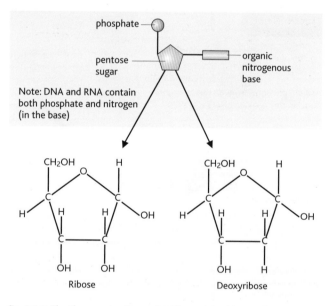

fig. 2.1.2 The three parts of a nucleotide are joined by condensation reactions. The arrangement of these molecules is crucial to the formation of DNA and RNA.

A phosphate group (PO_4^{3-}) is the third component of a nucleotide. Inorganic phosphate ions are present in the cytoplasm of every cell. This phosphate group makes the mononucleotides, and hence the nucleic acids, acidic.

The sugar, the base and the phosphate group are joined together by condensation reactions (with the elimination of two water molecules) to form the nucleotide (see fig. 2.1.2).

Building the polynucleotides

Mononucleotides are themselves linked together by condensation reactions to form polynucleotide strands (nucleic acids) which can be millions of units long. The sugar of one nucleotide bonds to the phosphate group of the next nucleotide so polynucleotides always have a hydroxyl group at one end and a phosphate group at the other.

To form DNA, nucleotides containing the bases C, G, A and T join together. RNA is made up of long chains of nucleotides containing C, G, A and U. Knowledge of how these units join together, and the three-dimensional structures that are produced in DNA in particular, is the basis of our understanding of molecular genetics.

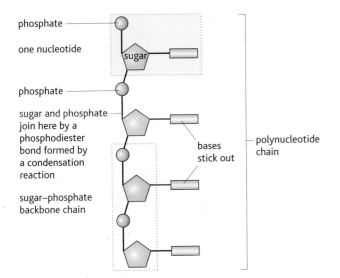

fig. 2.1.3 A polynucleotide strand like this makes up the basic structure of both DNA and RNA.

RNA molecules form single polynucleotide strands which may be folded into complex shapes or remain as long thread-like molecules. A DNA molecule is made up of two polynucleotide strands twisted around each other. The sugars and phosphates form the backbone of the molecule. Pointing inwards from this 'spine' are the bases, which pair up in specific ways. A purine always pairs with a pyrimidine – in DNA adenine pairs with thymine and cytosine with guanine. This results in the famous DNA **double helix**, a massive molecule that resembles a spiral staircase.

The two strands of the double helix are held together by hydrogen bonds between the **complementary base pairs** (fig. 2.1.4). There are ten of these pairs for each complete twist of the helix. The two strands are known as the 5' (5 prime) and 3' (3 prime) strand, named

according to the number of the carbon atom in the pentose sugar to which the phosphate group is attached in the first nucleotide of the chain. As you will see, these features of the structure of DNA and RNA are crucial to the way the molecules work within cells.

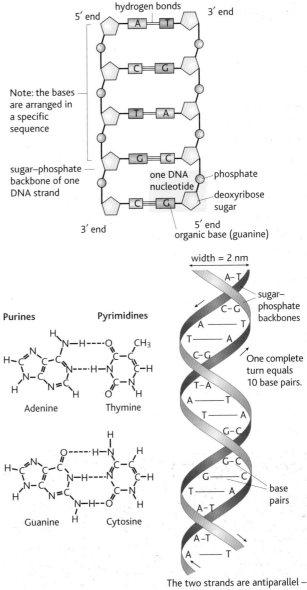

fig. 2.1.4 The detailed double helix structure of DNA depends on the hydrogen bonds which form between the base pairs.

Questions

1 Describe the structure of a mononucleotide.

2 Explain how complementary base pairing and hydrogen bonding are responsible for the double-helix structure of DNA.

The structure of DNA is considered common knowledge, yet the story of how it was worked out in very recent history reads like a combination of detective story and thriller.

The molecules of inheritance

Gregor Mendel (1822–84) introduced the concept of individual inherited 'particles' around 1865, without knowing what these might be. (These 'particles' were later called **genes**.) When scientists saw how the chromosomes duplicated and moved during cell division, chromosomes were quickly accepted as the means of transferring inherited information.

Analysis showed that chromosomes contain both DNA and proteins, but for many years there were arguments between biologists about which carried the genetic information. By 1900 the basic structure of DNA was known, including the fact that there were only four variations in the bases. Many scientists felt this wasn't enough to produce the 20 different amino acids found in human proteins. Proteins were the logical candidates for genes because of their greater variation.

In the late 1940s and early 1950s Oswald Avery, Colin MacLeod and Maclyn McCarty destroyed different types of molecules to see if this affected an inherited change (transformation) in bacteria. If the DNA was destroyed, no transformation took place. No other molecules – not even proteins – had any effect on transformation. Their elegant experiments convinced other scientists that DNA was the material of inheritance. But to understand the genetic code, the structure of DNA itself had to be understood.

Evidence about DNA

A key piece of information was discovered in 1951 by Erwin Chargaff (1905–2002), who analysed DNA from a wide range of species. He found that in every case the proportions of cytosine and guanine were the same. In the same way he found that the proportion of adenine was always the same as that of thymine. But there was no relationship between other combinations of bases.

Linus Pauling (1904–94), a Nobel prize-winning biochemist, showed that polypeptide chains are often held in the shape of an α-helix by hydrogen bonds that could be broken by moderate heating. He suggested a similar helical structure might explain changes that had been seen in DNA on heating.

At King's College London Maurice Wilkins (1916–2004) and Rosalind Franklin (1920–58) were working on the **X-ray diffraction** (also known as X-ray crystallography; see chapter 1.3) of DNA. It was very hard to get pure crystals of DNA to work with – DNA doesn't crystallise easily – and the pictures were so complicated that interpreting them proved very difficult. Wilkins and Franklin were supposed to be working together, but for a number of reasons this didn't happen and an intense rivalry developed.

Franklin developed methods of obtaining the highest quality X-ray diffraction pictures of the DNA molecule (see **fig. 2.1.5**) which allowed her to make accurate measurements. From these she could work out the positions of the groups in the DNA molecule relative to each other and the distances between them. This in turn allowed her to build up a two-stranded helical model for the structure of DNA. A summary of her findings – but not the crucial details – were shared with other scientists, including Watson and Crick, at a relatively informal meeting.

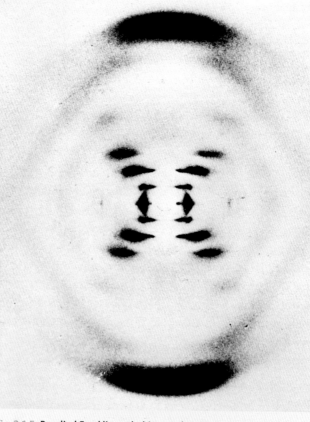

fig. 2.1.5 Rosalind Franklin took this complex picture and worked out a model of the large three-dimensional molecule that produced it.

Deducing the structure

Meanwhile at Cambridge James Watson (1928–) and Francis Crick (1916–2004) were trying a different approach. They gathered all the available information about DNA and kept trying to build a model that fitted with all the facts. They worked with space-filling models and also with simpler representations of the known components of DNA. Any model they produced had to explain all the available data about the structure of the molecule and how it behaved. By a process of assimilating information from other researchers, long discussions and hours of manipulating the models, an idea emerged which seemed to work. Without her knowledge or permission, Maurice Wilkins passed them detailed measurements which Rosalind Franklin had obtained. This gave them the final evidence they needed to confirm their ideas without ever actually carrying out an experiment!

What finally took shape was the now famous double helix. The patterns from the X-ray crystallography suggested a helix measuring 3.4 nm for every complete turn. The idea of a double or parallel helix emerged – but how was the structure maintained? Watson noticed that if in every case cytosine was paired with guanine, and thymine with adenine, hydrogen bonds would hold them together. The two sets of base pairs (cytosine/guanine and thymine/adenine) are roughly the same size and they fit within the measured dimensions of the molecule. Two purines would be too large to fit and two pyrimidines too small to be held together by hydrogen bonds.

The realisation that the bases are always paired in this way was a major breakthrough in understanding the structure of the DNA molecule. The base pairs occupied 0.34 nm, meaning that ten of them would neatly make up one complete twist of the helix (3.4 nm) as measured by Franklin at King's College London. This model explained Chargaff's results too. And best of all, the two complementary chains of DNA could 'open up' along the line of hydrogen bonds between the base pairs. They could then replicate to produce two identical double helices, thus explaining the vital role of DNA in reproduction.

fig. 2.1.6 (a) Frances Crick and James Watson, who along with Maurice Wilkins were awarded a Nobel prize for this masterly piece of molecular detection. (b) Rosalind Franklin had died before the prize was awarded.

Questions

1   How did Chargaff's data and Pauling's ideas support Watson and Crick's deductions on the structure of DNA?

2   Explain how the communication of science is important to developing new theories, in relation to the identification of the structure of DNA.

The double helix and how it works

How DNA makes copies of itself

The model of DNA you are probably most familiar with is something like the image in **fig. 2.1.1**. But it isn't easy to understand from this how DNA works as the means of inheritance. Looking at simple models makes it clearer how the DNA molecule works (see **fig. 2.1.7**). One of the most important features of the DNA molecule – and the one that allows it to pass on genetic information from one cell or generation to another – is that it can **replicate**, or copy itself, exactly.

Uncovering the mechanism of replication

Although we know now how DNA is replicated, it took some years for this mechanism to be established. Several years after Watson and Crick had produced their double helix model for the structure of the DNA molecule, there were two main ideas about how replication happens. One was known as conservative replication (see **fig. 2.1.8**).

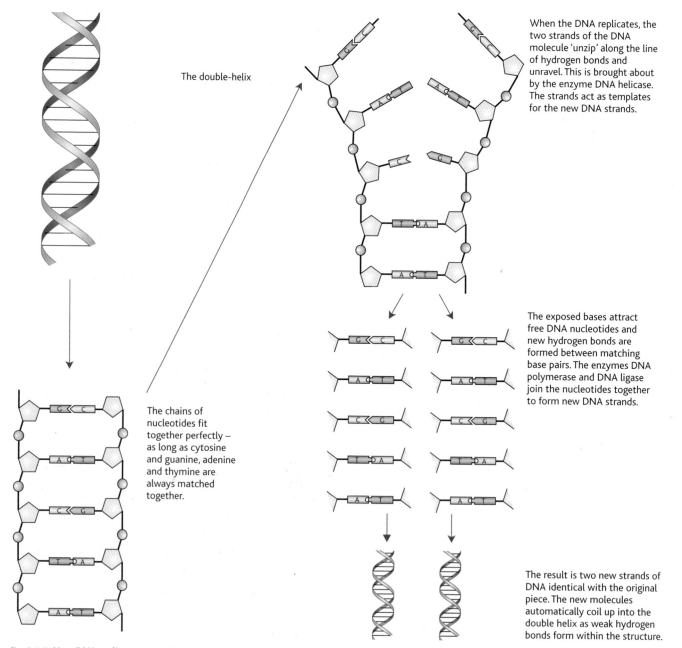

The double-helix

When the DNA replicates, the two strands of the DNA molecule 'unzip' along the line of hydrogen bonds and unravel. This is brought about by the enzyme DNA helicase. The strands act as templates for the new DNA strands.

The chains of nucleotides fit together perfectly – as long as cytosine and guanine, adenine and thymine are always matched together.

The exposed bases attract free DNA nucleotides and new hydrogen bonds are formed between matching base pairs. The enzymes DNA polymerase and DNA ligase join the nucleotides together to form new DNA strands.

The result is two new strands of DNA identical with the original piece. The new molecules automatically coil up into the double helix as weak hydrogen bonds form within the structure.

fig. 2.1.7 How DNA replicates in semiconservative replication

In this the original double helix remained intact and in some way instructed the formation of a new, identical double helix all made up of new material. The other was known as semiconservative replication. This assumed that the DNA 'unzipped' and new nucleotides aligned along each strand, so each new double helix contained one strand of the original DNA and one strand made up of new material.

As the result of a very elegant set of experiments by Matthew Meselson and Franklin Stahl, semiconservative replication became the accepted model.

Conservative replication, where the double helix remains intact and new strands form on the outside, would give:

heavy DNA

light DNA   heavy DNA

Replicates in medium containing only light nitrogen.

Half of the DNA molecules have 2 light strands and half have 2 heavy strands.

Semiconservative replication, where the double helix unzips and each strand replicates to produce a second, new strand would give:

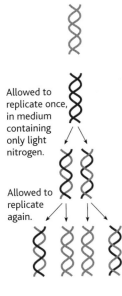

control with only light DNA

light DNA

Allowed to replicate once, in medium containing only light nitrogen.

Sample 1 with only heavy DNA

heavy DNA

Sample 2 first generation

hybrid DNA

Allowed to replicate again.

All of the DNA has one heavy strand and one light strand (hybrid).

Sample 3 second generation

light DNA   hybrid DNA

Half of the DNA molecules have light DNA and half are hybrid with one light and one heavy strand.

fig. 2.1.8 After Meselson and Stahl produced their evidence, support for the idea of conservative replication melted away.

HSW DNA replication: the experiments of Meselson and Stahl

The Watson and Crick hypothesis was that the DNA double helix would unzip, allowing semiconservative replication to take place. In the late 1950s Matthew Meselson (1930–) and Franklin Stahl (1929–) at the California Institute of Technology in the US performed a classic series of experiments which confirmed this hypothesis.

1 They grew several generations of the gut bacteria *E. coli* in a medium where their only source of nitrogen was the radioactive isotope ^{15}N from $^{15}NH_4Cl$. Atoms of ^{15}N are denser than those of the isotope usually found, ^{14}N. The bacteria grown on this medium took up the radioactive isotope to make the cell chemicals, including proteins and DNA. After several generations all the bacterial DNA was labelled with ^{15}N.

2 The bacteria were then moved to a medium containing normal $^{14}NH_4Cl$ as their only nitrogen source, and the density of their DNA was tested as they reproduced.

3 Meselson and Stahl predicted that if DNA reproduces by conservative replication, some of the DNA would have the density expected if it contained nothing but ^{15}N (the original strands), and some of it would have the density expected if it contained nothing but ^{14}N (the new strands). However, if DNA reproduces by semiconservative replication, then all of the DNA would have the same density, half-way between that of ^{15}N- and ^{14}N-containing DNA.

The DNA was all found to have the same density – and so DNA must replicate semiconservatively.

Questions

1 Explain how Meselson and Stahl's classic experiment supported the theory of semiconservative replication and refuted the idea of conservative replication.

2 Explain clearly how DNA is replicated.

How does DNA act as the genetic code?

Research has revealed the structure of DNA and the way in which it replicates itself. But the code needs to be translated into the parts of a living cell.

Proteins are vital components of almost all parts of a cell, including the enzymes that control the synthesis and biochemistry of everything in the cell and the larger organism. By controlling protein synthesis the DNA instructions control not only how the cell is built, but also how it works.

Proteins are made up of amino acids. There are only 20 naturally occurring amino acids that combine to make proteins, but joined together in countless combinations they make up an almost infinite variety of proteins. The amino acids are joined together to build proteins using the code from the DNA. This process of **translation** happens on the surface of the cell organelles known as **ribosomes**.

The genetic code

In a double helix of DNA, the components that vary along the structure are the bases. So scientists guessed that it was the arrangement of the bases that carries the genetic code – but how? There are only four bases, so if one base coded for one amino acid there could be only four amino acids, and we know this is not the case. Even two bases do not give enough amino acids – the possible arrangements of four bases into groups of two is $4 \times 4 = 16$. A **triplet code** of three bases gives $4 \times 4 \times 4 = 64$ possible combinations – more than enough for the 20 amino acids that are coded for.

Cracking the code

By the early 1960s it had been proved that a triplet code of bases was the cornerstone of the genetic code. Each sequence of three bases along a strand of DNA codes for something very specific. Most code for a particular amino acid, but some triplets instead signal the beginning or the end of one particular amino acid sequence.

A sequence of three base pairs on the DNA or RNA is known as a **codon**. The codons of the DNA are difficult to work out because the molecule is so large, so most of the work in this field was done on the codons of the smaller molecule mRNA. This mRNA is formed as a **complementary strand** to the DNA, so it is like a mirror image of the original base sequence. Once the RNA sequence is known, the DNA sequence is simplicity itself to deduce from the way bases always pair: T/U always with A, and G always with C. Sequencing tasks like this have become much easier in the twenty-first century as computer technology has advanced.

The result of all this work was a sort of dictionary of the genetic code, as shown in **table 2.1.1**. Much of the original work, done in the 1960s, used the gut bacteria *E. coli* – but all subsequent studies suggest that the code is identical throughout the living world. The genetic code is based on genes, and a **gene** is usually defined as a sequence of bases on a DNA molecule coding for a sequence of amino acids in a polypeptide chain.

DNA triplet	Amino acid	DNA triplet	Amino acid	DNA triplet	Amino acid	DNA triplet	Amino acid
AAA	Phe	GAA	Leu	TAA	Ile	CAA	Val
AAG	Phe	GAG	Leu	TAG	Ile	CAG	Val
AAT	Leu	GAT	Leu	TAT	Ile	CAT	Val
AAC	Leu	GAC	Leu	TAC	Met/start*	CAC	Val
AGA	Ser	GGA	Pro	TGA	Thr	CGA	Ala
AGG	Ser	GGG	Pro	TGG	Thr	CGG	Ala
AGT	Ser	GGT	Pro	TGT	Thr	CGT	Ala
AGC	Ser	GGC	Pro	TGC	Thr	CGC	Ala
ATA	Tyr	GTA	His	TTA	Asn	CTA	Asp
ATG	Tyr	GTG	His	TTG	Asn	CTG	Asp
ATT	stop	GTT	Gln	TTT	Lys	CTT	Glu
ATC	stop	GTC	Gln	TTC	Lys	CTC	Glu
ACA	Cys	GCA	Arg	TCA	Ser	CCA	Gly
ACG	Cys	GCG	Arg	TCG	Ser	CCG	Gly
ACT	stop	GCT	Arg	TCT	Arg	CCT	Gly
ACC	Trp	GCC	Arg	TCC	Arg	CCC	Gly

table 2.1.1 The triplet code which underpins all work on genetics. (*TAC codes for 'start' at the beginning of a gene, or methionine if within the gene.)

But a problem remains. The DNA is contained within the chromosomes in the nucleus of the cell. The ribosomes where proteins are synthesised are found in the cytoplasm and, as no nuclear DNA has ever been detected there, the message cannot be carried direct. The messages are relayed from the nuclear DNA to the active synthetic enzymes on the ribosomes by the **ribonucleic acids** (**RNAs**).

Protein synthesis

RNA is closely related to DNA but it does not form enormous and complex molecules like DNA. The sequence of bases along a strand of RNA is related to the sequence of bases on a small part of the DNA in the nucleus. Different types of RNA play different roles in the process of protein synthesis.

Messenger RNA (**mRNA**) carries information from the DNA in the nucleus out into the cytoplasm. It is formed when a small length of the DNA double helix unzips. The coding or **antisense strand** of the DNA acts as a template for the formation of the mRNA. The mRNA then moves out of the nucleus, transporting the instructions from the genes to the surface of the ribosomes which are the site of protein synthesis (see fig. 2.1.9).

Transfer RNA (**tRNA**) is found in the cytoplasm. It picks up particular amino acids from the vast numbers always free there. The tRNA molecules, each carrying an amino acid, line up alongside the mRNA on the surface of the ribosome, building up a long chain of amino acids. Peptide links are formed between the amino acids, joining them together to form a polypeptide chain which in turn can be used to form a larger protein.

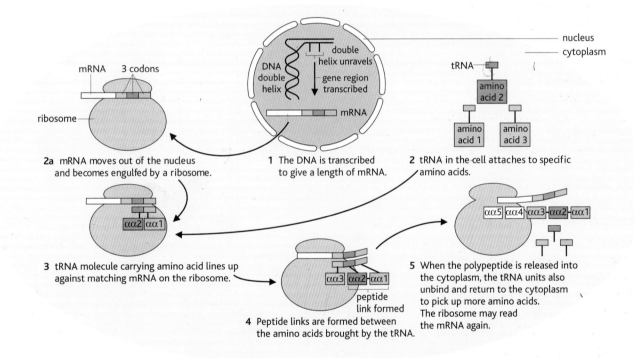

fig. 2.1.9 In protein synthesis the information held in the sequence of bases in a gene is translated into a sequence of amino acids in a polypeptide chain. This diagram is highly simplified. The mRNA strand and the amino acid chain will be hundreds or even thousands of units long.

Questions

1 In many organisms the DNA is contained within the nucleus of the cells. The proteins for which it codes are needed in the cytoplasm. Explain carefully the roles of the following in translating the genetic code into an active enzyme in the cytoplasm of a cell:

a DNA   b messenger RNA   c transfer RNA.

Mutation

Changes in the codons

The genetic code carried on the DNA is translated into living cellular material during protein synthesis. The nucleic acids are central to the process, as both the carriers and the translators of the genetic code. If a single codon is changed or misread during the process, then the amino acid for which it codes may be different and so the whole polypeptide chain and indeed the final protein may be altered. A change like this is known as **mutation**. Such a tiny alteration at this molecular level may have no noticeable effect at all – but equally it can have devastating effects on the whole organism. Many of the genetic diseases you are going to look at in the following chapters are the result of these random mutations in the genetic material of the **gametes** (sex cells). Examples include some human genetic diseases such as **thalassaemia**, in which the blood proteins are not manufactured correctly, or **cystic fibrosis**, in which a membrane protein does not function properly.

Mutations are changes in the arrangement of bases in an individual gene or in the structure of the chromosome, which change the arrangement of the genes on the chromosome. These changes can happen when gametes (sex cells) are formed, although they also occur during the division of **somatic** (body) **cells**. The chance of a mutation taking place during DNA replication is around $2–30 \times 10^{-7}$ per base, although estimates vary widely as it is very difficult to measure. Fortunately the body also has its own DNA repair systems. Specific enzymes cut out or repair any parts of the DNA strands that become broken or damaged. However, some mutations remain and are copied from the DNA when new proteins are made.

Some mutations are caused by the miscopying of just one or a small number of nucleotides. These are **point mutations**. If you think of the amino acids produced from each codon as the equivalent of the letters of the alphabet, the result of a point mutation is like changing a letter in one word. It may well still make an acceptable word, but the meaning will very probably be different. If the change is in a protein which plays an important role in many different types of cell, the effect can be catastrophic. There are a few cases where several different arrangements of nucleotides code for the same amino acid. In that case, a point mutation may have no effect.

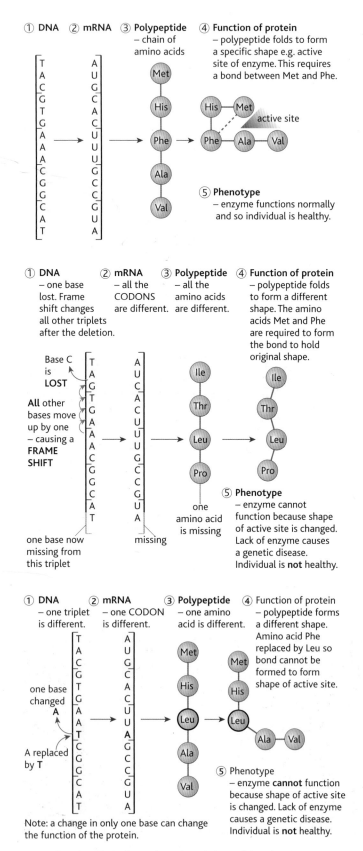

fig. 2.1.10 The effect of a single nucleotide being deleted or changed on the structure and function of the protein formed.

Chromosomal mutations involve changes in the positions of genes within the chromosomes. This is like rearranging the words within a sentence – if you are lucky they still make sense, but it will not mean the same as the original sentence.

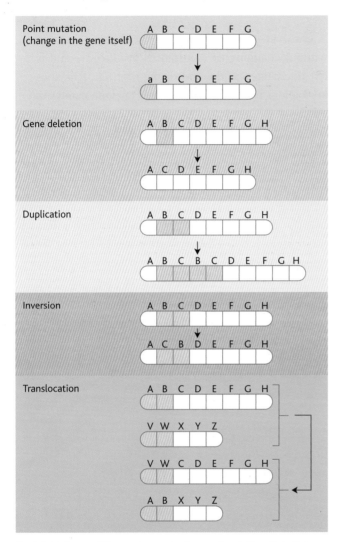

fig. 2.1.11 **Some of the most common types of mutation. Note the scale of genes to chromosomes is very much exaggerated.**

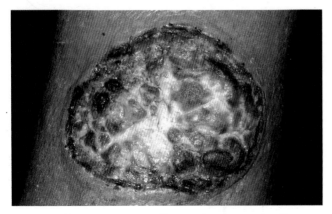

fig. 2.1.12 **Skin cancers are caused by mutations in somatic cells.**

Finally there are **whole-chromosome mutations**, where an entire chromosome is either lost during meiosis (cell division to form the sex cells) or duplicated in one cell by errors in the process. This is like the loss or repetition of a whole sentence. For example, Down's syndrome is caused by a whole-chromosome mutation at chromosome 21 – affected individuals have three copies of this chromosome instead of the usual two.

The effect of mutations on the organism

Mutations are a source of variation within an organism. Very occasionally a mutation occurs that results in the production of a new and superior protein. This may help the organism gain a reproductive advantage so that it leaves more offspring than other individuals of that species (you will learn more about this in chapter 4.3). Most mutations are neutral, meaning that they neither improve nor worsen the chances of survival. Some mutations cause great damage, disrupting the biochemistry of the entire organism.

Mutations can happen to any cell at any time, though they occur most commonly when the DNA is copied for cell division. Mutations in the cells of the body can cause serious problems such as cancer (fig. 2.1.12). But the most damaging mutations are the ones that occur in the gametes because they will be passed on to future offspring. These are the mutations that give rise to genetic diseases.

Exposure to **mutagens**, such as X-rays, ionising radiation and certain chemicals, increases the rate at which mutations occur. For this reason exposure to these mutagens is best kept to a minimum.

Questions

1 Some point mutations will have as big an impact on the way the body works as any chromosomal or whole-chromosome mutation. Others have no effect at all on the organism. Explain this.

2 Explain why, in terms of risk and benefit, doctors still use X-rays for diagnosis.

Enzymes

Under the conditions of temperature and pH found in living cells, most of the reactions that provide cells with energy and produce new biological material would take place very slowly – too slowly for life to exist. But within your cells is a group of molecules that make life possible by speeding up the chemical reactions without changing the conditions in the cytoplasm. These molecules are the enzymes, which are one group of proteins produced during protein synthesis.

What is an enzyme?

Within any cell many chemical reactions are going on at the same time. Those reactions that build up new chemicals are known as **anabolic reactions** ('ana' means up, as in 'build up'). Those that break substances down are **catabolic reactions** ('cata' means down). The combination of these two processes results in the complex array of biochemistry which we refer to as **metabolism**. Most of the reactions of metabolism occur not as single events but as part of a sequence of reactions known as a **metabolic chain** or **pathway**.

A **catalyst** is a substance that speeds up a reaction without changing the substances produced – and is left unchanged at the end of the reaction. Enzymes are biological catalysts, with each enzyme catalysing only a specific reaction or group of reactions. We say enzymes show great **specificity**. Enzymes are proteins. They have a very specific shape as a result of their tertiary and quaternary structures, and this means they can only catalyse specific reactions. Changes in temperature and pH affect the efficiency of an enzyme because they affect the intramolecular bonds within the protein which are responsible for the shape of the molecule. Each cell contains several hundred different enzymes to control the multitude of reactions going on inside.

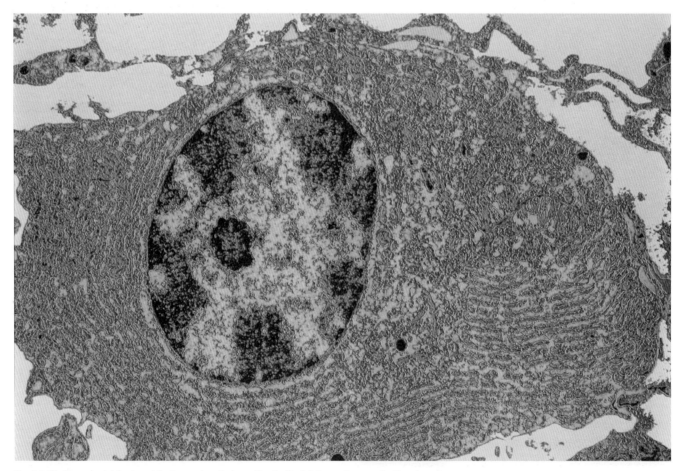

fig. 2.1.13 It can be tricky to get just one chemical reaction in the lab to work properly. Now imagine 100 or more happening in the tiny volume of a single cell! They are all under the control of enzymes, translating the instructions of the DNA in the nucleus into the running of the cell.

HSW A short history of enzyme discovery

In modern biology and medicine, knowledge of enzymes is very much taken for granted. But this knowledge has developed over a long period of time.

- In 1835 it was noticed that the breakdown of starch to sugars was brought about more effectively by malt (barley which has been allowed to sprout) than by sulfuric acid. This suggested that there was a substance present in the living malt that was more effective than the inorganic acid.

- For a long time it was suspected that there was a biological catalyst in yeast which brought about the fermentation of sugar to alcohol, but nobody could prove it. Initially called 'ferments', it was in 1877 that the name enzyme (literally 'in yeast') was introduced to describe these chemicals.

- In 1897 Eduard Buchner (1860–1917) extracted from yeast cells the enzyme responsible for fermenting sugar, and showed it could work outside the living cell structure.

- The first pure, crystalline enzyme was produced from jack beans in 1926 by James B. Sumner (1887–1955). It was the enzyme urease, which catalyses the breakdown of urea. Sumner showed that the crystals were protein

and concluded that enzymes must be proteins. Unfortunately no one believed him because Sumner was a young researcher claiming a great breakthrough! Senior scientists had tried for years to isolate a pure enzyme and concluded that enzymes contained no protein. Eventually Sumner was given the credit he deserved in the form of a Nobel prize in 1946.

- Animal hides used to be softened before being tanned to make leather, and one method was particularly revolting as it used a mixture of dog faeces and pigeon droppings. At the end of the nineteenth century the German chemist Otto Röhm (1876–1939) discovered that it was proteases (enzymes that break down protein) in the dog faeces that had the desired effect on the leather. By 1905 he worked out a method to extract proteases from the pancreas of cows and pigs to supply the enzymes needed for softening hides.

- In 1930–36 the protein nature of enzymes was finally firmly established when the protein-digesting enzymes pepsin, trypsin and chymotrypsin were extracted from the gut and crystallised in their pure form outside the body by John H. Northrop (1891–1987).

Naming enzymes

Many of the enzymes found in animals and plants work inside the cells. These are known as **intracellular enzymes**, eg DNA polymerase, DNA ligase. Other enzymes are secreted by cells to have an effect beyond the boundaries of the cell membrane. These are **extracellular enzymes**. The digestive enzymes and lysozyme, the enzyme in your tears, are well-known examples of these.

Most enzymes – both intracellular and extracellular – have several names:

- a relatively short recommended name, which is often the name of the molecule that the enzyme works on (the **substrate**) with '-ase' on the end, or the substrate with an indication of what it does, eg creatine kinase

- a longer systematic name describing the type of reaction being catalysed, eg ATP:creatine phosphotransferase

- a classification number, eg EC 2.7.3.2.

Some enzymes, such as urease, ribonuclease and lipase, are known by their recommended names. But there are still some enzymes that have distinctly uninformative names – like trypsin and pepsin for example. However, the names of most enzymes will give you useful information about what the enzyme does.

Questions

1 From which organisms were the first enzymes isolated?

2 What is the difference between an intracellular enzyme and an extracellular enzyme?

3 Investigate Sumner's work and discover which scientists were particularly against his ideas and why.

What does an enzyme do?

For a chemical reaction to take place, the reacting molecules must have enough energy to make or break bonds. To get started, it is as if the reaction has to get over an 'energy hill', known as the **activation energy**.

Raising the temperature increases the rate of a chemical reaction by giving more molecules sufficient energy to react. However, living cells could not survive the temperatures needed to make many cellular reactions fast enough – and the energy demands to produce the heat would be enormous. Enzymes solve the problem by lowering the activation energy needed for a reaction to take place (fig. 2.1.15).

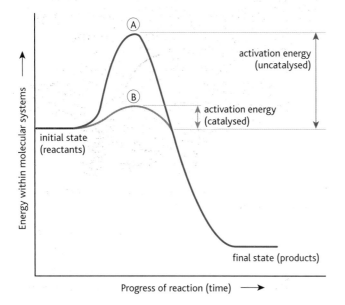

Ⓐ = Energy of transition state in uncatalysed reaction.
Ⓑ = Energy of transition state, ie enzyme/substrate complex, during catalysed reaction.

fig. 2.1.14 **This energy diagram shows the difference between a catalysed and an uncatalysed reaction**

How do enzymes work?

To lower the activation energy and catalyse a reaction, enzymes form a complex with the substrate or substrates of the reaction. A simple picture of enzyme action in a catabolic reaction is:

substrate + enzyme ⇌ enzyme/substrate complex ⇌ enzyme + products

Once the products of the reaction are formed they are released and the enzyme is free to form a new complex with more substrate.

How does this relate to the structure of the enzyme? The **'lock-and-key hypothesis'** gives us a model that helps us understand what happens (see fig. 2.1.15). Within the protein structure of each enzyme is an area known as the **active site** which has a very specific shape. Only one substrate or type of substrate will fit the shape of the gap, and it is this that gives each enzyme its specificity. Just as a key fits into a lock, so the enzyme and substrate slot together to form a complex. The formation of this complex lowers the activation energy of the reaction. In an anabolic reaction, the reacting substances are brought close together, making it easier for bonds to form between them. In catabolic reactions the active site affects the bonds in the substrate, making it easier for them to break. Once the reaction has been catalysed, the products are no longer the right shape to stay in the active site and the complex breaks up, releasing the products and freeing the enzyme for further catalytic action.

The lock-and-key hypothesis fits most of our evidence about enzyme characteristics (see below). However, it is now thought that the lock-and-key hypothesis is a simplification. Evidence from X-ray crystallography, chemical analysis of active sites and other techniques suggests that the active site of enzymes is not the rigid

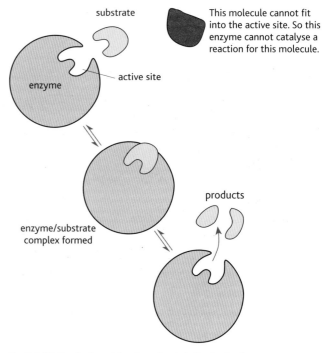

fig. 2.1.15 **The lock-and-key hypothesis is the basis of our understanding of enzyme action.**

shape that was once supposed. In the **induced-fit hypothesis**, generally accepted as the best current model, the active site is still thought of as having a very distinctive shape and arrangement, but a rather more flexible one. Once the substrate enters the active site, the shape of that site is modified around it to form the active complex. Once the products have left the complex the enzyme reverts to its inactive, relaxed form until another substrate molecule binds (see **fig. 2.1.16**).

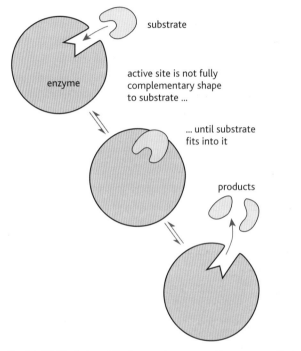

fig. 2.1.16 The induced-fit theory of enzyme action proposes that the catalytic groups of the active site are not brought into their most effective positions until a substrate molecule is bound to the site, inducing a change in shape.

How is the structure of an enzyme related to its functions?

What is the evidence for the structure of enzymes and the relationship between structure and function?

1   All enzymes are **globular proteins** (see chapter 1.3) which contain an active site that is vital to the functioning of the enzyme. The active site is a small depression on the surface of the molecule that has a specific shape because of the way the whole large molecule is folded. Anything affecting the shape of the protein molecule affects its ability to do its job, which indicates that the three-dimensional nature

of the molecule is important to the way it works. A change in shape changes the shape of the active site as well – and so the enzyme can no longer function.

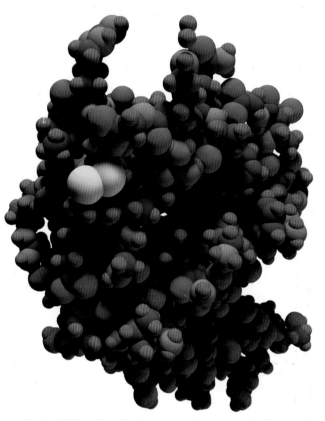

fig. 2.1.17 The active site of an enzyme depends on the three-dimensional structure of the molecule

2   Enzymes change only the rate of a reaction. They do not alter the end products that are formed, or affect the equilibrium of the reaction. They act purely as catalysts and not as modifying influences in any other way.

3   Enzymes are present in only very small amounts. They speed up reactions to such an extent that only minute amounts of them are needed to catalyse the reaction of many substrate molecules into products. This is described by the **molecular activity** or **turnover number**, which measures the number of substrate molecules transformed per minute by a single enzyme molecule. The number of molecules of hydrogen peroxide catalysed by the enzyme catalase extracted from liver cells is 6×10^6 in 1 min! A normal enzyme reaction would catalyse thousands of molecules per minute rather than millions – catalase is the fastest known enzyme.

Measuring the effect of an enzyme

A practical way of measuring the effect of an enzyme on the reaction rate is to measure the rate of the reaction without a catalyst. Using this method it has been shown that when urea breakdown is catalysed by **urease** extracted from the jack bean, the rate of the reaction is increased by a factor of 10^{14}. The efficiency of enzyme catalysis is such that reaction rates are generally increased by factors from 10^8 to 10^{26}. No wonder only tiny amounts are needed!

Measuring the initial rate of reaction

When investigating the way the substrate concentration affects the rate of reaction, biologists measure the initial rate of reaction. Every other factor – temperature, pressure, pH and the concentration of enzyme and of product – must be kept the same so that any changes are clearly seen to be the result of changing the substrate concentration. Measuring the initial rate only, right at the start of the reaction, means these other factors have not had time to change and influence the rate. You can investigate each of the other factors independently by measuring the initial rate.

4   Enzymes are very specific to the reaction that they catalyse. Inorganic catalysts such as platinum or iron filings are frequently used to catalyse a wide range of industrial reactions, although often only at extremes of temperature and pressure. In comparison, some enzymes are so specific that they will catalyse only one particular reaction. Others are specific to a particular group of molecules which are all of similar shape, or to a type of reaction which always involves the same groups. This suggests that there is a physical site within the enzyme with a particular shape into which a specific substrate will fit.

5   Enzymes are affected by the number of substrate molecules present (the concentration of the substrate). Take a simple reaction where substrate A is converted to product Z. If the concentration of A is gradually increased, the rate of the enzyme-catalysed reaction A ⇨ Z will increase – but only for so long. Then the enzyme becomes **saturated** – all of the active sites are occupied by substrate molecules – and a further increase in substrate concentration will not increase the rate of the reaction further (see **fig. 2.1.18**). Once all the

available active sites are involved in the reaction, further increases in substrate concentration will not increase the rate any further.

6   The rate of an enzyme-catalysed reaction is affected by temperature in a characteristic way. The effect of temperature on the rate of any reaction can be expressed as the **temperature coefficient, Q_{10}**. This is expressed as:

$$Q_{10} = \frac{\text{rate of reaction at } (x + 10)\ {}^{\circ}\text{C}}{\text{rate of reaction at } x\ {}^{\circ}\text{C}}$$

Between about 0 °C and 40 °C Q_{10} for any reaction is 2. In other words, in that temperature range every 10 °C rise in temperature produces a doubling of the rate of reaction. Outside this range, however, Q_{10} for enzyme-catalysed reactions decreases markedly whilst Q_{10} for other reactions changes only slowly. The rate of enzyme-catalysed reactions falls as the temperature rises, and at about 60 °C the reaction has stopped completely in most cases.

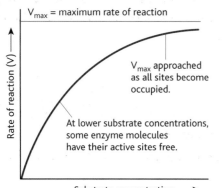

fig. 2.1.18 The effect of substrate concentration on an enzyme-catalysed reaction, showing how the enzyme becomes saturated with substrate.

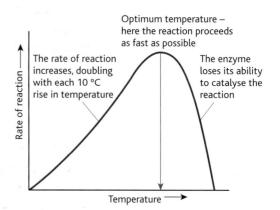

fig. 2.1.19 The effect of temperature on the rate of a typical enzyme-catalysed reaction. All other factors must be kept constant.

At temperatures over 40 °C most proteins, including enzymes, start to lose their tertiary and quaternary structures – they unravel or **denature**. When enzymes denature, they lose their ability to catalyse reactions. There are some exceptions to this rule. For example, the enzymes of **thermophilic** (heat-loving) bacteria, found in hot springs at temperatures of up to 85 °C, clearly function at very high temperatures. They are made of most unusual temperature-resistant proteins.

7 The pH has a major effect on enzyme activity by affecting the shape of protein molecules. Different enzymes work in different ranges of pH, because of the particular arrangement of weak bonds which hold their shape together (see **fig. 2.1.20**). Changes in pH affect the formation of the hydrogen bonds and sulfur bridges that hold together the three-dimensional structure of the protein.

Interestingly, the optimum pH for an enzyme is not always the same as the pH of its normal surroundings. It is thought that this could be one way in which cells control the effects of their intracellular enzymes, increasing or decreasing their activity by minute changes in the pH level.

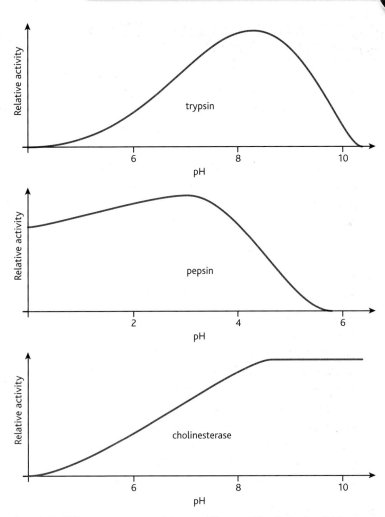

fig. 2.1.20 **Different enzymes work best at different pH levels. Again, all other factors must be kept constant.**

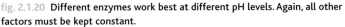

HSW Enzymes in medicine

Increasingly enzymes are being put to use in medical fields, particularly in the area of diagnostic tests. Glucose oxidase is the enzyme found on the 'dip-sticks' used for testing urine to see if it contains glucose. The enzyme is mixed with peroxidase and a blue dye. If the urine contains no glucose then the dye stays blue. However, if there is glucose present, the enzyme breaks it up releasing hydrogen peroxide, which in turn reacts with the other chemicals present to turn the dye from blue to green to brown depending on the glucose levels. Enzyme-linked immunosorbent assays (ELISAs) are also widely used to detect antibodies to particular infections.

Questions

1 a Summarise the characteristics of enzymes.

   b Explain how each characteristic of enzymes provides evidence for the induced-fit hypothesis.

2 Plan a practical investigation into the effect of temperature on enzyme activity.

2.2 The basics of genetics

In chapter 2.1 you saw how the genetic code is translated through the process of protein synthesis into the enzymes that control all the reactions of life. In this chapter you will be looking at the way this genetic information is inherited, and how you can predict the possible offspring of a known genetic cross.

The process of transcription and translation is extremely complex and sometimes things go wrong. These mistakes, known as **mutations**, can have a serious effect on the resulting organism and you will be looking at the ways in which they can be passed from one generation to another. You will be looking first at the general principles of genetics, and then in more detail at how they apply to humans. In the twenty-first century, biotechnology and medicine are increasingly offering us methods of overcoming problems that arise from genetic mutations. However, these advances raise many ethical issues, some of which you will consider at the end of this chapter.

The basis of inheritance

The physical and chemical characteristics that make up the appearance of an organism are known as its **phenotype** – for example, the shape of a cabbage, the colour of a flower or the shape of a nose. This phenotype is partly the result of the genetic information (the **genotype**) passed from parents to their offspring, and partly the effect of the environment in which the organism lives. For example, the shape of a cabbage will depend on levels of soil nutrients and sunlight as well as on the genetic make-up of the individual plant. Differences in the genotype between different organisms of a species are due in part to the shuffling of genes that occurs during the process of meiosis (see chapter 3.2), and in part to the inheritance of genes from two different individuals in sexual reproduction.

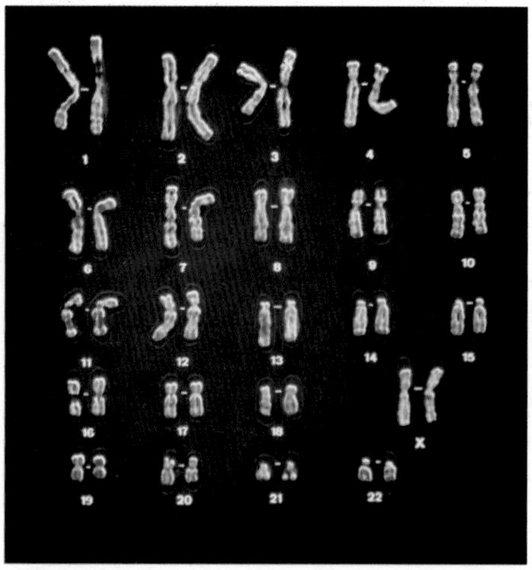

The cells of any individual organism contain a particular number of chromosomes that is characteristic of the species. So, for example, humans have 46 chromosomes. Half the chromosomes are inherited from the female parent and the other half come from the male. The two sets can be arranged as matching **homologous pairs** in a **karyotype** (see fig. 2.2.1). This karyotype is from a woman as you can see by the presence of the XX sex chromosomes.

fig. 2.2.1 The 23 homologous pairs of human chromosomes from a woman displayed in a karyotype.

Along each chromosome are thousands of **genes**, each gene being a different segment of DNA coding for a particular protein or polypeptide. The chromosomes in a homologous pair carry the same genes (the only exception being the sex chromosomes). The gene for a particular characteristic is always found in the same position or **locus**, which means that you usually carry two genes for each characteristic.

There may be slightly different versions of a gene, called **alleles**, that produce variations of the characteristic. For example, at the locus for the gene for the height of a pea plant the allele may code for a tall plant or for a dwarf plant (see **fig. 2.2.2**). And since the pea plant has two homologous chromosomes carrying this gene, it may have two alleles for the tall characteristic, two for the dwarf, or one of each. If both of the alleles coding for a particular characteristic are identical, then the organism is **homozygous** for that characteristic – it is a **homozygote** ('homo' means 'the same'). If the two alleles coding for a characteristic are different, the organism is **heterozygous** for that characteristic and is called a **heterozygote** ('hetero' means 'different').

Some alleles are described as **dominant**. This means that their effect is **expressed** or shown whether the individual is homozygous or not. As long as one dominant allele is present, it will be expressed in the phenotype even in the presence of another, **recessive**, allele. Recessive alleles are only expressed in the phenotype when there are two of them, when the organism is homozygous recessive. In genetic diagrams, dominant alleles are usually represented by a capital letter and recessive alleles by the lower-case version of the same letter.

Monohybrid crosses

Homozygotes are referred to as **true breeding**, because if you cross two individuals that are homozygous for the same characteristic, all the offspring of all the generations that follow will show this same characteristic in their phenotype (unless a mutation occurs). Heterozygotes are not true breeding. If two heterozygotes are crossed, then the offspring will include homozygous dominant, homozygous recessive and heterozygous types and at least two different phenotypes.

When one gene is considered at a time in a genetic cross, it is referred to as a **monohybrid cross**. Fig. 2.2.3 shows a cross between a pea plant homozygous for the dominant, round pea seed shape and a pea plant homozygous for the recessive wrinkled pea. The first generation of this cross is called the **F1** (**first filial generation**) and you can see that they all have the same genotype for the characteristic and they are heterozygous. They also all have the same phenotype – round pea shape because the round allele is dominant. There is no sign of the wrinkled pea allele.

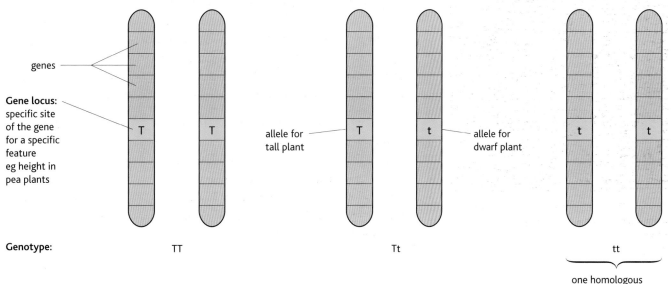

fig. 2.2.2 In sexual reproduction it is the different alleles passed on that determine the genotype and hence eventually the phenotype of the offspring.

If we cross individuals from the F1 generation we call the next generation the **F2** (**second filial generation**). In fig. 2.2.3 you can see that theory predicts the genotypes to be 1 homozygous round : 2 heterozygous round : 1 homozygous wrinkled. You would expect to see three round peas for every wrinkled one. The recessive trait of wrinkled pea has become visible again, after being 'hidden' in the F1 generation.

R = round
r = wrinkled

Parental phenotype	round-seeded	×	wrinkle-seeded
Parental genotype	RR		rr

Gametes

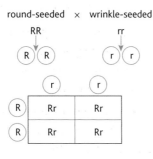

Offspring genotype (F₁ generation)	Rr
Offspring phenotype	round-seeded

If two F₁ offspring are crossed, we get the following results.

F₁ phenotype	round-seeded	×	round-seeded
F₁ genotype	Rr		Rr

Gametes

	R	r
R	RR	Rr
r	Rr	rr

Offspring genotypes (F₂ generation)	RR : 2Rr : rr
Offspring phenotypes	3 round-seeded : 1 wrinkle-seeded

fig. 2.2.3 A cross between a pea plant homozygous for the round pea allele and a plant homozygous for the wrinked pea allele, through the F1 and F2 generations.

HSW Test crosses

As can be seen from fig. 2.2.3, individuals that are homozygous dominant or heterozygous have identical phenotypes. For a plant or animal breeder this can present all sorts of difficulties. A breeder often needs to know that the stock will breed true – in other words, that it is homozygous for the desired feature. If the feature is inherited through a recessive allele, then any plant showing the feature in the phenotype must be homozygous recessive. However, if the required feature is inherited through a dominant allele the physical appearance does not show whether the organism is homo- or heterozygous. To find out which it is, the individual must be crossed with a homozygous recessive individual (see fig. 2.2.4). This type of cross, known as a **test cross**, reveals the parental genotype.

Y = yellow
y = green

If a homozygous yellow parent is crossed:

Parental phenotype	yellow seeds	×	green seeds
Parental genotype	YY		yy

Gametes

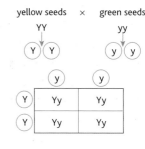

Offspring genotype (F₁ generation)	Yy
Offspring phenotype	yellow seeds

If a heterozygous yellow parent is crossed:

Parental phenotype	yellow seeds	×	green seeds
Parental genotype	Yy		yy

Gametes

	y	y
Y	Yy	Yy
y	yy	yy

Offspring genotypes	Yy : yy
Offspring phenotypes	1 yellow : 1 green

fig. 2.2.4 Test cross to show the genotype for seed colour of a parent pea plant.

Carrying out genetics experiments

Much of our understanding of human genetics comes from experiments on other organisms. A 'suitable' organism for a genetic experiment should, ideally, have the following features:

- be relatively easy and cheap to raise, to maximise the chance of successful breeding and minimise the cost of the experiment – for valid results, genetics investigations need lots of organisms

- have a short life cycle so that the results of crosses and/or mutations can be seen quickly

- produce large numbers of offspring so that the results of any crosses are statistically relevant (see below)

- have clear, easily distinguished characteristics such as tall or dwarf plants, the colour and shape of plant seeds or the length of the wings of an insect.

Organisms commonly used by scientists for genetic investigations include the fruit fly *Drosophila melanogaster*, pea plants, fungi such as *Aspergillus* and a variety of bacteria. Specially bred 'fast plants' which germinate, mature, flower and set seed in a matter of weeks are now being used increasingly, particularly in schools and colleges. Another advantage of these organisms is that there are fewer ethical concerns about using them than experimenting with larger organisms such as mammals.

fig. 2.2.5 **Drosophila are ideal for carrying out genetics experiments.**

HSW Sampling errors in genetic experiments

The theoretical ratios predicted by a genetic cross are usually seen (approximately!) in real genetic experiments, but the numbers are never precise. There are a number of reasons for this:

- some offspring dying before they can be sampled (eg seeds that do not germinate, embryos that miscarry or die at birth, etc)

- inefficient sampling techniques (eg it is very easy to let a few *Drosophila* escape!)

- chance plays a large role in reproduction – the joining of particular gametes (sex cells) is a completely random affair, unlike the theoretical diagrams that we draw.

So **sampling error** must be taken into account when you look at a real genetic cross, and the smaller the sample, the larger the potential sampling error. For example, if a 3 : 1 ratio of phenotypes is expected in the offspring of a cross, it is unlikely to show itself if only four offspring are produced. Looking at 400 offspring increases the likelihood of the expected ratio emerging, and 4000 offspring would be better still.

Questions

1 Define the terms homozygous, heterozygous, dominant and recessive.

2 In pea plants the gene for plant height has two alleles. The allele for 'tall' is dominant to the 'dwarf' allele. Choose suitable letters for the two alleles and draw genetic diagrams to show the following:

   a a cross between two homozygous 'tall' parents

   b a cross between heterozygous parents

   c the F1 and F2 generations of a cross between a homozygous 'tall' parent and homozygous 'dwarf' parent.

3 Explain the importance of a test cross to identify genotype.

4 Why are organisms such as *Drosophila*, *Aspergillus* and bacteria suitable as experimental organisms in genetics?

For centuries the theories developed to explain inheritance assumed that the characteristics of parents in some way blended or fused in the offspring. The birth of Gregor Mendel (1822–84) was to herald the end of these old ideas.

Johann Mendel was born in the Czech Republic into a very poor family. When he was 17 he became a monk at the Augustinian monastery at Brunn where he took the name Gregor. Here he taught maths, physics and statistics as well as botany, which may well explain the rigorous way he later analysed the results of his research.

Mendel was fascinated by the variety in the plants in the monastery gardens and wanted to find out more about them. He decided to investigate peas, because he could see that they had characteristics (their height and the colour and shape of the peas and the pods) that varied in a clear-cut way. What's more, they were easy to grow. Mendel realised that by using pea plants he could control which plants pollinated each other (see fig. 2.2.7). Between 1856 and 1863 Mendel grew and observed around 28 000 pea plants!

fig. 2.2.6 Gregor Mendel, the father of genetics, working with his peas. Pea plants were Mendel's main experimental organism, and by focusing on particular traits he was able to build up a mass of statistics about the various crosses he made.

Mendel was meticulous. He spent the first two years of his research making sure that the plants he was using were true breeding for the characteristics he was investigating. He then tried crossing pure-breeding parents that had two different forms of a chosen characteristic, and found the offspring were all the same. But when he crossed the first-generation offspring together, he found that both characteristics reappeared in predictable ratios. Fig. 2.2.7 and table 2.2.1 show some of his actual results.

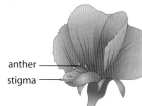

anther
stigma

Both the male and female parts of the pea flower are held within a hood-like petal so that self fertilisation frequently occurs. Mendel opened the bud of one flower before the pollen matured and fertilised the stigma with pollen from another chosen flower. In this way he could control the cross.

Mendel used seven clearly differentiated, pure-breeding traits of the pea plant for his experiments. They are shown here in both their dominant and recessive forms.

	Stems		Pods		Seeds/flowers		
Dominant trait	Tall	Axial flowers	Green	Inflated	Round	Yellow	Red flowers
Recessive trait	Short	Terminal flowers	Yellow	Pinched	Wrinkled	Green	White flowers

fig. 2.2.7 Mendel worked on clear-cut characteristics like these and built up a mass of experimental data.

Experiment	Characteristic	Results at second generation
1	round/wrinkled seed	336 round : 107 wrinkled
2	yellow/green pea	355 yellow : 123 green
3	purple/white flower	705 purple : 224 white
4	fat/shrunken pod	882 fat : 299 shrunken
5	green/yellow unripe pod	428 green : 152 yellow

table 2.2.1 Some of Mendel's results.

In 1865 Mendel presented his results to the Brunn Natural History Society. In 1866 his papers on the subject were published, describing the two fundamental laws of heredity he deduced from them. Sadly, he was a man ahead of his time and his work was poorly received and almost ignored. Remember that no one knew of the existence of chromosomes at this point, let alone genes – so Mendel's ideas were presented to a scientific community with no framework of ideas ready to receive them. Along with this, his statistical work seemed odd to many biologists of the time. Strange as it seems today, the idea of collecting lots of data and analysing them mathematically was very unusual in those days! As a result the full importance and impact of his work was not recognised until 16 years after his death.

By 1900 chromosomes had been discovered and meiosis had been observed and described. Hugo de Vries (1848–1935) in the Netherlands and Karl Correns (1864–1933) in Germany discovered Mendel's work and duplicated his results. To their credit, they gave him the recognition he deserved.

Then in 1902 it was suggested that Mendel's units of inheritance might be found on the chromosomes and in the early 1900s Thomas Morgan (1866–1945) and his team in America worked with *Drosophila* to gain evidence for this idea. From this point the study of genetics developed.

Mendel's practical experimental results showed ratios that coincided remarkably well with the theoretical ones. If you have tried doing genetics experiments yourself, you will realise how impressive this is! It has been suggested that Mendel worked out in advance what he expected to happen and then either he or his helpers 'managed' the actual results to achieve the desired ratios. Even if this was the case, the ideas were correct and the laws which Mendel came up with, although oversimplifying matters, still apply today.

The law of segregation: Mendel's first law

The first law that Mendel presented is known as the **law of segregation**. It was the result of work with monohybrid crosses. The law states that in a diploid organism (an organism with two sets of chromosomes in the nuclei of its somatic cells), one unit or allele for each trait is inherited from each parent to give a total of two alleles for each trait. The segregation (separation) of alleles in each pair takes place when the gametes are formed.

This idea of independent units of inheritance, some dominant over others and which are maintained throughout the life of an individual and do not fuse to form a homogeneous mass, was Mendel's real breakthrough.

The law of independent assortment: Mendel's second law

The **law of independent assortment** states that different traits are inherited independently of each other. This means that the inheritance of a dominant or recessive allele for one characteristic, such as grey or ebony bodies, has nothing to do with the inheritance of alleles for other characteristics such as wing length or eye colour.

Considering that Mendel formulated these laws so early in the history of our understanding of inheritance it is remarkable that they are still largely relevant today, although we now recognise that the second law in particular has many exceptions as a result of **gene linkage** and **polygenic inheritance**.

Questions

1 Summarise the aspects of Mendel's work that suggest that his results stood a good chance of being both accurate and valid.

2 Choose one of Mendel's crosses shown in table 2.2.1, give suitable genetic diagrams to work out the theoretical results and compare them with Mendel's experimental results.

3 If Mendel did 'tweak' his experimental evidence to get the results he predicted, does the end justify the means?

Understanding human genetic traits

There are many mutations that result in human genetic diseases. Scientists are building up an increasingly detailed picture of how the mutations are caused and how the genetic conditions are inherited.

Some human traits are inherited through a single gene. Others are **polygenic**, meaning they are controlled by several interacting genes. This is different from having several possible alleles for the same gene. For example, scientists know that eye colour in humans is controlled by at least three genes, but how they all interact is still being investigated.

Enzyme chains in cells also show polygenic inheritance. It takes many enzymes to release energy in the mitochondria or to make a particular protein in a cell. A large number of genes are involved – at least one for each enzyme in the chain. If mutation of any of these genes results in a missing enzyme, the process will not work.

Genetic diseases cannot be cured yet, although treatments for some of them are improving. Understanding the causes and inheritance of genetic conditions will help scientists find treatments and even cures. You will look at some of these in relation to cystic fibrosis in chapter 2.4. Here, though, we will look at two other examples.

Thalassaemia

Thalassaemia is not a single disorder – there are a range of thalassaemias and they all affect the polypeptide chains of the haemoglobin molecule, which carries oxygen in the blood. Haemoglobin is a huge molecule made up of 574 amino acids arranged in four polypeptide chains held together by sulfur bridges. There are two types of polypeptide chain in haemoglobin, the alpha (α) and beta (β) chains (**fig 1.3.19**). Each chain is arranged around an iron-containing haem group. Haemoglobin shows polygenic inheritance, with a different gene coding for each of the two different chains. Depending on the type of disease, thalassaemia prevents the formation of either the alpha or the beta haemoglobin chains. This means the haemoglobin cannot carry enough oxygen. People affected by the disease suffer a variety of symptoms but they all include the classic symptoms of anaemia – fatigue and lack of energy due to insufficient oxygen in the blood.

Alpha thalassaemia provides an example of polygenic inheritance. It is inherited from genes on two loci on chromosome 16 (a total of four genes, two on each chromosome). The most common cause of alpha thalassaemia is the deletion of one or more of these genes. The more genes that are deleted, the less alpha haemoglobin is made and the worse the symptoms are. If all four genes are deleted the condition is very serious and may be fatal. However, alpha thalassaemias are now the most common blood disorders in the world. This seems to be linked to the fact that people who have the milder forms of alpha

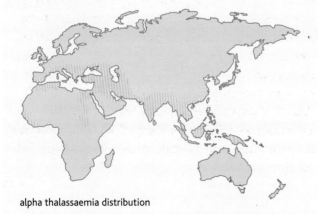

alpha thalassaemia distribution

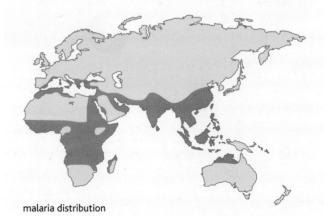

malaria distribution

fig. 2.2.8 The incidence of alpha thalassaemia is closely linked to the incidence of malaria, because the protection provided by the mild forms gives people with alpha thalassaemia a survival advantage

thalassaemia have low levels of anaemia but increased resistance to malaria. Alpha thalassaemia is particularly common in Mediterranean people and in areas of Asia, Africa and South America where malaria is most prevalent (**fig.2.2.8**). Scientists do not fully understand how the milder thalassaemias give protection against malaria – this is an area of active research.

Beta thalassaemia is caused by mutations in the HBB gene on chromosome 11. This gene is not usually deleted, but about 100 different mutations are known. The mutations vary in their effect, so some thalassaemias are more severe than others. Many of the mutations mean the number of beta haemoglobin chains made is reduced, which causes relatively mild symptoms of anaemia. The most severe form of beta thalassaemia is known as **thalassaemia major** or **Cooley's anaemia**. This mutation means the beta chains of the haemoglobin molecule cannot be made at all. Without treatment, beta thalassaemia can be life threatening.

The mutation that causes beta thalassaemia forms a recessive allele. This means that affected children are usually born to carrier parents who both have the beta thalassaemia allele (see **fig. 2.2.9**). These carriers usually appear normal or suffer a very mild form of the disease. They may have some anaemia because with some mutations the alleles show a degree of incomplete dominance – neither allele is completely dominant or recessive and so both affect the phenotype. In some areas – for example Cyprus – as many as one person in seven is a carrier for the disease.

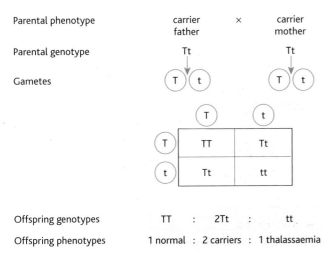

Parental phenotype	carrier father	×	carrier mother
Parental genotype	Tt		Tt
Gametes	T t		T t

	T	t
T	TT	Tt
t	Tt	tt

Offspring genotypes	TT	:	2Tt	:	tt
Offspring phenotypes	1 normal	:	2 carriers	:	1 thalassaemia

fig. 2.2.9 Thalassaemia is a crippling and sometimes fatal disease for people who inherit two recessive alleles

Thalassaemia cannot be cured. It can be treated by regular blood transfusions, but the body builds up an excess of iron, so drugs are needed to deal with this problem too. Treatment is expensive and has to be carried out frequently throughout life. Some countries where thalassaemias are very common have introduced screening programmes to try and reduce the numbers of affected children born. So for example, in Sardinia, the birth of children affected by thalassaemia major has fallen from 1 in 250 live births to 1 in 4000. In Cyprus the number of affected births has fallen to almost zero. This has been achieved largely by prenatal testing of embryos, which you will consider in more detail in chapter 2.4.

Genetic pedigree diagrams

A genetic diagram such as **fig. 2.2.9** shows how a trait can be passed on theoretically. A family tree or **genetic pedigree diagram** (**fig. 2.2.10**) shows what happens in reality. A pedigree diagram includes all the members of a family, indicating their sex and whether or not they have the disease. Because humans are not available as experimental animals, much of our understanding of human genes has come from the analysis of genetic pedigree diagrams. In families affected by conditions such as thalassaemia and cystic fibrosis, genetic pedigree diagrams can be useful in predicting which family members may be carriers of the genetic mutation, allowing them to consider their options before they conceive a child.

The albino trait

Albinism is a condition in many species in which the natural melanin pigment of the skin, eyes and hair does not form. There are several different forms of albinism, and mutations in several different genes can give similar results in the phenotype. One of the most common forms is due to a mutant allele which prevents the formation of a normal enzyme in the cells. The enzyme **tyrosinase**, normally active in the **melanocytes** (pigment-forming cells), is not formed correctly and so the reactions that make melanin cannot take place.

Albinism is inherited through a recessive allele. The parents may appear normal – in which case they are both carriers of the albino allele – or one or more of the parents may be an albino themselves (see **fig. 2.2.10**).

In the general human population, about 1 person in 30 000 is affected by albinism. As well as lacking pigment in the cells of their skin, hair and eyes, their vision is often poor and they are at higher risk of developing skin cancers because they do not have the natural protective pigment melanin. However, albinism is not life threatening and both difficulties can normally be overcome.

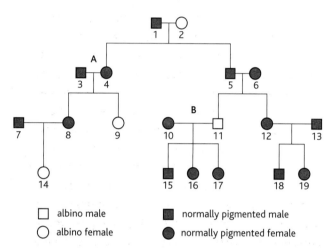

fig. 2.2.10 Albinism is another example of a genetic mutation where a small change in the DNA leads to a different protein being made – in this case an enzyme which no longer works. The genetic pedigree diagram shows you how the affected allele is inherited.

The biochemical pathway involved in albinism shows another example of polygenic inheritance. It involves the breakdown of the amino acid phenylalanine by a series of enzymes. Other genes involved in the pathway are also prone to mutations, with very different results as you can see in fig. 2.2.11.

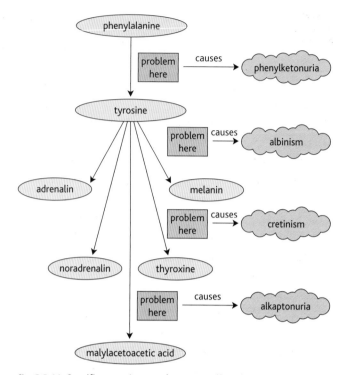

fig. 2.2.11 Specific mutations are known to affect three different enzymes in this biochemical pathway – each with different and possibly devastating results.

The future of genetic diseases

Our understanding of the genetic cause of inherited diseases has enabled some genes to be identified on the human genome. This opens up the possibility of treating these diseases using gene therapy, (discussed in more detail in chapter 2.4). Further opportunities have developed from the **Human Genome Project**, a massive research programme set up in the 1990s to map the entire human genome. In 265 laboratories in six countries, scientists worked towards determining the locations, DNA nucleotide sequences and functions of all the human genes – between 25 000 and 30 000 of them. The basic analysis of the whole genome was completed in April 2003. Improvements in technology enabled the project to be completed about two years ahead of schedule. In 2008, the new international 1000 Genomes Project was set up to develop a more detailed map of the human genome, based on DNA from at least 1000 people around the world. As a result, it should become possible to identify the gene for all single-gene genetic disorders, and provide screening tests for them. It should also clarify the situation in **multifactorial diseases** that result from a number of genes and lifestyle factors. All this knowledge will raise some tough issues for society to consider.

HSW Genetic choices

Knowing where to find a human gene makes it much easier to identify whether a person has it or not, but this knowledge can be a double-edged sword. Scientists have identified the mutation that causes **Huntington's disease**, a genetic condition that doesn't develop until middle age but which is fatal. In a family affected by a severe genetic condition like this, some people might find it unbearable to learn that they have inherited a completely untreatable, fatal disease. Others would want to know and plan for the future. Should the tests be carried out within a family or not?

If you are screened and it is found that you have an allele for a genetic disease, the information could remain confidential, to you. Then it would be up to each individual to decide whether to tell partners, parents and friends, whether to have children and indeed, whether to tell the insurance companies. However, some people would argue that society has a right to know. Do you think you have a right to know if the person with whom you plan to have children is knowingly going to pass on abnormal genes to them? Financial institutions might argue that they have a right to know if they are lending money or offering cheap life insurance to someone who is carrying a genetic time bomb. However, based on these arguments every one of us could be a bad risk – because no one has a set of completely perfect genes.

Some genetic conditions are rare – for example, most people do not carry the dominant allele for Huntington's disease. However, 1 in 25 people carries the gene for cystic fibrosis. Should this affect your choice of partner? This type of information is already becoming available, affecting our life choices, and perhaps we need to decide how to deal with it now rather than later.

The Human Genome Project, along with the 1000 Genomes Project, also offers the promise of a medical revolution, with drugs tailored to fit and interact with the way your genes work, so that your tendency to develop diseases such as some cancers is simply never turned on.

In another perturbing twist to the story, commercial companies are moving into the genetic market place. And whereas the aim of the international effort was to provide the human genome free to everyone who needs it, biotech companies and universities are using genes to earn money. A paper published in the journal *Science* in 2005 suggested that around 20% of the human genes had been patented for different uses, such as adding them to a particular virus to make genetically modified organisms. It costs a lot to develop new gene technology. The companies want to protect their developments so that they can make some money from their investment. People who are against this idea fear that any drugs produced in future using these genes will be unrealistically expensive, limiting their use to richer nations or individuals. So we now have to ask not only how society should use the information of the human genome, but also should the fundamental details of our DNA be for sale?

Questions

1 How is thalassaemia inherited and what are the symptoms in the individuals affected.

2 How can family trees be used to help understand the way in which a human genetic trait is inherited?

3 From the genetic pedigree in **fig. 2.2.10** show the possible genotypes of the couples labelled A and B and the possible offspring they could have produced. How many of these phenotypes actually appeared in their children?

4 Discuss the difficulties of investigating human genetics and comment on some of the methods used.

5 Now that the whole of the human genome has been worked out, it will be possible in future to analyse the genome of every newborn baby and keep the information on a database. Discuss both the advantages and the possible ethical objections to this plan.

6 Explain why we cannot use humans as experimental animals, and therefore why pedigree diagrams are so useful in understanding the inheritance of human characteristics.

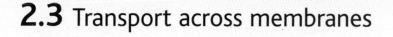

2.3 Transport across membranes

Membranes in cells

In this chapter you are going to look at some of the ways in which cells and organisms work, before moving on in chapter 2.4 to consider in detail the disruptive effect of a mutation in a single gene. Every process of life in a whole organism depends on reactions happening in its cells. Many of these basic cellular processes are affected by the **cell membrane**, so it is important to understand the best available model of this important structure.

There are many membranes within cells, such as those that surround **organelles** like the nucleus and mitochondria. But the most obvious membrane is the cell surface membrane (outer cell membrane) which forms the boundary of the cell – anything that leaves or enters the cell must pass through it. All membranes act as barriers, controlling what passes through them and allowing the fluids either side of them to have different compositions. This makes it possible to have the right conditions for a particular reaction in one part of a cell and different conditions to suit other reactions elsewhere in the cell.

fig. 2.3.1 Cells are the basis of all life. The cell membrane controls much of what goes on in the cell.

Membranes perform many other functions too. Many chemical processes take place on membrane surfaces, such as some reactions of respiration which happen on the mitochondrial membrane. Enzymes and any other factors needed to make the reactions go are held closely together so that the process can proceed smoothly from one reaction to the next. The cell surface membrane must also be flexible to allow the cell to change shape very slightly as its water content changes, or quite dramatically, eg when a white blood cell engulfs a bacterium (see fig. 2.3.1). Chemical secretions made by the cell are packaged into membrane bags known as **vesicles**, so some membranes must be capable of breaking and fusing together readily.

The structure of membranes

Our current model of the structure of membranes has been worked out over many years (see the box on page 103). The model developed as microscopy improved, from light to electron and then scanning electron microscopes. In time there may well be further refinements to the model presented here, but the overall picture seems unlikely to change dramatically. The membrane is made up mainly of two types of molecules – **lipids** and **proteins** – arranged in a very specific way.

The phospholipid bilayer

The lipids in the membrane are of a particular type called **polar lipids**. These are lipid molecules with one end joined to a polar group. Many of the polar lipids in the membrane are **phospholipids**, with a phosphate group forming the polar part of the molecule (fig. 2.3.2). The fatty acid chains of a phospholipid are neutral and insoluble in water. In contrast the phosphate head carries a negative charge and is soluble in water.

When these phospholipids come into contact with water the two parts of the molecule behave differently. The polar phosphate part is **hydrophilic** (water-loving) and dissolves readily in water. The lipid tails are **hydrophobic** (water-hating) and insoluble in water. If the molecules are tightly packed in water they either form either a **monolayer**, with the hydrophilic heads in the water and the hydrophobic lipid tails in the air, or clusters called **micelles**. In a micelle all the hydrophilic heads point outwards and all the hydrophobic tails are hidden inside (fig. 2.3.2).

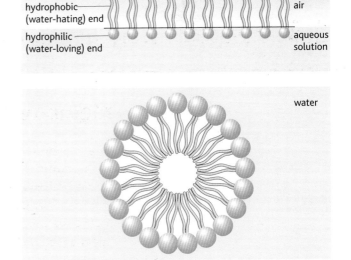

fig. 2.3.2 Phospholipids form a monolayer at an air/water interface. Sometimes they will form a micelle when submerged in water.

A monolayer may form at a surface between air and water, but this is a fairly rare situation in living cells where there are water-based solutions on either side of membranes. With water on each side the phospholipid molecules form a **bilayer** with the hydrophilic heads pointing into the water while the hydrophobic tails are protected in the middle (see fig. 2.3.3). This structure, the **unit membrane**, is the basis of all membranes. However, a simple lipid bilayer alone would not explain either the microscopic appearance of membranes or the way in which they behave.

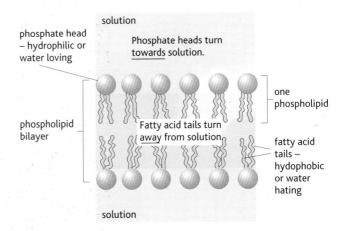

fig. 2.3.3 This bilayer structure is the 'backbone' of the cell membrane – the unit membrane.

A simple lipid bilayer allows fat-soluble organic molecules to pass through it, but many vital chemicals needed in cells are ionic. Whilst these dissolve in water they cannot dissolve in or pass through lipids, even polar lipids. They can enter cells because the membrane consists not only of lipids but also of proteins and other molecules too.

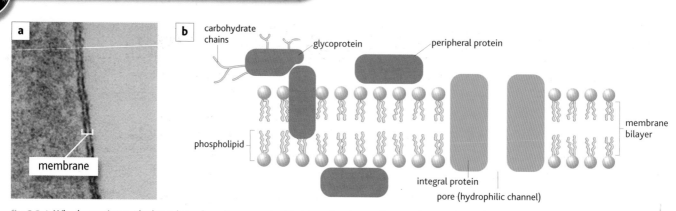

fig. 2.3.4 Whether acting as the boundary of a cell or as part of its internal make-up, the complex structure of the membrane is closely linked to its wide variety of functions. (a) Is a TEM of a membrane showing the protein layers as the two dark lines.

The membrane proteins

The best model of a membrane we have today sees the bilayer of lipid as a fluid system, with many proteins and other molecules floating within it like icebergs whilst others are fixed in place (see fig. 2.3.4). The proportion of phospholipids containing unsaturated fatty acids (see chapter 1.3) in the bilayer seems to affect how freely the moving proteins float about in the membrane. Many of the proteins have a hydrophobic part which is buried in the lipid bilayer and a hydrophilic part which can be involved in a variety of activities. Some proteins penetrate all the way through the lipid, while others span only part of the bilayer.

One of the main functions of the membrane proteins is to help substances move across the membrane. The proteins may form pores or channels – some permanent, some temporary – which allow specific molecules to move through the pores. Some of these channels can be open or shut, depending on conditions in the cell. These are known as gated channels. Some of the protein pores are active carrier systems using energy to move molecules, as you will see later. Others are simply gaps in the lipid bilayer which allow ionic substances to move through the membrane in both directions.

Proteins may act as specific receptor molecules – for example, making cells sensitive to a particular hormone. They may be enzymes, particularly on the internal cell membranes, to control reactions linked to that membrane. Some membrane proteins are **glycoproteins**, proteins with a carbohydrate part added to the molecule. These are very important on the surface of cells as part of the way cells recognise each other.

This model of the floating proteins in a lipid sea is known as the **fluid mosaic model** and was first proposed by S. Jonathan Singer and Garth Nicholson in 1972.

Investigating membrane permeability

Membrane structure can be investigated experimentally in a school laboratory by looking at the effect of alcohol concentration or temperature on the permeability of the membrane.

HSW Where does our picture of the membrane come from?

Our ideas about membrane structures have developed over many years from scientific observations.

- The first indications that lipids are important components of cell membranes came at the end of the nineteenth century when Charles Ernest Overton (1865–1933) made a series of observations on how easily substances passed through cell membranes. Because lipid-soluble substances entered more easily than any others, he concluded that a large part of the membrane structure must be lipid.

- Observations on the behaviour of cell surface membranes when cells join together, and the way in which most membranes seal themselves if they are punctured with a fine needle, led to the idea that cell membranes are not rigid but are fluid structures.

- In 1917 Irving Langmuir (1881–1957) demonstrated the lipid monolayer mentioned earlier, and developed a piece of equipment for collecting lipid monolayers known as the Langmuir–Blodgett trough.

- In 1925 two Dutch scientists, Evert Gorter and F. Grendel, set out to measure the total size of the monolayer film formed by lipids extracted from human red blood cells (erythrocytes). They also estimated the total surface area of an erythrocyte, and found that their measured area of monolayer was about twice the estimated surface area of the cell. As a result they concluded that the cell membrane was a lipid bilayer. We now know that their results were wrong on two counts – they didn't extract all the lipid, and they miscalculated the surface area of the erythrocytes because they thought they were flat rather than biconcave discs. In spite of this their conclusions were correct – by lucky coincidence the two errors cancelled each other out!

- By 1935 Hugh Davson and James Danielli had produced a further model of the membrane with a lipid centre coated on each side by protein. This is broadly the basis of our current ideas.

- The Davson–Danielli hypothesis was backed up in the 1950s by work on the electron microscope by James D. Robertson. He found ways of staining the membrane which showed it up as a three-layered structure – two distinct lines with a gap in the middle. When the membrane was treated with propanone (acetone) to extract the lipid, the two lines remained intact, suggesting that they were the protein layers.

More recently techniques such as X-ray diffraction and new electron microscopy methods have added to our knowledge of the structure of cell membranes, giving more detail of the layers, the pores and the carrier molecules.

fig. 2.3.5 The nuclear pores through which mRNA leaves the nucleus are clearly visible in this freeze-etched electron micrograph of the nuclear membrane.

Questions

1   Which kinds of molecules make up the structure of a membrane and how do their properties affect the properties of the membrane itself?

2   Explain why a membrane may be more fluid when it contains more unsaturated fatty acids. (You may need to refer back to chapter 1.3.)

3   Choose three different pieces of evidence that have been important in building up our models of the cell membrane. For each one explain how the evidence moved the model forward towards the fluid mosaic model.

How the membrane works

We know from scientific analysis that the concentration of substances either side of a membrane can be very different. This suggests that a membrane exercises control over the passage of substances across it. Some substances, particularly those that dissolve very easily in lipids, simply pass through the membrane in a process of **diffusion** as though the membrane was not there. Other very small molecules, such as the gases oxygen and carbon dioxide as well as water itself, also pass freely in and out of cells through the membrane.

Diffusion

In physical terms, diffusion is the movement of the molecules of a liquid or gas from an area where they are highly concentrated to an area where they are at a lower concentration. We say that they move down their **concentration gradient**. This occurs because of the random motion of molecules due to the energy they have, which is dependent on the temperature. If you have a large number of molecules tightly packed together, random motion will result in their spreading out and eventually reaching a uniform distribution. The molecules don't stop moving once they reach a uniform distribution; however, the movement no longer causes a net change in concentration because equal numbers are moving in all directions.

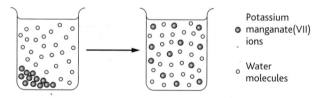

Potassium manganate(VII) ions

Water molecules

fig. 2.3.6 **Observing the process of diffusion. If the beaker is left to stand, the random motion of both the water and the purple manganate(VII) ions will ensure that they are eventually evenly mixed.**

For many small molecules, like oxygen and carbon dioxide, the membrane is no barrier and they can diffuse freely across it. Other molecules move by diffusion through protein-lined pores. This movement, by diffusion alone, is a form of **passive transport.** However, larger hydrophilic molecules and ions larger than carbon dioxide molecules cannot move across the membrane by simple diffusion.

Facilitated diffusion

Substances with a strong positive or negative charge and large molecules cannot cross cell membranes by simple diffusion. Nevertheless they may move into and out of the cell down a concentration gradient by a specialised form of diffusion. **Facilitated diffusion** involves proteins in the membrane which allow only specific substances to move through passively down their concentration gradient (see fig. 2.3.7). They may simply be **channel proteins** which form pores through the membrane. Each type of channel protein allows one particular type of molecule through, dependent on its shape – eg some are sodium ion channels and others form potassium ion channels. Some channels open only when a specific molecule is present or there is an electrical change across the membrane, such as during the passage of nervous impulses along neurones.

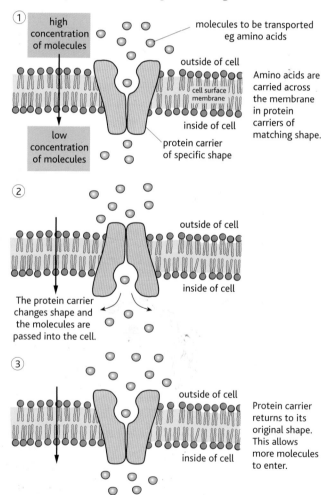

fig. 2.3.7 **Facilitated diffusion acts as a ferry across the lipid membrane sea. But this is a boat with no oars, sail or engine – it can only work when the tide (the concentration gradient) is in the right direction.**

Another form of facilitated diffusion depends on carrier molecules floating on the surface of the membrane. The carriers will be found on the *outside* surface of the membrane structure when a substance is to be moved *into* the cell or organelle, and on the *inside* for transport *out* of the cell or organelle. The protein carriers are specific for particular molecules or groups of molecules, depending on the shape of the protein carrier and the substance to be carried. Once a carrier has picked up a molecule it rotates through the membrane to the other side, carrying the molecule with it, and then releases the molecule. The movement through the membrane takes place because the carrier changes shape once it is actually carrying something. The process can only take place down a concentration gradient – from a high concentration of a molecule to a low one. It doesn't use energy, so is considered a form of diffusion. For example, red blood cells have a carrier to help glucose move into the cells rapidly.

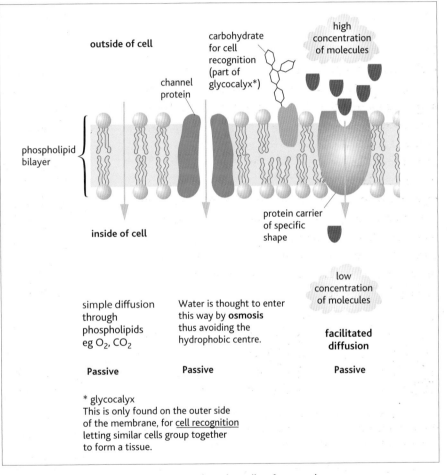

fig. 2.3.8 **Three of the main transport routes through a cell surface membrane.**

Questions

1   Describe the conditions needed for molecules to enter a cell passively.

2   Water and ions tend to enter the cell through protein pores. Why can't they pass through the lipid layer like oxygen and carbon dioxide do?

3   Explain the differences between simple diffusion and facilitated diffusion.

Osmosis – a special case of diffusion

As you have seen, diffusion takes place where molecules or ions can move freely. Free water molecules – water molecules which are not involved in hydrogen or any other type of bonding – can move easily through membranes. So, as a result of random motion, water molecules will tend to move across a membrane down their concentration gradient. The diffusion of solvents, including water, is given its own term: **osmosis**. Osmosis can be defined as the net movement of solvent molecules from a region where they are at a high concentration to a region where they are at a lower concentration through a partially permeable membrane. In living organisms the solvent is always water, and membranes in cells are generally partially permeable, in that they let some molecules through but not others.

So osmosis in cells involves the movement of water from a region of high concentration of water molecules (in other words, a dilute solution of the solute) to a region of lower concentration of water molecules (a more concentrated solution of the solute) across a partially permeable membrane such as the cell surface membrane or nuclear membrane. It may help you to think of osmosis in terms of water molecules moving from the side of a membrane where there is a more dilute solution to the side where there is a more concentrated solution.

If the solution bathing the outside of a cell has a lower concentration of dissolved substances (**solutes**) than the solution inside the cell, there will be a concentration gradient that encourages water molecules to move into the cell. If the opposite is true and the solution bathing the cell has a higher concentration of solutes

than the cell contents, water will move out of the cell. The **osmotic concentration** of a solution concerns only those solutes that have an osmotic effect. Many large insoluble molecules found in the cytoplasm don't affect the movement of water and so are ignored when considering osmotic concentration. Only soluble particles are considered.

Modelling diffusion and osmosis in cells

You can make a model cell using an artificial membrane that is permeable to some molecules – in particular water – and impermeable to others, such as sucrose. There are many experiments showing the movement of water in these circumstances, and one of the simplest is illustrated here.

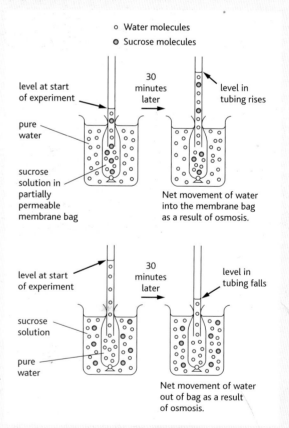

fig. 2.3.10 The artificial partially permeable membrane in this experiment allows water to pass through freely but not the solute molecules.

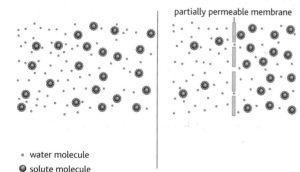

fig. 2.3.9 In diffusion the random movement of particles results in an even distribution of both solute and solvent particles. A partially permeable membrane allows only the solvent molecules and small solute molecules to move freely in osmosis.

In the context of living cells the movement of water by osmosis, and the control of this process, are very important. In animal cells in particular it is vital that water does not simply move continuously into the cells from a dilute external solution, because the end result would be that the cells swell up and burst.

Red blood cells

hypotonic solution

Red onion cells

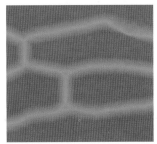

hypotonic solution

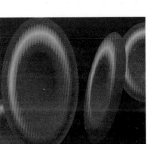

isotonic solution

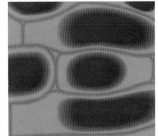

isotonic solution – incipient plasmolysis

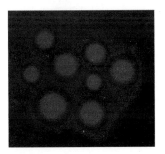

hypertonic solution

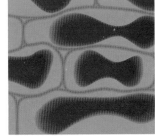

hypertonic solution – cells fully plasmolysed

fig. 2.3.11 The effects of osmosis on animal and plant cells in different solutions. These situations don't usually occur in living organisms because they have systems to control water balance and prevent damage to cells.

Describing concentrations

During osmotic experiments cells are often immersed in solutions of varying concentrations. An **isotonic** (sometimes also called **isosmotic**) solution has the same osmotic concentration as the cell. Isotonic sports drinks are designed to keep your body fluids at the same osmotic concentration as your cell contents

even when you are exercising hard. If the surrounding solution is **hypotonic**, the osmotic concentration of solutes is lower than that in the cytoplasm of the cells. If the solution surrounding your cells is **hypertonic**, then the osmotic concentration of solutes is higher than that in the cytoplasm.

Osmosis in living cells

Osmosis needs to be carefully controlled in animal cells, where the net movement of water in or out needs to be kept to a minimum. When too much water moves in the cells burst; too little and the cells shrivel as the concentrated cytoplasm loses its internal structure and the chemical reactions that normally take place in the cell stop working.

In plant cells the cellulose cell wall prevents cells bursting. If the surrounding fluid is hypotonic to the cytoplasm of a plant cell, water will enter the cell by osmosis – but not indefinitely. As the cytoplasm swells, the inward pressure of the cell wall on the cytoplasm increases until it cancels out the tendency for water molecules to move in. At this point the plant cell is rigid, in a state known as **turgor**.

The movement of water in the right direction is vital for the healthy functioning of cells. This movement depends largely on the concentrations of various solutes both in the cytoplasm and in the surrounding fluids. So cells control the movement of water by selecting which solute molecules move into and out of the cell. Many molecules move passively by diffusion, but cells have another way of selecting which molecules cross the membranes, as you will see next.

Questions

1 Explain the meaning of the term osmosis?

2 In an experiment, human cheek cells were placed in three solutions: an isotonic solution, a hypertonic solution and a hypotonic solution. Describe what you would expect to happen in each case. Explain your answers in terms of osmosis.

Active transport

Diffusion and facilitated diffusion are passive processes that allow small molecules to move across membranes. Cells can maintain steep concentration gradients by simply 'mopping up' the substance as soon as it arrives inside the cell. They can do this by immediately starting to **metabolise** the substance – by chemically changing it to something else – or by using a carrier molecule on the surface of an organelle to take it into the organelle.

Both diffusion and facilitated diffusion rely on a concentration gradient in the right direction to move a substance into the cell. However, cells have another system to move substances across membranes, which uses energy supplied by the cell. This system is called **active transport** and it makes it possible for cells to move substances against their concentration gradient.

How does active transport work?

Active transport involves a **carrier protein** which often spans the whole membrane (see **fig. 2.3.4**). It may be very specific, picking up only one type of ion or molecule, or it may work for several relatively similar substances which have to compete with each other for a place on the carrier.

The energy needed for active transport is provided by molecules of **adenosine triphosphate** (**ATP**). The ATP is produced during cellular respiration, so cells that carry out a lot of active transport generally have a lot of mitochondria too to carry out respiration and supply the ATP. The active transport carrier system in the membrane involves the enzyme **ATPase**. This enzyme catalyses the breakdown of ATP, removing a phosphate group to form **adenosine diphosphate** (**ADP**). Energy is released by the breaking of this bond, and this energy may be used to move the carrier system in the membrane or to release the transported substances and return the system to normal.

Active transport is a one-way system – the carriers will not transport substances back through the membrane. An active transport system moves substances only in the direction required by the cell. In some cases they will move out again through open channels, down the concentration gradient that has just been overcome, but active transport can move substances in faster than they can move out by diffusion.

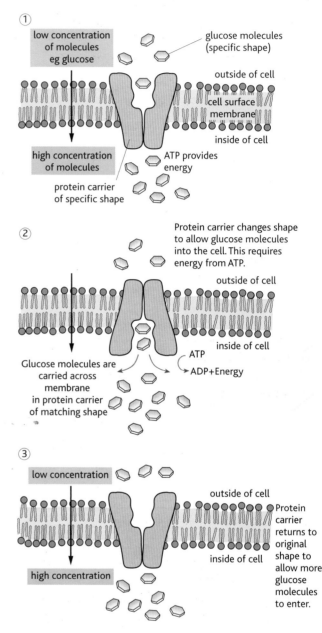

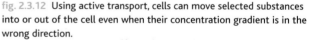

fig. 2.3.12 Using active transport, cells can move selected substances into or out of the cell even when their concentration gradient is in the wrong direction.

In active transport the movement of a substance is often linked with that of another particle, such as a sodium ion. One of the best known examples of active transport is the 'sodium pump' that actively moves potassium ions into the cell and sodium ions out. This pump is vital for the working of the nervous system – each nerve impulse depends on an influx of sodium ions through the axon membrane. These ions have to be actively pumped out of the neurone again afterwards so that another impulse can pass.

HSW Evidence for active transport

Active transport requires energy in the form of ATP produced during cellular respiration. Much of the evidence for active transport comes from linking these two processes together, showing that without ATP active transport cannot take place.

1 Active transport takes place only in living, respiring cells.

2 The rate of active transport depends on temperature and oxygen concentration. These affect the rate of respiration and so the rate of production of ATP.

3 Many cells that are known to carry out a lot of active transport contain very large numbers of mitochondria – the site of aerobic cellular respiration and ATP production.

4 Poisons which stop respiration or prevent ATPase from working also stop active transport. For example, **cyanide** prevents the synthesis of ATP during cellular respiration. It also stops active transport. However, if ATP is added artificially, active transport starts again.

The combination of diffusion, facilitated diffusion and active transport means that the cell surface membrane provides control over what moves into or out of the cell. The concentration of ions and molecules within the cell can be maintained at very different levels from those of the external fluids. In a similar way the membranes inside the cell, around the organelles and in the cytoplasm, provide a range of microenvironments within the cell itself, each suited to different functions, such as the protein-packaging systems in the Golgi body (see chapter 3.1).

Endocytosis and exocytosis

Diffusion and active transport allow the movement of small particles across membranes. However, there are times when larger particles need to enter or leave a cell, eg when white blood cells ingest bacteria or gland cells secrete hormones. Membrane transport systems cannot do this job, but the membrane has properties that make it possible to move larger particles into or out of the cell.

Materials can be surrounded by and taken up into membrane-lined vesicles in a process known as **endocytosis** (see fig. 2.3.13). This can occur at a relatively large scale, eg during bacteria ingestion, called **phagocytosis** (cell eating). It also happens at a microscopic level, when tiny amounts of the surrounding fluid are taken into minute vacuoles. This is known as **pinocytosis** (cell drinking). Electron microscope studies have shown that pinocytosis is very common as cells take in the extracellular fluid as a source of minerals and nutrients.

Exocytosis is the term for the emptying of a membrane-lined vesicle at the surface of the cell or elsewhere (fig. 2.3.13). For example, in cells producing hormones, vesicles containing the hormone fuse with the cell surface membrane to release their contents. These processes are made possible by the fluid mosaic nature of the membrane.

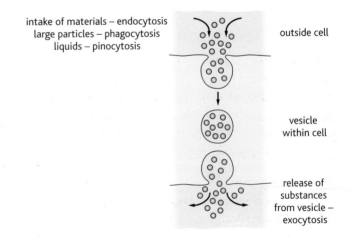

fig. 2.3.13 The properties of the membrane allow cells to take in large particles or release secretions.

Questions

1 Explain the importance of active transport in cells.

2 Explain how endocytosis and exocytosis provide evidence for the fluid mosaic model of membranes.

Diffusion and gaseous exchange

Many exchanges of substances by cells, such as the exchange of gases and other substances between the blood system and cells, occur by diffusion. In chapter 1.1 you saw how the decreasing surface area to volume ratio makes it increasingly difficult for diffusion to supply materials fast enough to cells as the size of the organism increases, and how this explains why larger organisms have circulatory systems. It also explains why they have a **respiratory system** to exchange gases with the environment. Examples of respiratory systems are the lungs in mammals and gills in fish. You are going to look at the adaptations of the human respiratory system to maximise the rate of diffusion, but first you need to understand its structure.

The human respiratory system

Most of the human respiratory system is found within the chest. It is linked with the outside world through the mouth or nose (see fig. 2.3.14). The nasal passages have a relatively large surface area, but no gaseous exchange takes place here. The passages have a good blood supply, and the lining secretes mucus and is covered in hairs. This means that the external air is prepared before entering the rest of the system. The hairs and mucus filter out and 'clean up' much of the dust and small particles such as bacteria that you breathe in. The moist surfaces increase the level of water vapour in the air and the rich blood supply raises the temperature of the air if this is necessary. This means that the air entering the lungs has as little effect as possible on the internal environment.

Factors affecting the rate of diffusion

The human respiratory system is specialised for the exchange of oxygen and carbon dioxide between the body and the environment. These gases are exchanged by simple diffusion. The rate of diffusion across a membrane is controlled by a number of factors:

- the surface area – the bigger the surface area the more particles can be exchanged at the same time

- the concentration gradient of the particles diffusing – the more there are on one side of a membrane compared with the other, the faster they move across, so maintaining the gradient (eg by transporting substances away once they have diffused) and making the diffusion faster

- the distance over which diffusion is taking place – the shorter the diffusion distance the faster diffusion can take place.

You can use this information to calculate the rate at which substances of a given size will diffuse at a known temperature. This relationship is known as **Fick's law**:

$$\text{rate of diffusion} \propto \frac{\text{surface area} \times \text{concentration difference}}{\text{thickness of exchange membrane or barrier}}$$

$\propto$ = proportional to

You will look at how Fick's law applies to the human respiratory system next.

Questions

1 Explain carefully why humans need a complex internal respiratory system.

2 Suggest why breathing through the nose is better for your body than breathing through the mouth.

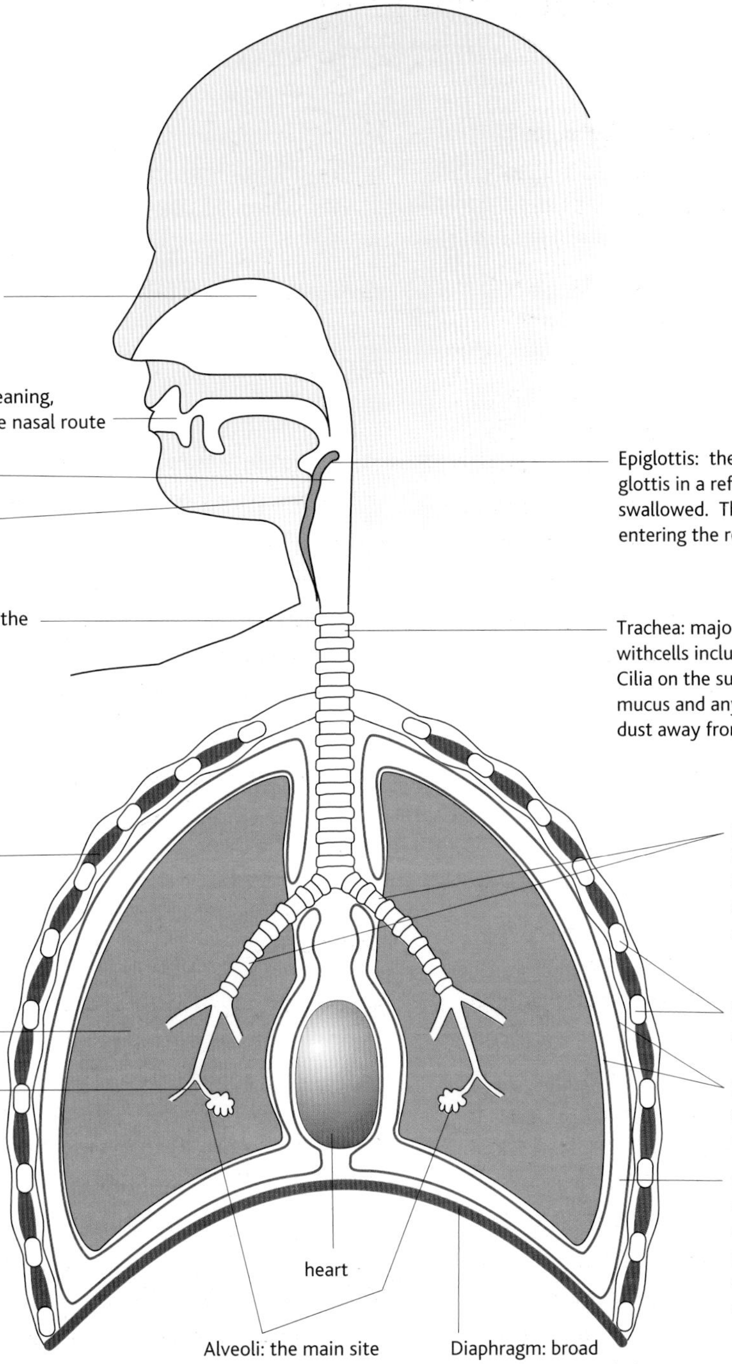

Nasal cavity: the main route by which air enters the respiratory system.

Mouth: air can enter the respiratory system here, but misses out on the cleaning, warming and moistening effects of the nasal route

glottis

Larynx: this uses the flow of air across it to produce sounds.

Incomplete rings of cartilage: prevent the trachea and bronchi from collapsing.

Intercostal muscles: found between the ribs and important in breathing.

lung

Bronchioles: small tubes which spread through the lungs and end in alveoli. The larger tubes have cartilage rings but once the diameter is 1 mm or less, there is no cartilage and they collapse quite easily. Their main function is still as an airway but a little gaseous exchange may occur.

Epiglottis: the epiglottis closes over the glottis in a reflex action when food is swallowed. This prevents food from entering the respiratory system.

Trachea: major airway to the bronchi, lined withcells including mucus-secreting cells. Cilia on the surface of the trachea move mucus and any trapped microorganisms and dust away from the lungs.

Left bronchus and right bronchus: tubes that are similar in structure to the trachea but are slightly narrower, which divide to form bronchioles.

Ribs: protective bony cage around the respiratory system.

Pleural membranes: surround the lungs and line the chest cavity.

Pleural cavity: space between the pleural membranes, usually filled with a thin layer of lubricating fluid which allows the membranes to slide easily with breathing movements.

heart

Alveoli: the main site of gaseous exchange in the lungs.

Diaphragm: broad sheet of tissue which forms the floor of the chest cavity, also important in breathing.

fig. 2.3.14 **The human respiratory system.**

Diffusion and the human respiratory system

Gas exchange in the alveoli

In the lungs most of the gaseous exchange occurs in tiny air sacs known as alveoli (single: **alveolus**) (see fig. 2.3.15). An alveolus is made of a single layer of flattened epithelial cells. The capillaries that run close to the alveoli also have a wall which is only one cell thick. Between the two is a layer of elastic connective tissue holding everything together. The elastic tissue helps to force air out of the lungs, which are stretched when you breathe in. This is known as the elastic recoil of the lungs.

Gaseous exchange occurs by a process of simple diffusion between the alveolar air and the deoxygenated blood in the capillaries. This blood has a relatively low oxygen content and a relatively high carbon dioxide content. The alveoli have a natural tendency to collapse, but this is prevented by a special phospholipid known as **lung surfactant** which coats the alveoli and makes breathing easier.

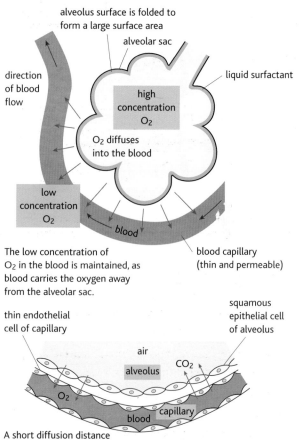

The low concentration of O_2 in the blood is maintained, as blood carries the oxygen away from the alveolar sac.

A short diffusion distance speeds up the rate of diffusion.

fig. 2.3.15 **The alveoli are the main respiratory surfaces of the lungs.**

Understanding Fick's law can help us explain how the adaptations of the human respiratory system optimise the exchange of gases in the lungs by diffusion.

Large surface area

The alveoli provide an enormous surface area for the exchange of gases in the human body. Recent calculations have shown that an average adult human has around 480–500 million alveoli in their lungs, which gives a surface area for gaseous exchange of around 100 m² packed into your chest!

Short diffusion distance

The walls of the alveoli are only one cell thick, as are the walls of the capillaries that run beside them. This means the distance that diffusing gases have to travel between them is only around 0.5–1.5 μm (micrometres, microns, 10^{-6} m).

Steep concentration gradient

Blood is continuously flowing through the capillaries past the alveoli, exchanging gases. The continuous flow of the blood maintains the concentration gradient on the capillary side. The air within the alveoli is constantly being refreshed with air from outside by breathing (see **table 2.3.1**). Movement of gases in and out of the alveoli is mainly by diffusion, but movement of air in and out of the lungs is a mass transport system.

	percentage of gas in:		
	Inspired air	Alveolar air	Expired air
Oxygen	20.70	13.2	14.5
Carbon dioxide	0.04	5.0	3.9
Nitrogen	78.00	75.6	75.4
Water vapour	1.26	6.2	6.2

table 2.3.1 **Composition of gases in breathing.**

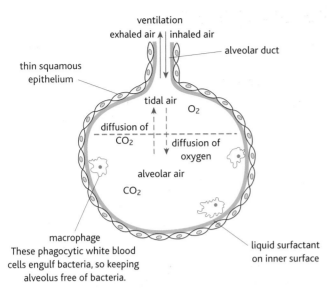

fig. 2.3.16 **Diffusion across the alveolar surfaces provides the blood with oxygen and disposes of carbon dioxide.**

Breathing

Although the exchange of gases at the alveolar surfaces in the lungs happens by passive diffusion alone, moving air between the lungs and the external environment is an active process known as **breathing**. There are two parts to the process of breathing – taking air into the chest (**inhalation**) and breathing air out again (**exhalation**). The chest cavity is effectively a sealed unit for air, with only one way in or out – through the trachea. Breathing involves a series of pressure changes in the chest cavity which in turn bring about movements of the air.

Inhalation is an active, energy-using process. The muscles around the diaphragm contract and as a result it is lowered and flattened. The intercostal muscles between the ribs also contract, raising the rib cage upwards and outwards. These movements result in the volume of the chest cavity increasing, which reduces the pressure in the cavity. The pressure within the chest cavity is now lower than the pressure of the atmospheric air outside, so air moves in through the trachea, bronchi and bronchioles into the lungs to equalise the pressure inside and out.

Normal exhalation is a passive process. The muscles surrounding the diaphragm relax so that it moves up into its resting dome shape. The intercostal muscles also relax so that the ribs move down and in, and the elastic fibres around the alveoli of the lungs return to their normal length. As a result, the volume of the chest cavity decreases, causing an increase in pressure. The

pressure in the chest cavity is now greater than that of the outside air, so air moves out of the lungs, through the bronchioles, bronchi and trachea to the outside air (**fig. 2.3.17**).

If you need to, you can force air out of your lungs more rapidly than passive exhalation allows. The internal intercostal muscles contract, pulling the ribs down and in. The abdominal muscles contract forcing the diaphragm upwards. This increases the pressure in the chest cavity, causing exhalation. This is known as forced exhalation. Singers use this to achieve a powerful voice and to maintain long notes, and free divers do it before a dive so they can fill their lungs with as much air as possible afterwards. Coughing is an exaggerated form of forced exhalation which is used to force mucus out from the respiratory system

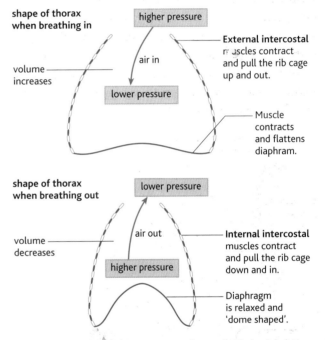

fig. 2.3.17 **You can feel the movements of your ribs during inhalation and exhalation, although the movements of your diaphragm are less obvious.**

Questions

1 Define Fick's law and explain how it applies to the human respiratory system.

2 Explain why breathing is important in maintaining concentration gradients in the alveoli.

3 Compare the last two columns in table 2.3.1. Explain the difference in oxygen and carbon dioxide percentages in expired air compared with alveolar air?

Transport proteins in action

Protecting the lungs

Your respiratory system carries out gaseous exchange. As well as gases, the air you breathe in also carries lots of tiny particles such as dust, pollen grains and smoke particles which could block the tiny alveoli. It also carries microscopic organisms like bacteria and viruses. Some of these microscopic organisms are **pathogens** (disease-causing organisms) and the respiratory system provides a potential route inside the body where they can cause infection.

To reduce the chances of damage happening to your lungs and of infection, your respiratory system produces lots of mucus which lines your airways and traps these tiny particles and organisms. The mucus is usually very runny, so that it is easily moved up the airways by cilia that sweep upwards to the back of your throat (see fig. 2.3.18). Here the majority of the mucus is swallowed without you even noticing it. The acid in your stomach and your digestive enzymes digest the mucus and everything carried with it.

fig. 2.3.18 Cilia beat constantly to move mucus with its load of pathogens and dirt out of your respiratory system.

The role of osmosis

The mucus is kept runny by various transport systems that move water out of the cells of your airways by osmosis. This is a very complex process that is described simply below. The numbers relate to the numbers in fig. 2.3.19.

1 Chloride ions (Cl^-) are actively transported into the epithelial cells that line the respiratory tubes from the tissue fluid surrounding them. This transport system is known as a chloride pump. Chloride pumps are found in many cells of the body, including those lining the digestive and respiratory systems.

2 As a result of this pump, the chloride ion concentration in the epithelial cells lining the airways is high. This creates a concentration gradient between the cell contents and the fluid on the surface of the epithelium inside the airway.

3 Chloride ions diffuse out of the cells into the fluid in the airways. They pass through chloride channels in the membrane that lines the lumen of the airway. The proteins that form these channels are known as the **cystic fibrosis transmembrane regulatory channel proteins** – or more simply **CFTR** channel proteins. The CFTR channel proteins form a gated channel. This allows the free diffusion of chloride ions when they are open – but they are only open in the presence of ATP. Although it takes energy to open the channel, this isn't active transport because the ions move by simple diffusion.

4 The CFTR channel protein also controls the sodium ions moving into the cell. When the CFTR channel is working, it inhibits the movement of sodium ions into the cell through the sodium channels. As a result, the concentrations of sodium and chloride ions in the fluid lining the airways are higher than the concentrations of the solutes in the cytoplasm of the epithelial cells. This means that water moves out of the cells into the fluid lining the airways by osmosis. More water moves into the epithelial cells by osmosis from the tissues and tissue fluid on the other side.

5 The water that moves out of the epithelial cells by osmosis mixes with the mucus which is also produced by cells in the epithelial layer and keeps it runny so that it can be moved easily by the cilia.

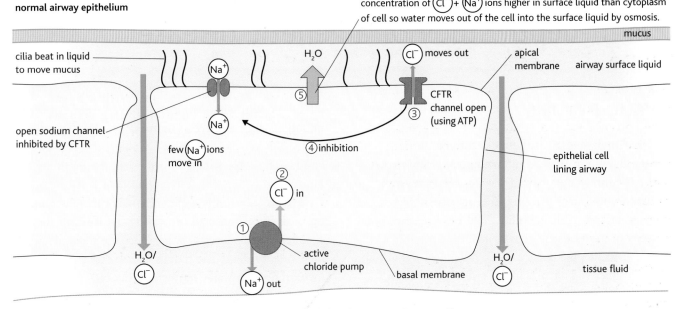

normal airway epithelium

concentration of Cl⁻ + Na⁺ ions higher in surface liquid than cytoplasm of cell so water moves out of the cell into the surface liquid by osmosis.

mucus

cilia beat in liquid to move mucus

H₂O

Cl⁻ moves out

apical membrane   airway surface liquid

Na⁺

⑤

Na⁺

open sodium channel inhibited by CFTR

few Na⁺ ions move in

④ inhibition

CFTR channel open (using ATP)

③

epithelial cell lining airway

② Cl⁻ in

① active chloride pump

basal membrane

H₂O/ Cl⁻

H₂O/ Cl⁻

Na⁺ out

tissue fluid

fig. 2.3.19 This complex process keeps the mucus produced by the epithelium lining your airways really runny. This helps your body prevent infection and also keeps the airways from getting blocked.

A similar process takes place in the epithelial cells that line your gut and your reproductive system, whether you are male or female. Runny mucus is important in both. It keeps open the narrow ducts and tubes along which enzymes pass from the pancreas into your gut, for example. It also makes the movement of sex cells in the reproductive system possible.

All of the biological systems described in this chapter are under the control of your genes. In the next chapter you will be looking at how a single mutation can cause devastation in the body because of the effect it has on this system for keeping mucus runny.

Questions

1   Explain why it is important to keep the mucus lining the airways very runny.

2   The process for keeping the mucus lining your respiratory system runny is very complex. Explain why each step is important and how it works.

2.4 In-depth study of a genetic disease

Cystic fibrosis

Cystic fibrosis (CF) is the most common serious genetic disease in the UK, affecting over 7500 people. It is a life-threatening condition that causes severe respiratory and digestive problems as well as very salty sweat and often infertility. The chlorine transport systems of the exocrine glands, including the mucus-secreting glands of the airways of the lungs, the gut and the reproductive system and the sweat glands, don't function properly. As a result, thick sticky mucus and very salty sweat are formed. It is the thick mucus which causes most of the symptoms.

Fig. 2.3.19 shows how the cystic fibrosis transmembrane regulatory (CFTR) channel protein lines the channels through which chloride ions leave the epithelial cells and move into the fluid outside the cells in the lungs. CFTR is an enormous protein containing 1480 amino acids. The gene that codes for it is also large, and it is found on chromosome 7. A mutation in any part of the gene can affect the CFTR protein and so cause cystic fibrosis. Around 1000 different mutations in this gene have been discovered, all of them coding for a faulty

CFTR protein, although some are very rare. All of the mutations are recessive. The most common mutation on the CF gene is known as DF508 and it is found in 75% of people with CF in the UK. People who inherit two copies of the faulty allele lack effective CFTR proteins. This means that chloride ions build up in their cells instead of moving out through the channels and as a result water does not move out of their cells to dilute the mucus on the surfaces of their membranes. So water moves *into* the cells by osmosis from the fluid surrounding the cells, making the mucus even more thick and sticky (**fig.2.4.1**).

The inheritance of cystic fibrosis

In the UK about 1 person in 25 carries a faulty CF allele – that's between 2 and 3 million people – and cystic fibrosis occurs in about 1 in 2400–2500 babies born to white Europeans. However it is much less common in other ethnic groups. Because cystic fibrosis is caused by a recessive allele, many people carry the mutation without knowing it. These carriers are phenotypically normal and usually have no idea that they are carrying the CF mutation. It is only if

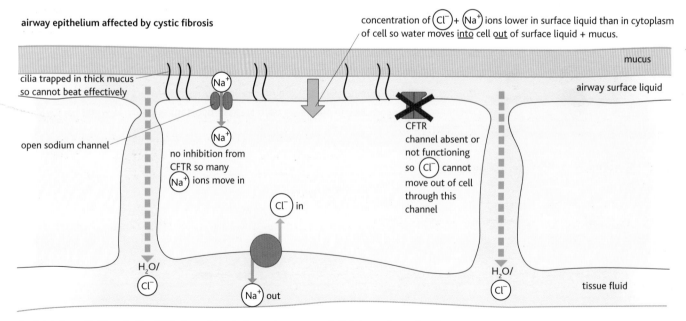

fig. 2.4.1 In cystic fibrosis, the CFTR channel is missing or does not work. This has a dramatic effect on the water balance of the cell and on the liquid and mucus lining the airways.

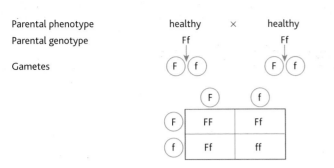

Offspring genotypes   FF : 2Ff : ff
Offspring phenotypes   3 healthy : 1 cystic fibrosis

fig. 2.4.2 **Two carrier parents have a 1 in 4 chance of having a child who will have cystic fibrosis.**

two carriers have children together that the problems may become apparent (see **fig. 2.4.2**). Even then, because the allele is recessive, there is only a 1 in 4 chance that any child of these parents will develop CF. Once the disease has appeared in a family, other family members become aware they may carry the faulty allele (see **fig. 2.4.3**). They will often be offered genetic counselling before they have children.

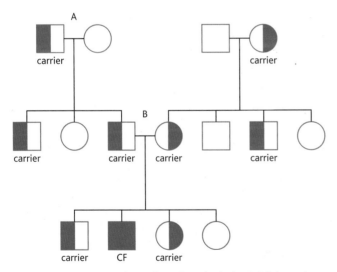

fig. 2.4.3 **This genetic pedigree shows how the faulty CF allele can be passed on through families until by chance two carriers have a family.**

Symptoms of cystic fibrosis

If someone inherits two alleles for cystic fibrosis it will affect many body systems because of its effect on the mucus that lines the tubes in these systems.

The respiratory system

The thick, sticky mucus typical of cystic fibrosis builds up in the tiny airways of the lungs and reduces the flow of air into the alveoli. It often blocks the smaller bronchioles completely and greatly reduces the surface area available in the lungs for gaseous exchange. This means affected individuals often have severe coughing fits as their body attempts to get rid of the mucus. They also feel breathless and are often short of oxygen making them feel tired and lacking in energy. In addition, the mucus is so thick and sticky that the cilia lining the airways cannot move it along and out of the system. Without treatment the lungs gradually fill up with mucus, making them less and less effective for gaseous exchange.

Bacteria and other pathogens that are breathed in and trapped in the mucus cannot be moved out of the respiratory system. As a result thick, pathogen-laden mucus builds up in the lungs, and this provides the bacteria with ideal conditions in which to grow.

Your body normally secretes antibodies into the mucus which inactivate pathogens. The normal chemical balance of the mucus is changed as water is moved out of the mucus into the cells. The solutes become more concentrated. Current research shows that the dehydrated surfaces of cells affected by CF lose their natural antibacterial properties, because the white blood cells and their antibodies cannot function effectively in the thickened mucus. However, these can be restored if the surface of the cell can be rehydrated.

The digestive system

The gut is badly affected in 85–90% of people with cystic fibrosis. Your digestive system makes enzymes that break down the large complex food molecules (carbohydrates, proteins and fats) into smaller molecules.

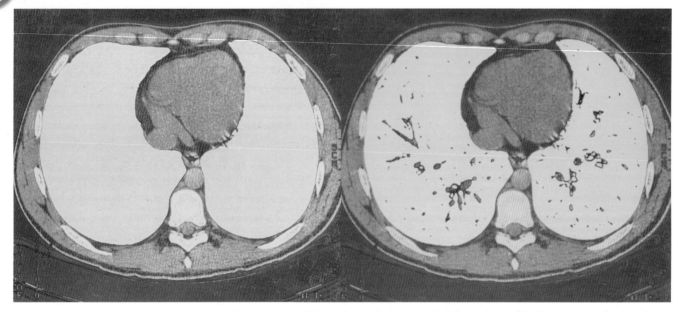

fig. 2.4.4 Scans of human lungs. The lungs on the left are clear and healthy. Those on the right show the blocked mucus-filled airways of someone affected by cystic fibrosis.

These can then be absorbed into your blood through the lining of your small intestine, which is covered in finger-like projections called **villi** that provide a large surface area for absorption to take place. Some digestive enzymes are made directly by glands in the lining of your gut. Some are made by associated organs such as the pancreas.

The enzymes from the pancreas are very important in the breakdown of carbohydrates, proteins and fats in the top part of your small intestine (the **duodenum**). The enzymes pass from the pancreas into your duodenum along a tube known as the **pancreatic duct**. Thin mucus is produced by the cells lining this tube in the same way as it is produced in the airways of the lungs. A faulty CFTR protein means the mucus produced in the pancreatic duct is also very thick and sticky. It often blocks the pancreatic duct, so that the enzymes do not reach the duodenum (see fig. 2.4.4). This has two damaging effects. If the digestive enzymes do not reach your gut, you cannot digest your food properly. This means you do not get enough nutrients from the food. Also, the digestive enzymes trapped in the pancreas may actually start to digest and damage the cells of the pancreas. If they affect the cells which make the hormone **insulin**, then the person may end up with diabetes.

Not only does the thick mucus stop enzymes getting to the gut to digest food, it also makes it more difficult for any digested food to be absorbed into the blood. Mucus is secreted throughout the gut to protect the delicate lining from damage by the digestive enzymes and to act as a lubricant. But, when this mucus is very thick and sticky, it forms a barrier between the contents of the gut and the lining of the intestine and clogs up the villi, reducing the surface area for absorption. These two effects put CF patients at severe risk of malnutrition and they often struggle to maintain their body mass. In fact one of the symptoms that can suggest a baby or small child may be affected by cystic fibrosis is a 'failure to thrive' – in other words, a failure to gain weight and grow as expected.

The reproductive system

The thick, sticky mucus produced in cystic fibrosis can have a damaging effect on the reproductive system. In women the mucus in the reproductive system normally changes through the menstrual cycle. When the woman is fertile it becomes thinner to help the sperm get through the cervix and along the oviducts. Women with cystic fibrosis usually produce fertile eggs, but the thick mucus can block the cervix so sperm cannot reach them. It can also block the oviducts, making fertilisation even less likely. In men the secretions of the reproductive system carry the sperm. Men with cystic fibrosis are often infertile. They may lack the tube that carries sperm out from the testis into the semen (the **vas deferens**). If the vas deferens is present, it may be partly or completely blocked by thick, sticky mucus so that only a reduced number of sperm (or no sperm at all) can leave the testis (see fig. 2.4.5).

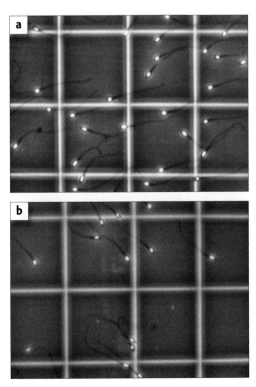

fig. 2.4.5 **(a)** A sample of normal semen contains vast numbers of sperm. **(b)** A man affected by cystic fibrosis may produce only few, or none if the tubes connecting the testes to the rest of the reproductive system are absent.

The sweat glands

The faulty CFTR protein means that people with cystic fibrosis usually have sweat that is more concentrated and salty than normal. Sweat is mainly salty water that is produced in your sweat glands. Normally, as the sweat passes along the duct of a sweat gland salt (sodium chloride) is reabsorbed, largely as a result of the CFTR protein moving chloride ions *into* the cells. Sodium ions follow along a concentration gradient. This reabsorption of salt prevents you losing too much salt in the sweat. So in sweat glands the chloride pump works in the opposite direction to that in the mucus-producing glands, where chloride ions are moved *out of* the epithelial cells.

Without functioning CFTR proteins, the chloride ions remain in the sweat, and so do the sodium ions. As a result the sweat is very salty – in fact this is one of the tests used to diagnose cystic fibrosis. The loss of sodium and chloride ions causes health problems linked to the balance of ions in the body. Levels of sodium and chloride ions are very important for the proper functioning of many body systems, including the nervous system and the heart. If too much is lost in the sweat, the concentration of the body fluids changes, which can affect the heart.

Babies and children affected by cystic fibrosis have very salty sweat. The condition is often noticed when parents comment that their baby tastes salty when they kiss it. Along with a 'failure to thrive' as a result of the gut complications that come with CF, this salty sweat is an early warning sign to health professionals that something may be wrong.

Questions

1 Nuala and her parents are both apparently healthy, but Nuala has a child with cystic fibrosis. Is it possible to decide the chances that Nuala's brother and sister might be carriers of the disease? Explain your answer, using genetic diagrams.

2 Using fig. 2.4.2, show genetic diagrams for couple A and couple B, giving all possible genotypes and phenotypes of the offspring. Comment on the ratio in the offspring that were actually born.

3 Someone with cystic fibrosis may often feel tired and lacking in energy. Explain why, with reference to the effect of CF on:
a the respiratory system
b the digestive system.

4 Explain why the loss of extra sodium and chloride ions in the sweat of someone with CF can affect all the other cells in the body.

Treating cystic fibrosis now

Cystic fibrosis is a life-threatening condition. In the past, affected children almost always died before reaching adulthood. However, the treatments available have improved enormously in recent years, and the average life expectancy for affected people is increasing all the time. The parents of babies and very young children who are diagnosed with the condition now are told that their child may well have a near-normal life expectancy.

At the moment there are no cures for cystic fibrosis. Current treatments aim to reduce the symptoms and allow the body systems to work as effectively as possible. In the UK, most people with CF will be looked after by a team of health professionals including doctors, nurses, physiotherapists, dieticians and others. There is a range of treatments that tackle different symptoms.

Physiotherapy

Physiotherapy is very important for removing as much of the thick, sticky mucus from the lungs as possible. This makes it easier to breathe and take in enough oxygen. It also reduces the risk of serious lung infections. Adults usually carry out the physiotherapy for babies and small children. Teenagers and adults with CF can often do most of their own physiotherapy unless they are producing excessive amounts of mucus. Physiotherapy is usually done a couple of times a day, though it may be more or less than this depending on the individual and their health at any particular time.

New devices which change the pressure in the airways and help people to clear their mucus without using more vigorous physiotherapy are being introduced – these include the 'Flutter', the positive expiratory pressure (PEP) valve and the ThAIRapy bronchial drainage system.

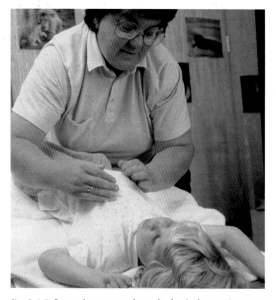

fig. 2.4.6 **Removing mucus through physiotherapy improves breathing and reduces the risk of lung infections taking hold.**

Diet and enzymes

Because the thick mucus of CF can block the pancreatic duct and coat the lining of the gut, digesting and absorbing food is difficult for most people with cystic fibrosis. To help overcome the effect of the blocked pancreatic duct, people with CF may take enzymes when they have a meal. These enzymes help to replace the missing pancreatic enzymes, so more of the food can be digested.

People with CF need to eat a carefully balanced diet, and also to eat more than other people to make up for what they cannot digest. They need more high-energy foods, such as high-fat and high-carbohydrate foods, and an adult with CF needs twice as much protein as someone unaffected by the condition.

Drug therapies

Most people in the UK who have cystic fibrosis take a cocktail of medicines to improve their health and protect them against the symptoms of the disease. These drugs include:

- Antibiotics – these destroy many of the bacteria that could potentially cause life-threatening lung infections. It can be difficult getting antibiotics to the tissue where they are needed, so aerosols (inhalers) are used to breathe them deep into the lungs.

- Vaccines – not only are all the childhood vaccines very important, but CF patients also need a 'flu vaccine every year to protect them against the most common forms of 'flu virus. A vaccine against pneumonia is important too, since prevention is better than cure.

- Mucolytics – these are drugs to make the mucus more runny and so easier to move.

- Asthma drugs, eg salbutamol and steroids – these are used to open the airways and to reduce inflammation in the lungs.

- DNAase enzymes – these make the mucus thinner and easier to cough up.

- Insulin – if problems with the pancreas lead to diabetes, insulin will need to be given regularly to control the blood sugar concentration.

Transplant surgery

In some cases the damage that cystic fibrosis causes to the lungs is so severe that the lungs cannot function properly. The heart may also be affected. In this case the only solution is a lung, or heart and lung, transplant. The new organs will not be affected by cystic fibrosis, although the rest of the body will still have the other problems associated with the disease. After a transplant, the person has to take immunosuppressant drugs for the rest of their life, to prevent their body reacting against the new tissue and rejecting it. This suppresses the immune system, making it harder to fight infections. However, people with CF who have lung transplants usually do very well.

Infertility treatments

Increasingly, women with cystic fibrosis are having babies of their own. Some need to use fertility treatment such as IVF, while for others their reproductive system is not badly affected. Men are also now being helped, with pioneering techniques taking sperm from their testes and using it to fertilise eggs *in vitro* (outside the body). Some of the resulting embryos are then returned to the mother's uterus to develop normally.

fig. 2.4.7 James Burke has cystic fibrosis, but it didn't stop him and six other people affected by CF completing the London Marathon in 2007. This shows what can be possible with modern treatments and a lot of determination!

Questions

1 Why is physiotherapy an important part of the treatment of cystic fibrosis?

2 Explain why someone with CF needs to eat more to maintain their body weight than other people.

3 Suggest why the combination of drugs needed for CF varies from person to person and over time for an individual.

Gene therapy for cystic fibrosis

Much current research into treatments for cystic fibrosis aims to remove the symptoms entirely or even cure the disease. The hope is that faulty alleles might be replaced by healthy ones in a process known as **gene therapy**. Gene therapy involves taking a copy of the healthy gene and finding an effective way of getting it into the cells that need it, so that they can produce the correct protein.

Copying the healthy gene

Copying genes and inserting them into other organisms is known as **genetic engineering** or **genetic modification**, and this is the first stage in gene therapy.

The best-known approach is to use special enzymes called **restriction endonucleases** to chop up healthy DNA strands, cutting them at specific sites. These enzymes cut the DNA into small pieces which can be handled more easily. Each type of endonuclease will only cut DNA at specific (restricted) sites within a particular DNA sequence, hence the name.

Some restriction endonucleases can cut the DNA strands in a way that leaves a few base pairs longer on one strand than the other, forming a **sticky end**. Sticky ends make it easier to attach new pieces of DNA to them. **DNA ligases** are also used as 'genetic glue' to join pieces of DNA together, which is what happens next as you will see below.

Artificial copies of the healthy gene can also be made by taking an mRNA molecule transcribed from the gene and using it to produce the correct DNA sequence – effectively reversing the transcription process, using the enzyme **reverse transcriptase**. DNA made like this is known as **complementary DNA** or **cDNA** and it can act as an artificial gene.

From research on human genes, including the Human Genome Project (see chapter 2.2), we know the base pair sequence in the gene that produces the normal CFTR protein. This knowledge can be used to build artificial genes from base pairs in the lab.

Vectors

The next stage is to attach the isolated gene to another piece of DNA, known as a **vector**, which will carry it into the target cell. **Plasmids**, the circular strands of DNA found in bacteria, are often used as vectors (see fig. 2.4.8). Once the plasmid gets into the host nucleus it can combine with the cell's DNA to form **recombinant DNA**. Plasmids are particularly useful in the formation of genetically modified bacteria.

Other vectors are needed to carry new DNA into human cells. Vectors that have been tested for the treatment of cystic fibrosis include harmless viruses and liposomes (spheres formed from a lipid bilayer). The viruses infect lung epithelial cells and insert the viral DNA (including the added gene) into the cell's DNA. Liposomes fuse with the cell membrane and can pass through it to deliver the new DNA into the cytoplasm. Viruses are much better at getting DNA into the

Stage 1. Isolate the required gene

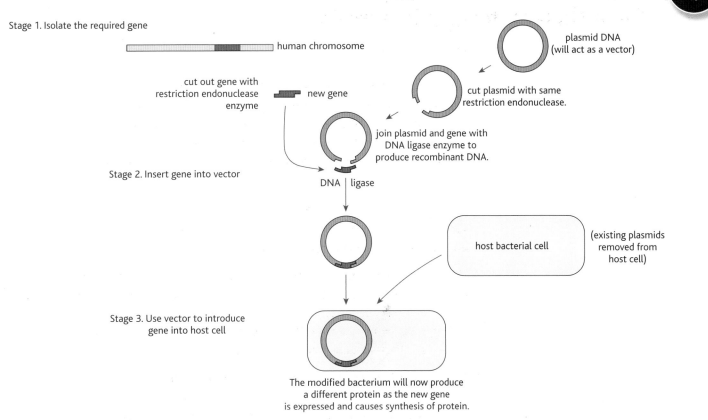

Stage 2. Insert gene into vector

Stage 3. Use vector to introduce gene into host cell

fig. 2.4.8 **The main stages involved in inserting a new gene into a plasmid.**

nucleus at the moment, but they can cause an immune reaction in some people. So a lot of research is focusing on non-viral vectors such as liposomes even though they are not as effective.

Specific marker genes can be added with the new gene to make it easier to see when the insertion of the gene into a cell has been successful (**fig. 2.4.9**).

Once the healthy new DNA is inside the lung epithelial cells, the healthy genes should be transcribed and translated, producing the normal, active CFTR protein and relieving the symptoms (**fig. 2.4.10**). This sounds good in theory, but the practice has shown that it's not quite that easy.

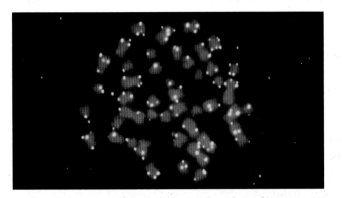

fig. 2.4.9 **Marker genes that fluoresce under ultraviolet light show which pieces of DNA have the new gene inserted.**

Gene therapy in CF

The position of the mutated gene on chromosome 7 which causes 70–75% of cystic fibrosis was isolated by scientists in August 1989. By 1990 the normal gene had been copied and added to cells affected by cystic fibrosis in the lab, where it proved effective in correcting the faulty biochemical pathway. This has now been repeated many times by different teams. Encouraged by these results, scientists were keen to begin human studies. First they tried to find out if it was possible to get a new, healthy CFTR protein allele into affected human cells in a living person – both viruses and liposomes were tried and both had very limited success. There was evidence that some effective CFTR protein was produced, but it only lasted a few days and was not enough to relieve the symptoms of the disease.

By 1999, techniques had improved so that about 25% of the normal chloride ion movement out of the cells could be restored, but sodium ions did not seem to be moving properly. Another problem was that epithelial cells are continually shed and replaced by new cells made by the body. These new cells only contain the faulty gene so the new DNA is lost. Other difficulties have appeared in testing.

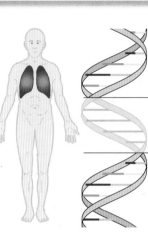

1 cystic fibrosis – occurs in people born missing a critical gene sequence

correct gene sequence

2 the missing genes are placed inside a specially modified virus

3 the virus carries the missing gene sequence into a lung cell

virus

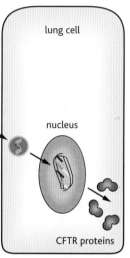

lung cell

nucleus

CFTR proteins

4 Hopefully the cell will start producing the correct CFTR proteins and so remove the symptoms of cystic fibrosis

fig. 2.4.10 **How gene therapy for cystic fibrosis works in theory.**

Extracellular barriers

There are many barriers to overcome before the new DNA in the vectors can reach the epithelial cells of the airways. This is partly because cystic fibrosis blocks the airways, making it difficult for sprays to penetrate the lungs, and partly because the surface is coated with thick, sticky mucus. New and better ways of dilating the airways and making the mucus more liquid (using mucolytics and DNAase) are one focus of research. Some research groups are trying to get the vectors to the lung cells by a completely different route – through the blood.

Intracellular barriers

Once the vector is inside the cell, the next challenge is to get the new DNA into the right place. This is proving very difficult, particularly when liposomes are used as the vectors. It has been estimated that only about 1 in every 1000 plasmids that enter a cell in a liposome get into the nucleus to be transcribed. Research is looking at modifying the vectors to get the new genes into the cells and then into the nuclei more effectively. So far viruses seem to be giving the best results.

Keeping the gene expression going

The final problem is that even when the new gene is taken into the nucleus and transcribed, it doesn't work for long – about two weeks at best. Research teams are looking at ways to extend this and even make it permanent, by using different vectors. They are also looking at ways of repairing the faulty allele on the original DNA. If stem cells could be used (see chapter 3.2) this might lead to a permanent cure.

The hope for gene therapy is that people will not need treatment after the initial medication. Their quality of life and life expectancy will be much improved with no respiratory complications. There would be a big financial saving without lifelong treatment. It may even be possible to treat the effects of the disease on the gut in the same way. It is hoped that by 2009–2010 human trials will be underway again, possibly on a larger scale than ever before.

Questions

1 Explain the role of genetic engineering in gene therapy.

2 Suggest why plasmids, viruses and liposomes are used as vectors, and explain why they have had limited success so far in treating cystic fibrosis.

3 What are the hopes for gene therapy in cystic fibrosis? How far has it lived up to these expectations?

4 Discuss the advantages and disadvantages of germ-line therapy (see opposite).

HSW Mouse models of CF

To find a cure for a disease like cystic fibrosis it is important to understand exactly how it affects the body and what aspects of the disease can best be targeted for treatment. However, there is a limit to the investigations that scientists and doctors can carry out on people. Some experiments can be done using cells in culture but sometimes there is no substitute for a living organism. This is where animal models play a vital role.

In the early 1990s David Porteous and Julia Dorin at the Medical Research Council Human Genetics Unit in Edinburgh used genetic manipulation techniques to develop mice with a faulty CFTR gene. These mice have similar chloride ion transport problems to humans affected by cystic fibrosis. They do not have all the same clinical symptoms but they have been very useful both in helping understand the disease and in trialling new treatments, including some early gene therapy work using DNA and liposomes. They demonstrated that this gene therapy was safe but did not completely correct the chloride ion transport abnormality associated with the disease. The same results were later seen in human trials.

Some people consider it unethical to manipulate and use animals like mice for scientific research. However, without this our understanding of this disease and the development of potential new treatments would take much longer.

fig. 2.4.11 The black mouse in this picture has a homozygous defect in the CFTR gene. Mice like this gave scientists an animal model for cystic fibrosis in people.

HSW The ethics of gene therapy

The gene therapy under development for cystic fibrosis is carried out on normal body cells (somatic cells) and so it is known as somatic cell gene therapy. Cystic fibrosis could be the first of many human conditions to be relieved in this way. However, gene therapy does not alter the fact that if someone with CF has a child they will pass on their faulty alleles. Their child may even have CF, depending on the genotype of the other parent. A potential solution to this problem is to alter the germ cells – the reproductive cells of the body – so that the faulty genes are no longer passed on. This could be done in the very early embryo immediately after *in vitro* fertilisation. The individual who developed would then be free of the disease, and would not risk passing it on to their own offspring.

This may sound like a good idea, and some people argue that when the technology is available it should be used. But many other people are very concerned. No one is yet sure of the effect on an early embryo of such an invasive intervention – and the impact might not become clear until years into the life of the individual. What is more, such manipulation of the human genome is seen as a Pandora's box. Whilst attempting to remove the risk of genetic diseases seems a very positive aim, it could be difficult to know where to draw the line – and some people would be prepared to pay a great deal of money to have the line moved a little. If it became possible to manipulate the genes not simply to remedy disease but to enhance longevity, change skin colour, increase adult height or intelligence or reduce adult weight, then some people would be tempted to try it.

At the moment most aspects of germ-line research are banned in the UK, most European countries and the US. But sooner or later, someone somewhere will try this and society needs to think carefully about how we will deal with that.

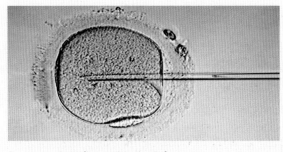
fig. 2.4.12 Germ-line engineering is for ever.

Genetic screening

In the future it may be possible to cure genetic diseases such as cystic fibrosis. At the moment it isn't. For individuals born with a genetic disease it can improve their chances of survival and their general state of health if their condition is diagnosed as early as possible. The screening may be carried out during pregnancy with the option of terminating the pregnancy if the fetus is affected by a severe genetic disorder, or it may be carried out on newborn babies to identify any problems and give the best treatment as early as possible.

For some genetic diseases whole populations are tested. This is known as **genetic screening**. For example, 1 in 15 000 babies born in the UK has the genetic condition **phenylketonuria** (**PKU**) (see chapter 2.2, **fig. 2.2.14**). In the late 1970s it was estimated that it would cost £20 000 to screen 15 000 babies and supply the predicted one affected baby with a special diet until adulthood. However, the lifetime treatment of a PKU baby who is not treated and is therefore severely handicapped, needing institutional care throughout its approximately 45-year lifespan, was calculated then as £126 000. It was decided that the cost of screening all newborn infants was well justified and as a result all newborn babies in the UK are now screened for PKU.

When the same calculation was done for cystic fibrosis in 1989, the benefit/cost ratio of whole-population screening was found to be much smaller. The cost of screening 2000 fetuses to find and abort the average of one with cystic fibrosis was £80 000. The cost of treating a child with cystic fibrosis for its relatively short life was £125 000, so the difference wasn't great. However, in the twenty-first century people with cystic fibrosis live far longer, so it is much more expensive to treat them for a lifetime. As a result, screening of all newborn infants is being introduced (**fig. 2.4.13**). This is a big step forward, because the sooner treatment is started, the better the chances of avoiding serious lung damage. The cost of screening is outweighed by the health benefits and lower costs of early treatment.

For families where the pedigree shows a history of an inherited disease, different screening tests can be offered to couples before they consider having a child. These tests offer help and hope – but also lead to some very difficult choices.

fig. 2.4.13 *All newborn babies are now tested for several genetic diseases. In most cases, early treatment can avoid serious problems later on.*

Identifying carriers

If one member of a family is born with cystic fibrosis, other members of the family will be offered genetic testing. It is possible to detect the CF allele in a carrier who has no symptoms. A sample of blood, or some cells from the inside of the mouth, can be used to carry out a simple test which identifies the allele. If one partner in a couple knows they are a carrier, the other partner is advised to be tested as well, because if two carriers have a baby there is a 1 in 4 risk that it will be affected by cystic fibrosis.

Prenatal screening

Couples who find they are at risk of having children with a genetic condition such as cystic fibrosis have several options open to them. They can go ahead and have a family as usual, hoping that in the genetic lottery they are lucky and their children inherit healthy genes, but being prepared to support and take care of them if they don't. Or they may decide not to have children at all, to prevent passing on a faulty gene even in a carrier.

The third option is to go ahead with pregnancies but to have each pregnancy screened (**prenatal screening**).

Ideally prenatal screening is used to try and discover whether a fetus is affected by cystic fibrosis early enough in the pregnancy for the parents to be offered a termination (abortion) if they do not wish to continue with the pregnancy. This option can be very traumatic for a couple to consider, and is not open to some individuals because of their beliefs about abortion.

However, it is effective at preventing the birth of children with genetic diseases such as cystic fibrosis. It can also identify individuals who will be carriers.

To find out if a fetus has cystic fibrosis, tests need to be carried out on some fetal cells. The fetal tissue can be obtained in one of two ways – amniocentesis or chorionic villus sampling (**fig. 2.4.14**).

Amniocentesis involves removing about 20 cm^3 of the amniotic fluid which surrounds the fetus using a needle and syringe. This is done at about the 16th week of pregnancy. Fetal epithelial cells and blood cells can be recovered from the fluid after spinning it in a centrifuge. After the cells have been cultured for 2–3 weeks a number of genetic defects as well as the sex of the baby can be determined from examination of the chromosomes.

Amniocentesis has the following disadvantages:

- It can only be carried out relatively late in the pregnancy, so that should termination of the pregnancy be necessary it is more traumatic.

- The results are not available until 2–3 weeks after the test.

- It carries about a 1% risk of spontaneous abortion after the procedure, regardless of the genetic status of the fetus.

In **chorionic villus sampling**, a small sample of embryonic tissue is taken from the developing placenta. This makes a much bigger sample of fetal tissue available for examination. The cells can be tested for a wide range of genetic abnormalities. This diagnostic technique can be carried out much earlier in the pregnancy, so that if a termination is necessary it is physically less traumatic for the mother. The results are also available more rapidly than for amniocentesis.

The disadvantages of chorionic villus sampling are:

- There is a 2.5–4.8% risk that the embryo may spontaneously abort after the tissue sample is taken, though the risk of miscarriage at this stage of pregnancy is much higher anyway.

- All paternal X chromosomes are inactivated in fetal placental cells so any problems in the genes on that chromosome cannot be detected by this technique.

Amniocentesis
- remove about 20 cm³ amniotic fluid at about 16th week of pregnancy
- cells from fluid cultured for several weeks before analysis

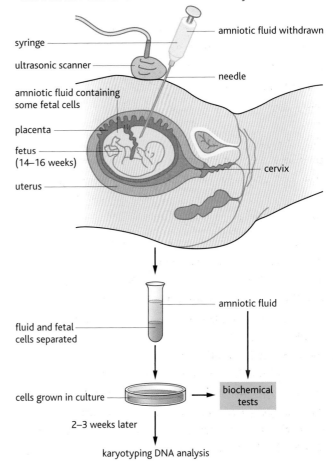

Choronic villus sampling
- small sample embryonic tissue taken from placenta at 8–10 weeks of pregnancy
- larger sample than amniocentesis, so cell culture not needed before analysis

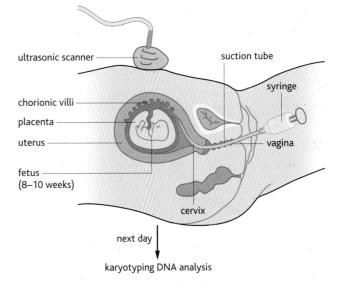

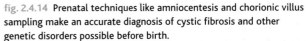

fig. 2.4.14 Prenatal techniques like amniocentesis and chorionic villus sampling make an accurate diagnosis of cystic fibrosis and other genetic disorders possible before birth.

Preimplantation genetic diagnosis

If parents have had a child affected by a disease such as cystic fibrosis, it is known that they are both carriers. If someone in the family has had a child affected by cystic fibrosis a couple may choose to be tested and find out if they are both carriers. Can parents who know they are carriers avoid having a child with cystic fibrosis without going through amniocentesis or chorionic villus sampling and a possible termination?

With major developments in human infertility treatments over the last 30 years, some sophisticated ways of genetically screening an early embryo before it is implanted in the uterus have been introduced. This technique, called **preimplantation genetic diagnosis**, is based on the technique of IVF (*in vitro* fertilisation) (see **fig. 2.4.15**). In this technique, the egg and sperm are fertilised outside the body. After a few cell divisions, a single cell is removed from each embryo. Amazingly, all the evidence so far suggests that this causes no harm to the development of the embryo. The genetic make-up is checked and only those embryos free of the problem alleles are placed in the mother's uterus to implant and grow. This avoids implanting not only embryos with the potential to develop a genetic disease but also carrier embryos as well, thus removing the faulty allele from the gene pool. In the case of genetic conditions found only in boys such as haemophilia, only female embryos would be replaced.

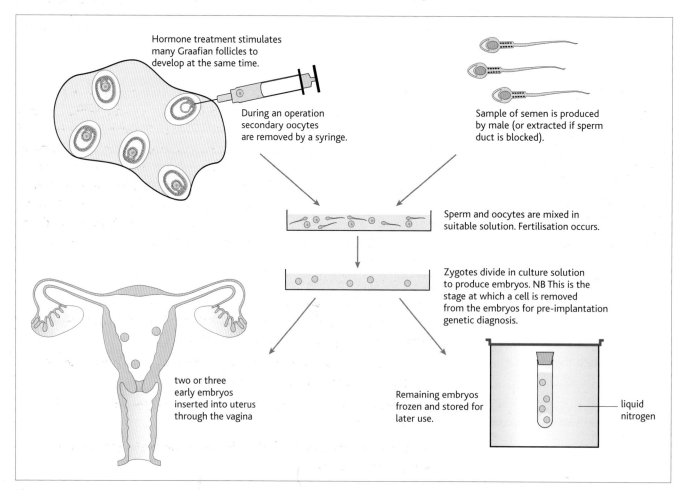

Hormone treatment stimulates many Graafian follicles to develop at the same time.

During an operation secondary oocytes are removed by a syringe.

Sample of semen is produced by male (or extracted if sperm duct is blocked).

Sperm and oocytes are mixed in suitable solution. Fertilisation occurs.

Zygotes divide in culture solution to produce embryos. NB This is the stage at which a cell is removed from the embryos for pre-implantation genetic diagnosis.

two or three early embryos inserted into uterus through the vagina

Remaining embryos frozen and stored for later use.

liquid nitrogen

fig. 2.4.15 **The main steps in the process of in vitro fertilisation.**

It is very difficult for any parent to come to terms with the fact that the child they are expecting will not be the normal, healthy baby they had hoped for. Some parents who find that their fetus is affected by a serious genetic disease decide to terminate the pregnancy rather than condemn the child to a life of constant medical treatment and themselves to a life of caring for the child. Other parents either do not accept abortion as a reasonable option or positively choose to carry their baby to term in spite of its genetic condition. For these couples too the testing can be valuable. It allows them time to grieve for the loss of the normal healthy baby they hoped to have and to come to terms with and welcome the child they actually have.

The nature of the amniocentesis test means that knowledge of a genetic condition such as cystic fibrosis often comes around half-way through a pregnancy. This means that if the parents choose to terminate the pregnancy, it involves a late and therefore relatively traumatic abortion. However, the option is still there, although most people now use either chorionic villus sampling or preimplantation diagnosis. Some people feel that the view of society (rather than individual parents), that the only acceptable baby is a perfect baby, is a sad indictment of very materialistic times.

Another issue to consider is this. A fetus which carries one recessive allele will be perfectly healthy, but could pass the allele on to other generations. Would anyone propose that such a fetus be terminated? Studies suggest most parents do not want to know if the fetus carries the faulty allele – only if it is affected by a genetic disease.

Genetic counselling

fig. 2.4.16 **Genetic counsellors can help couples to understand the problems they or their children may face from a genetic disease.**

For most people finding out about genetic diseases in their family is very traumatic. All of the issues discussed above – decisions to do with having children or not, aborting affected pregnancies, who to tell – are suddenly of immediate and personal relevance. Genetic counsellors are trained to help people to understand and come to terms with the situation of carrying a faulty allele that can cause a genetic disease. They will assess the statistical risk of a couple producing an affected child and help couples recognise the options they have. They will then work with the parents to choose what they believe is right for them within their own framework of moral, family, religious and social beliefs and traditions.

1   Why has testing of newborn babies for cystic fibrosis been introduced only recently, even though the test was available 20 years ago?

2   Why is chorionic villus sampling becoming more popular than amniocentesis? Why is amniocentesis still necessary as an option?

3   What are the advantages of preimplantation genetic diagnosis over prenatal techniques such as amniocentesis?

4   Summarise the ethical issues raised by the new techniques for prenatal and preimplantation testing.

Examzone: Topic 2 practice questions

1 Copy the table below which refers to four membrane transport processes: diffusion, facilitated diffusion, osmosis and active transport. If the statement is correct, place a tick (✓) in the appropriate box and if the statement is incorrect, place a cross (✗) in the appropriate box. **(Total 4 marks)**

Process	Takes place against a concentration gradient	Requires energy in the form of ATP
Diffusion		
Facilitated diffusion		
Osmosis		
Active transport		

2 a State *three* characteristic features of gas exchange surfaces. (3)

b Describe how the process of inspiration (breathing in) takes place in mammals. (3)

(Total 6 marks)

3 Proteins are chains of amino acids and have a wide range of functions in living organisms.

a i The diagram below shows part of the general structure of two different amino acids.

Copy and complete the diagram to show how these two amino acids can be joined together. (1)

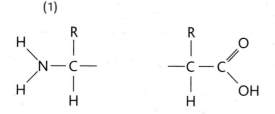

ii Name the bond that is formed between the two amino acids. (1)

b Explain how the specific sequence of amino acids in a protein determines its three-dimensional structure. (2)

c Enzymes are proteins that speed up chemical reactions within living organisms. The graph below shows the effect of changing enzyme concentration on the rate of reaction.

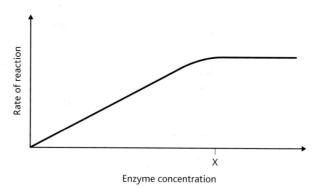

i Explain why increasing the enzyme concentration above point X on the graph does not increase the rate of the reaction further. (2)

ii Outline the practical procedures that could have been used to obtain the results shown on the graph. (4)

(Total 10 marks)

4 The diagram below shows the sequence of bases in one strand of the DNA from part of a gene. The base sequence is read from left to right.

DNA base sequence

A	C	C	C	C	A	T	T	T	C	A	T	C	C	A

The table below shows the anticodons of some tRNA molecules and the specific amino acids each would carry.

Amino acid	tRNA anticodon
Alanine	CGA
Glycine	CCA
Lysine	UUU
Proline	GGA
Tryptophan	ACC
Valine	CAU

a Using this information write down the amino acid sequence coded for in this part of the gene. **(2)**

b The diagram below shows the same length of DNA after it has undergone a mutation.

A	C	C	T	C	A	T	T	T	C	A	T	C	C	A

  i Name the type of mutation that has occurred.**(1)**

  ii Suggest how this mutation might affect the protein produced. **(3)**

(Total 6 marks)

5 Cystic fibrosis is a common inherited disorder amongst Europeans. It is caused by a single gene.

The diagram below is part of a family pedigree showing the inheritance of this disorder.

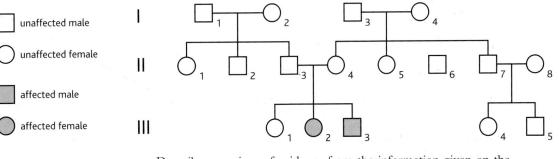

□ unaffected male

○ unaffected female

■ affected male

● affected female

a Describe one piece of evidence from the information given on the diagram which suggests that the allele causing this inherited disorder is recessive and one piece of evidence that it is autosomal. **(2)**

b A person whose family includes people with cystic fibrosis may wish to know if they or their partner are carriers of this inherited disease, before deciding to have children. Suggest how an individual with the heterozygous genotype for this disorder could be identified. **(2)**

c Suggest *two* reasons why chorionic villus sampling is more acceptable than amniocentesis in detecting genetic disorders in a fetus. **(2)**

(Total 6 marks)

Topic 3 The voice of the genome

This topic deals with cells, their structure, their division and how unspecialised stem cells in the embryo differentiate to form the many different kinds of cell in a multicellular organism. Recent developments in the research of undifferentiated stem cells offer the potential for treating many illnesses, but also raise difficult social and ethical questions. New ways of altering cells may overcome these problems.

What are the theories?

Eukaryotic cells have a complex internal structure that enables them to carry out many different processes at the same time, and these processes vary depending on the tissues and organs that the cells are in.

After meiosis and fertilisation, the cells of the fertilised egg divide by mitosis to produce a new individual, with cells that all contain the same genetic information but are differentiated to carry out different tasks. The cells in the early embryo are totipotent, able to differentiate into any kind of cell, but most cells lose this ability gradually over many cell divisions. Cell differentiation is the result of genes in the cell being switched on and off.

What is the evidence?

You will see that the technological development of microscopes, and the development of practical techniques such as the use of chemical markers, has gradually led to our current understanding of cell structure and the way cells work. We are still learning how cells differentiate during growth and development. There will be opportunities to carry out your own practical investigations, such as observing mitosis and using plant tissue culture to study totipotency in plant cells.

What are the implications?

The understanding of how stem cells can differentiate into specialised cells offers the possibility to treat many conditions caused by faulty cells, but could be extended to achieve other results such as cloning humans. This raises important social and ethical questions. Is it right that while treatments are being tested in the lab, people are dying who might have been saved by those treatments? How do we control research in these areas and who decides what is the right research and what should be stopped? In the rush for scientific progress, the system can fail and scientists may even be driven to publish false results.

The map opposite shows you all the knowledge and skills you need to have by the end of this topic. The colour in each box shows which chapter they are covered in and the numbers refer to the Edexcel specification.

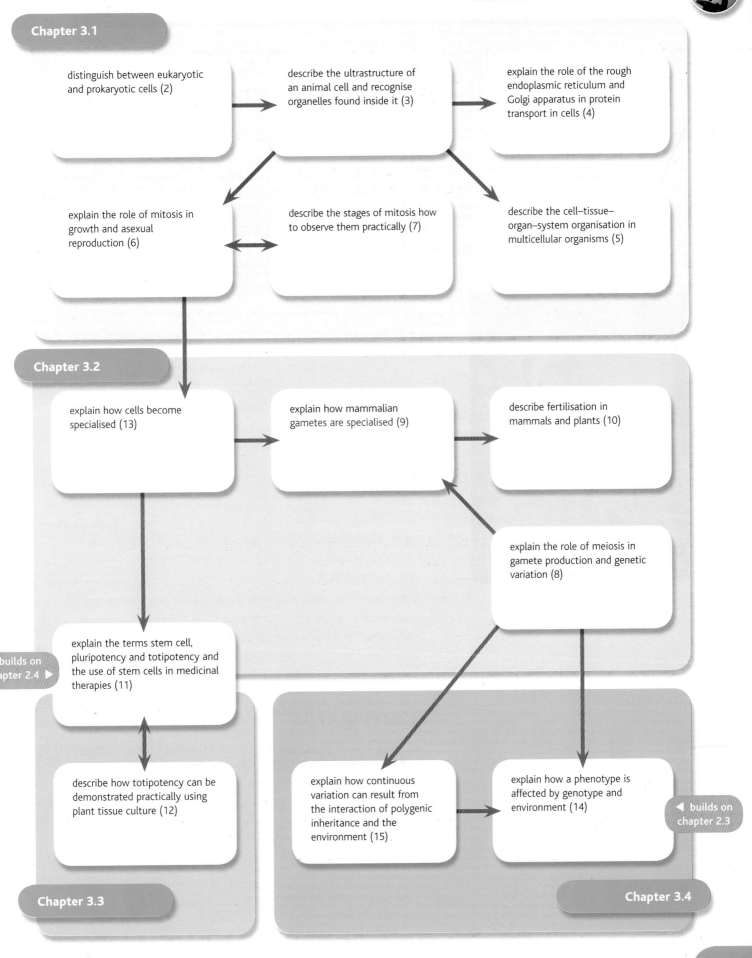

Chapter 3.1

distinguish between eukaryotic and prokaryotic cells (2)

describe the ultrastructure of an animal cell and recognise organelles found inside it (3)

explain the role of the rough endoplasmic reticulum and Golgi apparatus in protein transport in cells (4)

explain the role of mitosis in growth and asexual reproduction (6)

describe the stages of mitosis how to observe them practically (7)

describe the cell–tissue–organ–system organisation in multicellular organisms (5)

Chapter 3.2

explain how cells become specialised (13)

explain how mammalian gametes are specialised (9)

describe fertilisation in mammals and plants (10)

explain the role of meiosis in gamete production and genetic variation (8)

builds on chapter 2.4 ▶

explain the terms stem cell, pluripotency and totipotency and the use of stem cells in medicinal therapies (11)

describe how totipotency can be demonstrated practically using plant tissue culture (12)

explain how continuous variation can result from the interaction of polygenic inheritance and the environment (15)

explain how a phenotype is affected by genotype and environment (14)

◀ builds on chapter 2.3

Chapter 3.3

Chapter 3.4

3.1 Animal cells and asexual reproduction

Looking at cells

The idea of cells is familiar to most people. Cells are discussed in the media on an almost daily basis in relation to topics such as cancer, test-tube babies and DNA testing. But, in spite of this, most people have only a vague idea of what a cell really is.

fig. 3.1.1 The work of the early pioneers of the microscope helped to reveal the structure of living cells.

HSW Discovering cells

Cells were first seen over 300 years ago. In 1665 Robert Hooke, an English scientist, designed and put together one of the first working optical microscopes. Amongst the many objects he examined were thin sections of cork. Hooke saw that these sections were made up of tiny, regular compartments which he called cells.

It took many years of further work for the full significance of Hooke's work to emerge. In 1676 Anton van Leeuwenhoek, a Dutch draper who ground lenses in his spare time, used his lenses to observe a wide variety of living unicellular organisms in drops of water, which he called 'animalcules'. At the same time the English plant scientist Nehemiah Grew was publishing drawings of 'tissues'. His work with microscopes led him to believe living material was made up of just a woven mass of fibres. By the 1840s it was recognised that cells are the basic units of life, an idea that was first expressed by Matthias Schleiden and Theodore Schwann in their **cell theory** of 1839.

In the years since 1839 knowledge about cells has progressed a long way. Improvements in the quality of lenses and the techniques used to prepare material for microscopy have allowed us to see cells in increasing detail and so develop our understanding of how they work.

Observing cells

Some cells can be seen easily with the naked eye, eg the ovum in an unfertilised bird's egg is a single cell. But most cells cannot be seen without some kind of magnification. Ever since it was first developed, the **light** or **optical microscope** has been the main tool for observing cells. In spite of the development of the electron microscope, the light microscope is still widely used. A good light microscope can magnify to 1500 times and still give a clear image. At this magnification an average person would appear to be 2.5 km tall. An electron microscope can give a magnification of up to 50 000 times, making your average person appear over 830 km tall!

Electron microscopes provide the highest levels of magnification, so you might assume that electron micrographs are always the best option for a scientist. However, there are some clear advantages and disadvantages to both light and electron microscopy, and the choice is not always as clear cut as it might seem.

Advantages of the light microscope

- Living plants and animals or parts of them can be seen directly. This is useful in itself and allows you to compare prepared slides with living tissue.
- Light microscopes are relatively cheap so are available in schools and universities, hospitals, industrial labs and research labs. They can be transported and used almost anywhere in the world.

Disadvantages of the light microscope

- Preservation and staining the tissue can produce artefacts in the tissues being observed, so what is seen may be the result of preparation rather than real.
- Light microscopes have limited powers of resolution and magnification.

Advantages of the electron microscope

- Electron microscopes have huge powers of magnification and resolution. Many details of cell structure have been seen for the first time since EMs were developed.

Disadvantages of the electron microscope

- All specimens are examined in a vacuum – air would scatter the electrons and make the image of the tissue fuzzy – so it is impossible to look at living material.
- Specimens undergo severe treatment that is likely to result in artefacts.
- Electron microscopes are extremely expensive. They are large, have to be kept at a constant temperature and pressure and need to maintain an internal vacuum. Relatively few scientists outside research laboratories have easy access to such equipment.

Eukaryotic and prokaryotic cells

Living organisms are made up of cells. Most of the most familiar organisms have the same sort of cells. Animals, plants, protoctists, eg algae, and many fungi have cells that contain membrane-bounded organelles such as a nucleus, mitochondria and chloroplasts.

These organisms are **eukaryotes** and are made up of eukaryotic cells.

But there is another ancient group of organisms that do not have eukaryotic cells. They include the bacteria and blue-green algae. They have cells of a very different type called prokaryotic cells, and are known as **prokaryotes**. Prokaryotic cells lack much of the structure and organisation of the eukaryotic cells. They do not have a membrane-bound nucleus – the genetic material is a single strand coiled up in the centre to form the **nucleoid** or loop and sometimes there are small additional bits of genetic material within the cell called **plasmids**. The cytoplasm contains enzymes, ribosomes and food-storage granules but lacks other features of eukaryotic cells such as endoplasmic reticulum, Golgi body, mitochondria and chloroplasts. Respiration takes place on a special piece of the cell membrane called a **mesosome** and those prokaryotes that can photosynthesise have a form of chlorophyll but no chloroplasts to hold it.

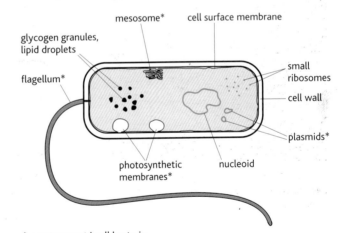

* = not present in all bacteria

fig. 3.1.2 A typical prokaryotic cell.

Questions

1 Knowledge about the structure of cells developed as light microscopes improved. Why were the two developments so closely linked?

2 Models of the internal structure of a cell have changed dramatically since electron microscopes were introduced. Some people have questioned the validity of the current models of the cell. Why do you think this is, and how would you justify relying on these models?

The characteristics of eukaryotic cells

Most microscopes images, apart from those of living material or from a scanning electron microscope, make cells appear flat and two-dimensional. But cells are actually spheres, cylinders or asymmetrical three-dimensional shapes – so try to use your imagination when you look at cells and see them in three dimensions.

In animals and plants there is a very wide range of different types of cells, each with a different function. But in both animals and plants there are certain cell features that turn up again and again, and we can put these together as a 'typical' plant or animal cell. Remember that this typical cell does not really exist, but it acts as a useful guide to what to look for in any eukaryotic cell. In this chapter you will be concentrating on a typical animal cell.

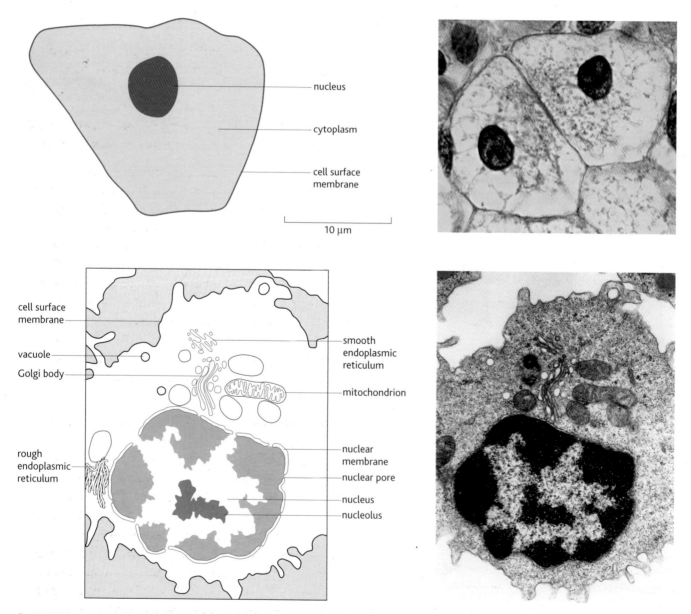

fig. 3.1.3 Many regions of an animal cell which appeared to have no particular features, or to be blurred areas when observed using a light microscope become complex structures when revealed by electron microscopy.

The typical animal cell

A typical animal cell is surrounded by a membrane known as the cell surface **membrane** (see chapter 2.3). Inside this membrane is a jelly-like liquid called **cytoplasm**, containing a **nucleus** – the two together are known as **protoplasm**. The cytoplasm contains much which is needed to carry out the day-to-day tasks of living, whilst the nucleus is vital to the long-term survival of the cell. This basic pattern gives rise to an enormous number of variations suited for the different functions that arise within the animal kingdom.

The various parts of the cell have complex and detailed structures, which are seen more clearly when an electron microscope is used. The structure of each part of the cell is closely related to its function – the job it has to do.

Membranes

Membranes in a cell are important both as an outer boundary to the cell and in the multitude of internal (**intracellular**) membranes. In chapter 2.3 you looked at the importance of cell membranes for controlling the movement of substances, but membranes inside the cell also have other functions, as you will see in the mitochondria and the large, internal membrane structure called the endoplasmic reticulum described below.

The protoplasm

When the light microscope was the only tool biologists had to observe cells, they thought that the cytoplasm was a relatively structureless, clear jelly. But the electron microscope revealed the cytoplasm to be full of all manner of structures, known as **organelles**. This detailed organisation is known as the **ultrastructure** of the cell, described below.

The nucleus

The nucleus is usually the largest organelle in the cell (10–20 μm) and can be seen with the light microscope. Electron micrographs show that the nucleus, which is usually spherical in shape, is surrounded by a double nuclear membrane containing holes or pores. Chemicals can pass in and out of the nucleus through these pores so that the nucleus can control events in the cytoplasm. Inside the nuclear membrane, or envelope as it is sometimes called, are two main substances, **nucleic acids** and **proteins**. The nucleic acids are **deoxyribonucleic acid** (**DNA**) and **ribonucleic acid** (**RNA**) (see chapter 2.1). When the cell is not actively dividing, the DNA is bonded to the protein to form **chromatin**, which looks like tiny granules. In the nucleus there is at least one **nucleolus** – an extra-dense area of almost pure DNA and protein. The nucleolus is involved in the production of ribosomes. Recent research also suggests that the nucleolus plays a part in the control of cell growth and division.

Mitochondria

The name **mitochondrion** simply means 'thread granule' and describes the tiny rod-like structures (1 μm wide by up to 10 μm long) in the cytoplasm of almost all cells as seen under the light microscope. In recent years we have been able to understand not only their complex structure but also their vital functions.

The mitochondria are the 'powerhouses' of the cell. Here, in a series of complicated biochemical reactions, energy is released from food by respiration using oxygen. This energy is in the form of ATP (see chapter 2.3) which can be used to drive the other functions of the cell and indeed the organism. The number of mitochondria present can give you useful information about the functions of a cell. Cells that require very little energy, eg fat storage cells, have very few mitochondria. Any cell with an energy-demanding function, eg muscle cells or cells that carry out a lot of active transport, will contain large numbers of mitochondria.

Mitochondria are surrounded by an outer and inner membrane. They also contain their own genetic material, so that when a cell divides the mitochondria replicate themselves under the control of the nucleus. They have an internal arrangement that is well adapted for their function (see **fig. 3.1.4**). The inner membrane is folded to form cristae surrounded by a fluid matrix. Scientists think that mitochondria (and chloroplasts) originated as symbiotic **eubacteria** living inside early cells. Over millions of years of evolution they have become an integral part of the cell.

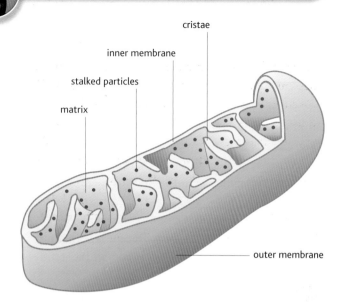

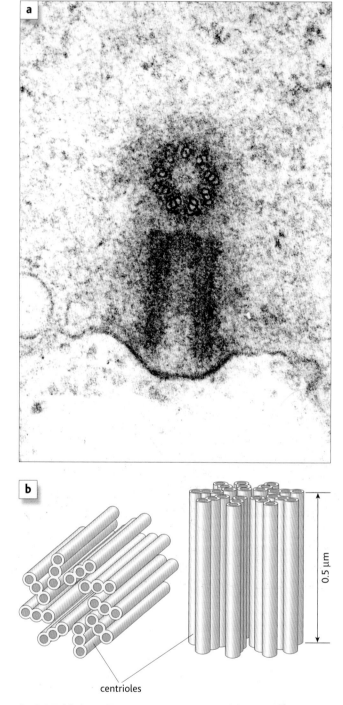

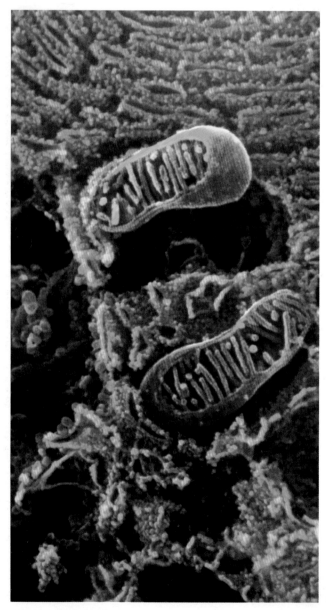

fig. 3.1.4 In the mitochondrion structure is closely related to the vital function of respiration.

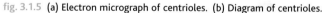

centrioles

fig. 3.1.5 (a) Electron micrograph of centrioles. (b) Diagram of centrioles.

The centrioles

In each cell there is usually a pair of **centrioles** near the nucleus (**fig. 3.1.5**). Each centriole is made up of a bundle of nine tubules and is about $0.5\,\mu$m long by $0.2\,\mu$m wide. The centrioles are involved in cell division. When a cell divides the centrioles pull apart to produce a **spindle** of microtubules which are involved in the movement of the chromosomes, as you will see later in this chapter.

The cytoskeleton

A cellular skeleton may seem a contradiction in terms, yet work in recent years has shown that a **cytoskeleton** is a feature of all eukaryotic cells. It is a dynamic, three-dimensional web-like structure that fills the cytoplasm (**fig. 3.1.6**). It is made up of **microfilaments**, which are protein fibres, and **microtubules**, tiny protein tubes about 20 nm in diameter. Microtubules are found, both singly and in bundles, throughout the cytoplasm. These microtubules are largely made up of the globular protein tubulin (see **fig. 3.1.6**). The cytoskeleton performs several functions. It gives the cytoplasm structure and keeps the organelles in place. Many of the proteins in the microfilaments are related to actin and myosin, the contractile proteins in muscle, and the cytoskeleton is closely linked with cell movements and transport within cells.

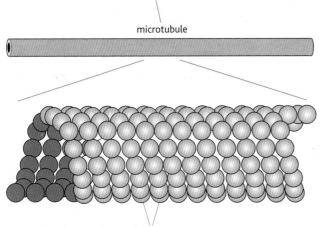

microtubule

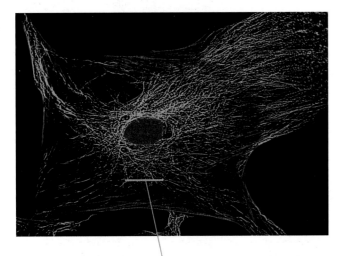

tubulin sub-units

fig. 3.1.6 The cytoskeleton forms a tangled web of structural and contractile fibres that hold the organelles in place and enable cell movement to occur.

Vacuoles

Vacuoles are not a permanent feature in animal cells. These membrane-lined enclosures are formed and lost as needed. Many simple animals make food vacuoles around the prey they engulf. White blood cells in higher animals form similar vacuoles around engulfed pathogens. **Contractile vacuoles** are an important feature in simple animals that live in fresh water because they allow the water content of the cytoplasm to be controlled. But in spite of these examples, vacuoles are not a major feature of animal cells and permanent vacuoles are never seen.

HSW The right tool for the job

Different types of microscopy provide very different types of information, as the images of a mitochondrion in **fig. 3.1.4** and centrioles in **fig. 3.1.5** show. The scanning EM shows the intact organelle and allows detailed measurements of the outer dimensions to be taken. The transmission EM provides clear images of the internal structures of the mitochondrion. This technique, combined with ways of marking molecules, helps scientists to work out exactly where the different reactions of cellular respiration take place.

Questions

1 What is the role of the cytoskeleton in the cytoplasm and why has its importance only recently been recognised?

2 Explain the importance of 'packaging' different parts of eukaryotic cells into organelles.

3 Look at the different images that result from transmission and scanning electron microscopes in **fig. 3.1.4** and **fig. 3.1.5** and describe how they differ. Suggest the value of each type of image and give examples where each would be more appropriate to use.

Protein transport in cells

Endoplasmic reticulum

The **endoplasmic reticulum** (**ER**) spreads through the whole cytoplasm. It is a three-dimensional network of cavities, some sac-like and some tubular, bounded by membranes. The ER network links with the membrane around the nucleus, and it is now known that it makes up a large part of the transport system within a cell as well as being the site of synthesis of many important chemicals.

HSW Finding out about the endoplasmic reticulum

The extensive presence of the endoplasmic reticulum in cells was only really revealed by the electron microscope. It has since been calculated that 1 cm³ of liver tissue contains about 11 m² of endoplasmic reticulum! Electron microscopes also helped scientists to work out the functions of the endoplasmic reticulum, by showing up the different forms – the rough and the smooth endoplasmic reticulum.

Another useful technique is to provide cells with radioactively labelled chemicals that are building blocks, eg labelled amino acids for the synthesis of proteins, and then find out where they appear in the cell. The labelled products can be tracked using microscopy. Another method of locating them is to break the cells open and then spin the contents in a centrifuge. The different parts of the cell can be separated and the regions containing the radioactively labelled substances identified.

Rough and smooth endoplasmic reticulum

Electron micrographs show that much of the outside of the endoplasmic reticulum membrane is covered with granules called **ribosomes**, so this is known as **rough endoplasmic reticulum** (**RER**) (fig. 3.1.7). The function of the ribosomes is to make proteins, and the RER isolates and transports these proteins once they have been made. Some proteins, such as digestive enzymes and hormones, are not used inside the cell that makes them, so they have to be secreted, that is moved out of the cell without interfering with the cell's activities. This is an example of exocytosis. Many other proteins are needed within the cell. The RER has a large surface area for the synthesis of all these proteins, and it stores and transports them both within the cell and from the inside to the outside. Cells that secrete materials, such as those producing the digestive enzymes in the lining of the gut, have a large amount of RER.

Not all endoplasmic reticulum is covered in ribosomes (fig. 3.1.7). **Smooth endoplasmic reticulum** (**SER**) is also involved in synthesis and transport, but in this case of the fatty molecules known as steroids and lipids. For example, lots of SER is found in the testes, which make the steroid hormone testosterone, and in the liver, which metabolises cholesterol amongst other lipids. The amount and type of endoplasmic reticulum in a cell give an idea of the type of job the cell does.

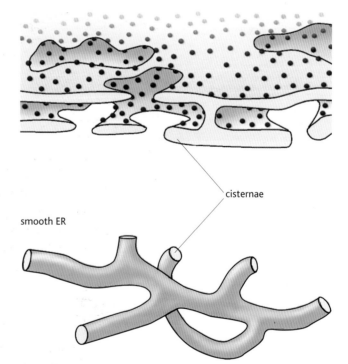

fig. 3.1.7 Rough and smooth endoplasmic reticulum. Apart from lacking encrusting ribosomes, smooth ER is more tubular than rough ER.

The Golgi body

Under the light microscope the **Golgi body** looks like a rather dense area of cytoplasm. An electron microscope reveals that it is made up of stacks of parallel, flattened membrane pockets called cisternae, formed by vesicles from the endoplasmic reticulum fusing together. The Golgi body has a close link with, but is not joined to, the rough endoplasmic reticulum. It has taken scientists a long time to discover exactly what the Golgi body does. Materials have been radioactively labelled and tracked through the cell to try and find out exactly what goes on inside it (see **fig. 3.1.8**).

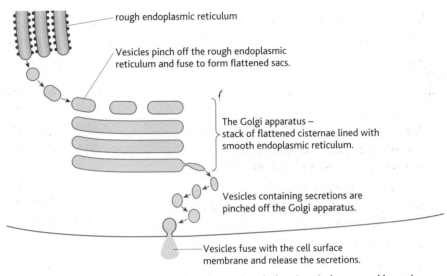

rough endoplasmic reticulum

Vesicles pinch off the rough endoplasmic reticulum and fuse to form flattened sacs.

The Golgi apparatus – stack of flattened cisternae lined with smooth endoplasmic reticulum.

Vesicles containing secretions are pinched off the Golgi apparatus.

Vesicles fuse with the cell surface membrane and release the secretions.

fig. 3.1.8 The Golgi body takes proteins from the rough endoplasmic reticulum, assembles and packages them and transports them to where they are needed. This may be the surface of the cell or different regions inside it.

Proteins are brought to the Golgi body in vesicles which have pinched off from the RER where they were made. The vesicles fuse with the membrane sacs of the Golgi body and the protein enters the Golgi stacks. As the proteins travel through the Golgi body they are modified in various ways. Carbohydrate is added to some proteins to form glycoproteins such as mucus. The Golgi body also seems to be involved in producing materials for plant cell walls and insect cuticles. Some proteins in the Golgi body are digestive enzymes. These may be enclosed in vesicles to form an organelle known as a **lysosome**. Alternatively enzymes may be transported through the Golgi body and then in vesicles to the cell surface membrane where the vesicles fuse with the membrane to release extracellular digestive enzymes.

HSW Discovering the Golgi

The Golgi body was first reported over 100 years ago in April 1898. The flattened stack of membranes was observed by the Italian scientist Camillo Golgi through a light microscope. For more than 50 years scientists argued over its function. Some thought it was an artefact from the process of fixing and staining during tissue preparation. The arrival of the electron microscope in the 1950s allowed the detailed structure of the Golgi body to be seen clearly.

The electron microscope has been central in showing details of the internal structure of the Golgi body. In addition a number of techniques have been developed which have allowed more detailed understanding. The most important of these has been the process of labelling specific enzymes so they can be seen using the electron microscope. The inner areas of the Golgi body, nearer to the RER, have been shown to be very rich in enzymes that modify proteins in various ways. This is where most enzymes or membrane proteins are converted into the finished product. In contrast, in the outer regions of the Golgi body you find lots of finished protein products, but few of the enzymes that make them.

The movement of cell membrane proteins through the Golgi body is very complex. Areas of the protein that need to be on the outside of the cell membrane, such as receptor binding sites, are orientated by the Golgi body so that when they arrive at the membrane they are inserted facing in the right direction.

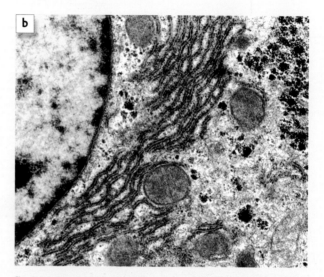

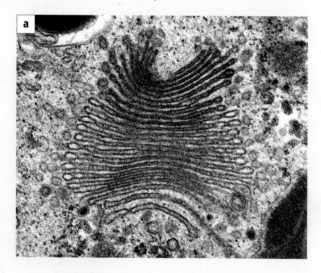

fig. 3.1.9 Scientists have used evidence from electron micrographs of the Golgi body to work out just how proteins are moved through the system. However, interpretation of electron micrographs is a skill in itself. Photograph (a) is a Golgi body. Even the scientists who made photograph (b) are not sure if it is a Golgi, or RER, and these micrographs don't have the added complication of labelled proteins.

Lysosomes

Food taken into the cell of a single-celled animal such as *Amoeba* must be broken down into simple chemicals that can then be used. Organelles in the cells of your body that are worn out need to be destroyed. These jobs are the function of the lysosomes. The word lysis from which they get their name means 'breaking down'.

Lysosomes appear as dark, spherical bodies in the cytoplasm of most cells and they contain a powerful mix of digestive enzymes. They frequently fuse with each other and with a membrane-bound vacuole containing either food or an obsolete organelle. Their enzymes then break down the contents into molecules that can be reused. A lysosome may fuse with the outer cell membrane to release its enzymes outside the cell as extracellular enzymes, eg to destroy bacteria or in digestion.

Lysosomes can also self-destruct. If an entire cell is damaged or wearing out, its lysosomes may rupture, releasing their enzymes to destroy the entire contents of the cell. This is known as **apoptosis**. Problems can arise if this starts to happen when it shouldn't.

HSW Discovering lysosomes

No one knew about lysosomes before the 1950s. The first clue to their existence came from an investigation into phosphatase enzymes in the cell by Christian de Duve (a Nobel prize winner in 1974). Working on rat liver cells, he always extracted his enzymes by breaking cells open in a very fast blender. One day he tried a more gentle process in a centrifuge. To his surprise, he found far fewer phosphatase enzymes than he expected. If the mixture was then blended fast or put in the fridge for a few days, the enzyme activity returned. De Duve hypothesised that the enzymes might be enclosed in a membrane which was damaged by the fast blending.

In 1955 Alex Novikoff and his team in the US were also looking at rat liver enzymes, using an electron microscope to examine the different layers (fractions) of enzyme-containing liquid after spinning in a centrifuge. They found what we now know as lysosomes in the fraction with lots of phosphatase activity. Soon afterwards a method of staining phosphatase enzymes was developed which worked for both light and electron microscopes.

In June 1958 de Duve presented his ideas for the possible role of lysosomes in intracellular digestion at a meeting of the Society of General Physiologists in the US. Lysosomes were on the map!

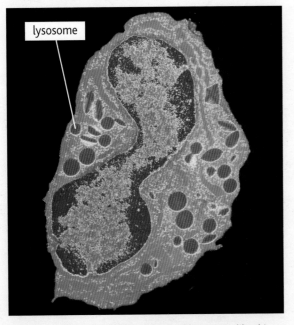

lysosome

fig. 3.1.10 **Clear microscopic evidence of lysosomes like this helped explain the role of these organelles in cells.**

HSW Apoptosis and disease

Apoptosis or 'cell suicide' is vital to the maintenance of a healthy body. Lysosomes rupture and their enzymes are released to kill cells that are old and coming to the end of their healthy life. Lysosomes may also destroy damaged cells or cells in which the DNA replication system is not functioning properly. But if apoptosis stops working properly – if too many cells are destroyed, or not enough lysosomes rupture so that cell death no longer takes place – this can have serious consequences for your health. For example, cancer is often thought of as a disease of uncontrolled cell growth (see below). But scientists are increasingly convinced that uncontrolled growth is not the whole story. Cancer cells also fail to die by apoptosis. As a result they propagate the genetic mutations that allow them to reproduce uncontrollably.

Excessive apoptosis also causes problems. It leads to the damage seen in the heart after a heart attack, and is linked to the death of killer T cells in HIV/AIDS (this is covered in more detail in the A2 course). The excessive rupturing of lysosomes may also be involved in autoimmune diseases such as rheumatoid arthritis, when cartilage tissue in joints self-destructs, and possibly in other conditions such as osteoporosis and retinitis pigmentosa.

Questions

1   What type of questions would scientists ask when they set out to investigate the functions of the endoplasmic reticulum, and how might they have set about finding the answers?

2   Describe the role of the RER and the Golgi body in the production of both intracellular and extracellular enzymes, and explain the importance of packaging products within a cell.

3   Make a flow chart to explain how scientists discovered lysosomes, including the questions that moved them on to each new stage of discovery.

4   Why is apoptosis so important in the body?

The organisation of cells

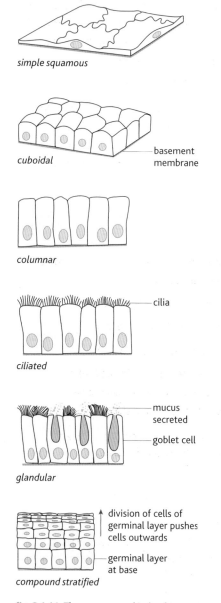

simple squamous

cuboidal — basement membrane

columnar

ciliated — cilia

glandular — mucus secreted — goblet cell

compound stratified — division of cells of germinal layer pushes cells outwards — germinal layer at base

fig. 3.1.11 There are many kinds of epithelial tissues in a human body.

Multicellular organisms are made up of specialised cells but these cells do not operate on their own. The specialised cells are organised into groups of cells known as **tissues**. These tissues consist of one or more types of cells all carrying out a particular function in the body. However, tissues do not operate in isolation. Many tissues are further organised into **organs**.

Tissues

Tissues are groups of similar cells that all develop from the same kind of cell. Although there are many different types of specialised cells, there are only four main tissue types in the human body – epithelial tissue, connective tissue, muscle tissue and nervous tissue. Modified versions of these tissue types containing different specialised cells carry out all the functions of the body. **Fig. 3.1.11** shows some different **epithelial tissues**, which are tissues that form the lining of surfaces both inside and outside of the body. Although some epithelial tissues consist of more than one kind of cell, they all originate from the basement membrane. Cells in epithelial tissues usually sit tightly together and form a smooth surface that protects the cells and tissues below.

Squamous epithelium is commonly found lining the surfaces of blood vessels, and forms the walls of capillaries and the lining of the alveoli. Cuboidal and columnar cells line many other tubes in the body. Ciliated epithelia often contain goblet cells that produce mucus. These epithelia form the surfaces of tubes in the lungs and the oviducts where the regular waving of the cilia from side to side moves materials along inside the tubes. Compound epithelia are found where the surface is continually scratched and abraded, such as the skin. The thickness of the tissue protects what lies beneath as new cells continue to grow from the basement membrane.

There are many other tissues in the body, including muscle tissue, nervous tissue, the collagen tissue and elastin tissue found in artery walls, glandular tissue that secretes substances from inside the cells. Connective tissue is the main supporting tissue in the body, and includes bone tissue and cartilage tissue as well as packing tissue that supports and protects some of the organs. Some of these tissues are shown in **fig. 3.1.12**.

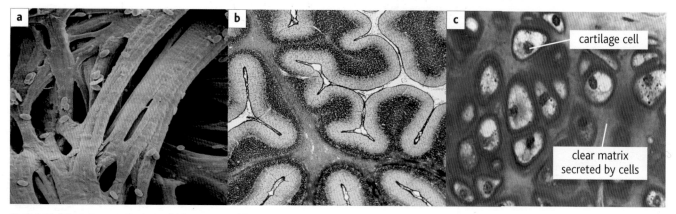

fig. 3.1.12 (a) cardiac muscle tissue, (b) brain tissue, (c) cartilage tissue.

Organs

An organ is made up of a group of tissues that are grouped into a structure so that they can work effectively together. There are many organs in the human body, some of which are shown in fig. 3.1.13.

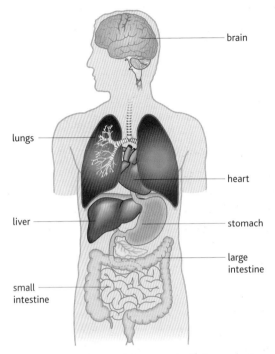

fig. 3.1.13 **The position of some of the organs in the human body.**

Plants also have cells grouped into tissues and organs. For example the leaf is an organ that is composed of vascular tissue, epithelial tissue and mesophyll tissues as shown in fig. 3.1.14.

Systems

In animals, in many cases a number of organs work together as a **system** to carry out large-scale functions in the body. For example the digestive system includes the organs of the stomach, pancreas, small and large intestines, and the nervous system includes the brain, spinal cord and all peripheral nerves.

Most of the cells in tissues, organs and systems have differentiated during development so that they are capable of carrying out their specific function. You will find out more about how this process happens in chapter 3.2.

Questions

1 Explain how the structure of the following tissues is related to their function.

  a squamous epithelium lining an alveolus

  b ciliated epithelium lining a bronchus

  c muscle tissue in the biceps muscle

2 a Choose one of the systems in the human body and describe briefly the cells, tissues and organs found within that system.

  b Explain why this grouping enables the system to carry out its function effectively.

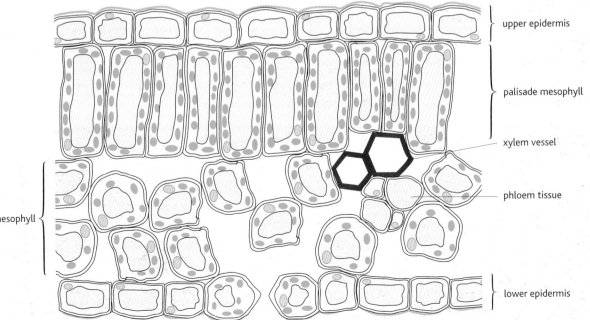

fig. 3.1.14 **The tissues in the organ of a plant leaf.**

Cell division

One of the most awe-inspiring processes of life is the way in which organisms reproduce. Like begets like – buttercups produce new buttercups, *Amoeba* produce more *Amoeba* and liver cells generate more liver cells. Most new biological material comes about as a result of the process of nuclear division known as **mitosis**, followed by the rest of the cell dividing. **Asexual reproduction** – the production of genetically identical offspring from a single parent cell or organism – and growth are both the result of mitotic cell division. The production of offspring by sexual reproduction is also largely dependent on mitosis to produce new cells after the gametes have fused. In mitosis the chromosomes of a cell are duplicated and the genetic information is then equally shared out between the two daughter cells that result. The formation of the sex cells involves a different process of nuclear division called **meiosis** (see chapter 3.2).

What are chromosomes?

A chromosome is made up of a mass of coiled threads of DNA and proteins. If a chromosome were as long as five consecutive letters on this page, the DNA molecule it contained would stretch the length of a football pitch or more. In a cell that is not actively dividing the chromosomes cannot easily be identified as individual structures. They are translucent to both light and electrons so they cannot easily be seen. When the cell starts to actively divide the chromosomes condense – they become much shorter and denser. They then take up stains very readily (this is the basis of the name 'chromosome' or 'coloured body'), and individual chromosomes can be identified.

When the DNA molecules condense, they have to be packaged very efficiently. This is achieved with the help of positively charged proteins called **histones**. The DNA winds around the histones to form dense clusters known as **nucleosomes** (**fig. 3.1.15**). These then interact to produce more coiling and then supercoiling to form the dense chromosome structures you can see through the microscope in the nucleus of a dividing cell. In the supercoiled areas the genes are not available to be copied to make proteins. This is one way in which cells of different types are produced (see chapter 3.2).

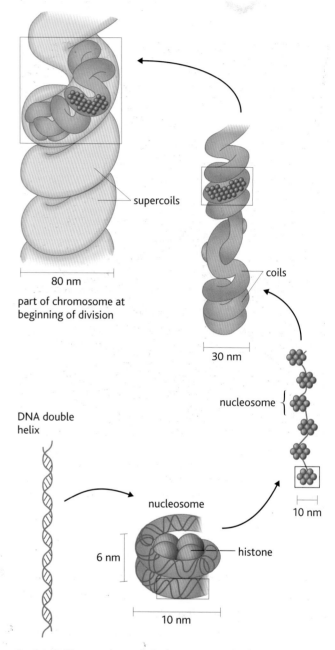

supercoils

80 nm

part of chromosome at beginning of division

coils

30 nm

nucleosome

DNA double helix

nucleosome

10 nm

6 nm

histone

10 nm

fig. 3.1.15 Histones play a role in the organisation of DNA into orderly chromosomes which can be replicated.

The cells of every different species possess a characteristic number of chromosomes – in humans this is 46. These chromosomes occur in matching pairs, one of each pair originating from each parent. In mitosis the two cells that result from the division must both receive a full set of chromosomes. So before a cell divides it must duplicate the original set of chromosomes. During mitosis these chromosomes are divided equally between the two new cells so that each has a complete and identical set of genetic information.

During the active phases of cell division the chromosomes become very coiled and condensed. In this state they are relatively easy to photograph and a special display or **karyotype** can be made (see **fig. 3.1.16**).

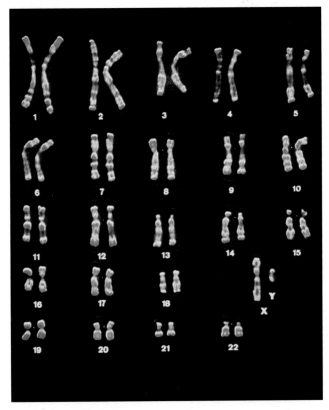

fig. 3.1.16 Human karyotypes organise the chromosomes into the 22 pairs of autosomes and 1 pair of sex chromosomes making up the 23 pairs of chromosomes found in every healthy human cell except the eggs and the sperm. This is the karyotype of a male; a karyotype for a female was shown in fig. 2.2.1.

The cell cycle

Cells divide on a regular basis to bring about growth and asexual reproduction. They divide in a sequence of events known as the **cell cycle** which involves several different phases (see **fig. 3.1.17**). There is a period of active division which is mitosis, when an increase in the number of cells takes place. This is followed by a period of non-division known as **interphase**, which is when the cells increase in mass and size, replicate their DNA and carry out normal cellular activities. The length of the cell cycle is variable. It can be very rapid, taking 24 hours or less, or it can take a few years.

In multicellular organisms the cell cycle is repeated very frequently in almost all cells during development. However, once the organism is mature, it may slow down or stop completely in some tissues. The cell cycle is controlled by a number of chemical signals made in response to different genes. This control is brought about at a number of checkpoints where the cell cycle moves from one phase to the next. The control chemicals are small proteins called **cyclins**. These build up and attach to enzymes called **cyclin-dependent kinases** (**CDKs**). The **cyclin/CDK complex** that is formed adds a phosphate group to (phosphorylates) other proteins, changing their shape and bringing about the next stage in the cell cycle. So, for example, when the chromatin in the nucleus is phosphorylated the chromosomes become denser, while when some nuclear membrane proteins are phosphorylated the structure of the membrane breaks down.

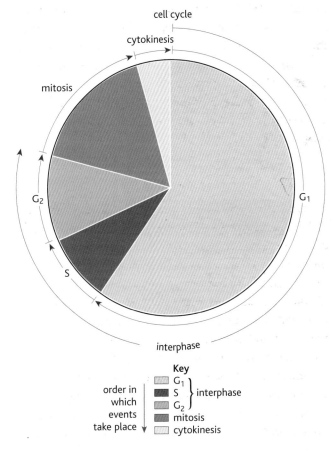

Key

order in which events take place	G_1 ⎫
	S ⎬ interphase
	G_2 ⎭
	mitosis
	cytokinesis

fig. 3.1.17 Phases of the cell cycle. In very actively dividing tissue the cycle is repeated as fast as possible, whilst in other tisues the time between successive divisions may be years.

Questions

1 Why do chromosomes only become visible as a cell goes into mitosis?

2 If a culture of cells is dividing every 48 hours, how long would you expect the different stages of the cycle to take?

Mitosis

During the process of cell division the chromosomes are duplicated. Then they, along with the remaining contents of the cell, are divided up in such a way that two identical daughter cells are formed. Walther Flemming (1843–1905), a German cytologist, was the first to describe what is sometimes called the 'dance of the chromosomes'. It refers to the complex series of movements that occur during cell division as the chromosomes jostle for space in the middle of the nucleus and then pull apart to opposite ends of the cell. The events of mitosis are continuous, but as in the case of so many biological processes it is easier to describe what is happening by breaking events down into phases. These are known as **prophase**, **metaphase**, **anaphase** and **telophase** (see **fig. 3.1.18**).

A cell is in interphase for much of the time. This used to be called the **resting phase**, but nothing could be further from the truth. During interphase the normal metabolic processes of the cell continue and new DNA is produced as the chromosomes replicate. Sufficient new proteins, cytoplasm and cell organelles are synthesised so that the cell is prepared for the production of two new cells. ATP production is also stepped up at times to provide the extra energy needed as the cells divide. Once all that is needed is present and the parent cell is large enough, interphase ends and mitosis begins.

Prophase

Before mitosis begins the genetic material has been replicated to produce exact copies of the original chromosomes. By the beginning of prophase both the originals and the copies are referred to as **chromatids**. In prophase the chromosomes coil up, take up stain and become visible. Each chromosome at this point consists of two daughter chromatids which are attached to each other in a region known as the **centromere**. The nucleolus breaks down. The centrioles begin to pull apart to form the spindle.

HSW Evidence for mitosis

The discovery of mitosis depended on the development of the microscope. Walther Flemming published his work on mitosis in 1882. Flemming had also discovered the presence of chromosomes in the cell using dyes which were taken up by the genetic material. A Belgian scientist, Edouard van Beneden, discovered chromosomes at much the same time. Flemming had not come across Mendel's work on inheritance and so he did not make the connection between what he was seeing and genetic inheritance. In spite of this, Flemming's discoveries are widely regarded as some of the most important work in cell biology.

 You can observe mitosis relatively easily in the cells of rapidly dividing tissues such as a growing root tip. Using an appropriate dye (such as acetic orcein) which shows up the chromosomes, tissue squashes can be produced which show all the stages of mitosis.

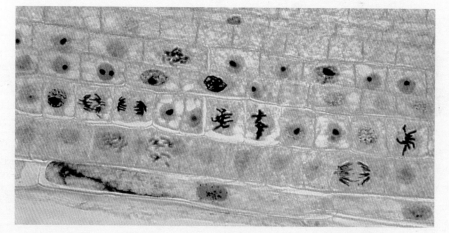

fig. 3.1.18 Stained section of a root tip squash showing cells at different stages of mitosis

Living tissue can be observed too, and dramatic recordings of the activity of the chromosomes have been made using time-lapse photography. This has moved our understanding forward considerably. It is impossible to show this on the printed page, but a viewing of the movements of the cell contents during mitosis shows why it is called the 'dance of the chromosomes'.

The events of mitosis

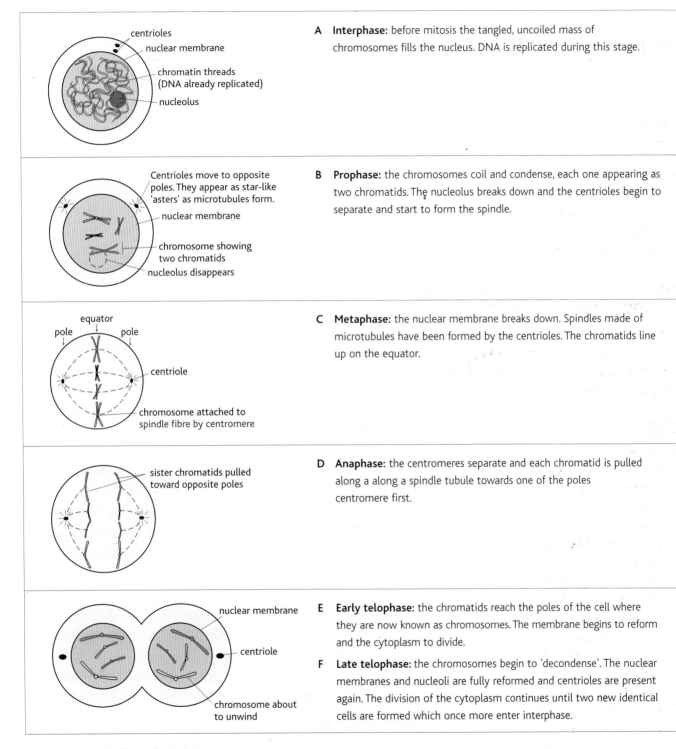

A **Interphase:** before mitosis the tangled, uncoiled mass of chromosomes fills the nucleus. DNA is replicated during this stage.

B **Prophase:** the chromosomes coil and condense, each one appearing as two chromatids. The nucleolus breaks down and the centrioles begin to separate and start to form the spindle.

C **Metaphase:** the nuclear membrane breaks down. Spindles made of microtubules have been formed by the centrioles. The chromatids line up on the equator.

D **Anaphase:** the centromeres separate and each chromatid is pulled along a along a spindle tubule towards one of the poles centromere first.

E **Early telophase:** the chromatids reach the poles of the cell where they are now known as chromosomes. The membrane begins to reform and the cytoplasm to divide.

F **Late telophase:** the chromosomes begin to 'decondense'. The nuclear membranes and nucleoli are fully reformed and centrioles are present again. The division of the cytoplasm continues until two new identical cells are formed which once more enter interphase.

fig. 3.1.19 **The main phases of mitosis in a simplified cell which has only two pairs of chromosomes.**

Metaphase

The nuclear membrane has broken down and the centrioles have moved to opposite poles of the cell, forming a set of microtubules between them which is known as the **spindle**. The chromatids appear to jostle about for position on the **metaphase plate** or **equator** of the spindle during metaphase. They eventually line up along this plate, with each centromere associated with a microtubule of the spindle.

Anaphase

The centromeres that have linked the two identical chromatids split, and from then on the chromatids act as completely separate entities. They effectively become new chromosomes. The chromatids from each pair are drawn (centromere first) towards opposite poles of the cell. This separation occurs quickly, taking only a matter of minutes. At the end of anaphase the two sets of chromatids have been separated to opposite ends of the cell.

The chromatids cannot move on their own. They rely on the microtubules of the spindle to allow them to move. The spindle was for many years envisaged as a structure running from one end of the cell to the other. It is now known to be made up of overlapping microtubules containing contractile fibres which are similar to those in animal muscles. Contraction of the overlapping fibres causes the movement of the chromatids. This is an energy-using process, and the energy is supplied by cell respiration.

Telophase

During telophase the spindle fibres break down and nuclear envelopes form around the two sets of chromosomes. The nucleoli and centrioles are also re-formed. The chromosomes begin to unravel and become less dense and harder to see.

The final phase is the division of the cytoplasm, sometimes referred to as **cytokinesis**. In animal cells a ring of contractile fibres tightens around the centre of the cell rather like a belt tightening around a sack of flour. These fibres seem to be the same as those found in animal muscles. They continue to contract until the two cells have been separated. In plant cells the division of the cell occurs rather differently, with a cellulose cell wall building up from the inside of the cell outwards. In both cases the end result is the same – two identical daughter cells are formed which then enter interphase and begin to prepare for the next cycle of division.

Mitosis is the source of all the new cells needed for organisms to grow and to replace worn out cells. It is also the method by which organisms undergo asexual reproduction.

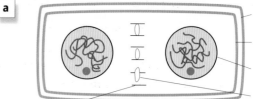

cellulose cell wall
cell surface membrane
nucleus in telophase
Golgi vesicle
remaining spindle fibre

Some spindle fibres remain and guide Golgi vesicles to the equator of the cell.

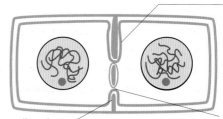

large vacuoles formed from Golgi vesicles
Golgi vesicle fuses with cell surface membrane

The vesicles enlarge and fuse together, forming a cell plate.

cell surface membrane formed from Golgi vesicle membrane
cellulose walls with middle lamella formed between cells
plasmodesma

The basic structure of the cell walls forms within each vesicle, and the vesicles fuse to join the cell wall together. Small gaps left between the vesicles form plasmodesmata.

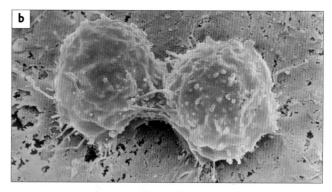

fig. 3.1.20 The final stages of mitosis (a) in a plant cell and (b) in an animal cell.

Questions

1 Summarise the stages of mitotic cell division in animal cells.

2 Explain why root tips are particularly suitable material to use for preparing slides to show mitosis.

Asexual reproduction and natural cloning

As you have already seen, mitosis is the basis of asexual reproduction. Asexual reproduction involves only one parent individual – plant or animal – and it results in genetically identical individuals or **clones**. It has many advantages for an organism. It is safe, does not rely on finding a mate and can give rise to large numbers of offspring very rapidly. It also has one large disadvantage – the offspring are almost all genetically identical to the parent organism. This becomes a problem when living conditions change in some way. Changes such as the introduction of a new disease to an environment, a change in the temperature or human intervention can cause the total destruction of a group of genetically identical organisms, because if one can't cope neither can all the others.

In the plant world many species undergo both sexual and asexual reproduction as a matter of course. They reproduce sexually by flowering, but they also produce bulbs, corms, tubers, rhizomes, runners or suckers to name but a few asexual methods of reproduction in plants.

Strategies for asexual reproduction

There are a variety of strategies for asexual reproduction, some of which are outlined below.

Binary fission

Fission involves mitosis followed by the splitting of an individual. It is found as a method of reproduction in only the simplest groups of organisms (**fig. 3.1.21**). Two new individuals are usually formed, so it is known as binary fission. Bacteria and protoctists such as *Amoeba* undergo this form of asexual reproduction. Bacteria are capable of enormous increases in numbers under ideal conditions, when they may divide every 20 minutes – one of the shortest known cell cycles.

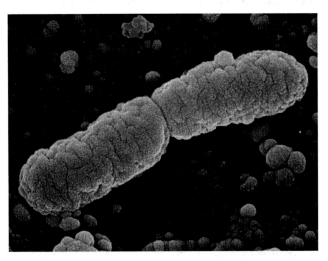

fig. 3.1.21 Fission as a reproductive strategy is very effective, but only works for the simplest living organism such as this bacterium.

Although fission is limited as a reproductive strategy in the world of whole organisms, a similar method is used in cell reproduction for growth and repair in all living things.

Producing spores

Sporulation involves mitosis and the production of asexual **spores** which are capable of growing into new individuals. These spores can usually survive adverse conditions, and are also easily spread over great distances. This form of asexual reproduction is most common in fungi and plants such as mosses and ferns.

Regeneration

Fragmentation and **regeneration** constitute a very dramatic form of asexual reproduction, occurring when organisms replace parts of the body which have been lost. For example, many lizards shed their tails when attacked and then grow another. This is known as regeneration. Some organisms manage an even more spectacular form of regeneration – they can reproduce themselves asexually from fragments of their original body. For example, certain starfish attack and eat oysters. To protect oyster beds from destruction oyster fishermen have attempted to destroy the starfish, often by chopping them up and throwing them back into the sea. This failed dismally as each fragment can regenerate to form another starfish hungry for oysters!

This type of cloning occurs naturally – some members of groups as diverse as fungi, flatworms, filamentous algae and sponges fragment and then regenerate as a regular method of reproducing. An adaptation of this ability has been developed to allow artificial cloning of plants, which you will look at in more detail in chapter 3.3.

Producing buds

Budding in a reproductive sense does not mean the production of buds by flowers. In reproductive budding there is an outgrowth from the parent organism which produces a smaller but identical individual, produced purely by mitotic cell division. This 'bud' eventually becomes detached from the parent and has an independent existence. Yeast cells reproduce by budding. In single-celled organisms like these the only recognisable difference between budding and binary fission is that in budding the parent cell is larger than the bud. Budding is relatively rare in the animal kingdom. A good example of an animal budding is in *Hydra*, seen in **fig. 3.1.22**. Asexual budding is only part of the reproductive strategy of *Hydra* – they reproduce sexually as well.

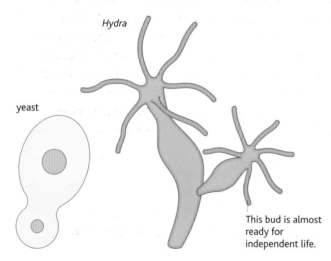

Hydra

yeast

This bud is almost ready for independent life.

fig. 3.1.22 Budding in yeast cells and in *Hydra*. Notice the asymmetry of the yeast cell and the bud which is what distinguishes this process from binary fission in single-celled organisms.

New plant structures

Vegetative propagation is in some ways a more sophisticated version of reproductive budding and occurs in flowering plants. A plant forms a structure which develops into a fully differentiated new plant, identical to the parent, and it eventually becomes independent. The new plant may be propagated from the stem, leaf, bud or root of the parent, depending on the type of plant. It involves only mitotic cell division. Vegetative propagation often involves **perennating organs**. These contain stored food from photosynthesis and can remain dormant in the soil to survive adverse conditions. They are often not only a means of asexual reproduction, but also a way of surviving from one growing season to the next. Examples include bulbs, corms, runners, suckers, rhizomes, stem tubers and root tubers.

Vegetative propagation is easily exploited by human gardeners to produce new plants. Splitting daffodil bulbs, removing new strawberry plants from their runners (**fig. 3.1.23**) and cutting up rhizomes are all easy ways of increasing plant numbers cheaply. As an added advantage the new plants are all clones so they will have exactly the same characteristics as their parents, eg they will be the same colour or produce fruit that is just as good.

Gardeners and farmers take asexual reproduction in plants one step further when they take cuttings. They induce fragmentation artificially. This involves taking a small piece of a plant – often part of the shoot – and planting it to grow on and develop by mitosis into another entire (identical) plant.

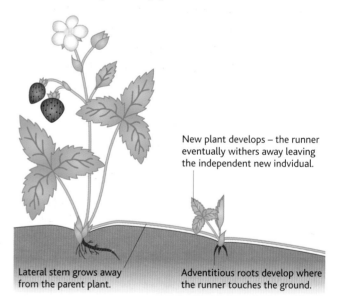

New plant develops – the runner eventually withers away leaving the independent new indvidual.

Lateral stem grows away from the parent plant.

Adventitious roots develop where the runner touches the ground.

fig. 3.1.23 Asexual reproduction in strawberry plants results in identical clones.

Asexual reproduction is common even in complex plants which maintain areas of unspecialised dividing cells throughout their life. In more complex animals, where the cells tend to become specialised much earlier, asexual reproduction is much less common.

HSW Komodo dragons and virgin births

Most vertebrates do not reproduce asexually. When they do it is known as **parthenogenesis**. It has been seen in about 70 species of vertebrates, including snakes, fish, a monitor lizard and a turkey. In spite of this, parthenogenesis has always been thought to be very rare in vertebrates.

Yet in the UK in 2006 new evidence appeared that suggests that parthenogenesis is not always rare in vertebrates. Komodo dragons are the largest land lizards on Earth. Their natural habitat is Indonesia where there are fewer than 4000 animals remaining. So the Komodo dragons in captivity are part of an important breeding programme. However, two different zoos in the UK (Chester Zoo and London Zoo) have reported that live, apparently healthy young dragons have hatched from eggs laid by females in the complete absence of any male dragons. The females were not related in any way, so it was not a rare family mutation. The young dragons have been DNA tested and their genetic make-up shows they come only from their mother. Both females have also bred sexually.

Female Komodo dragons have one W and one Z chromosome, while males have two Zs. Each egg carries either a W or a Z. When parthenogenesis takes place, the single chromosome is duplicated. Any eggs with WW will not develop (it makes genetic nonsense) but ZZ eggs can develop into normal male baby lizards.

fig. 3.1.24 **When this baby Komodo dragon hatched, scientists had to re-examine their ideas about parthenogenesis in reptiles.**

Richard Gibson is in charge of reptiles at London Zoo and an international expert in Komodo dragons. Richard feels that the arrival of these parthenogenic dragons should lead scientists to rethink their ideas on how common the process of parthenogenesis is, at least in reptiles. The ability to reproduce asexually as well as sexually could have evolved so that animals stranded in an isolated situation can nevertheless breed. If so it would be a very useful adaptation indeed – and therefore it would not be surprising if it was relatively common.

Questions

1 Suggest factors that would affect the rate of the cell cycle in dividing bacteria.

2 More living organisms result from asexual reproduction than sexual reproduction. Do you think this statement is accurate? Explain your response.

3 Why have no female Komodo dragons resulted from parthenogenesis?

4 Why is it important that the Komodo dragon mothers have also produced offspring sexually?

5 In the space of a few months new observations have changed a long-held scientific idea. Why have the Komodo dragon hatchlings forced scientists to rethink their ideas about asexual reproduction in vertebrates?

Growth

Mitosis isn't just about asexual reproduction. It plays a vital role in growth as well. Everyone is familiar with the concept of growth, but defining it in a biological sense isn't so easy! Growth is a permanent increase in the number of cells, or in the mass or size of an organism.

There are three distinct aspects of growth. They are cell division, **assimilation** and cell expansion. Cell division or mitosis is the basis of growth. Once cells have divided they usually get larger before dividing again. The resources needed to produce new cell material come from photosynthesis in plant cells and from feeding in animal cells. This is what is meant by assimilation, and when these materials are incorporated into cells the result is cell expansion. Cells can expand in other ways, eg by taking in water, but this increase in size may be only temporary. So growth is defined as involving a *permanent* increase in cell number, size or mass – or all three!

How is growth measured?

The measurement of growth is important both scientifically and medically. Growth may be affected by factors such as the availability of food, temperature and light intensity as well as the genetic make-up of the organism. Unfortunately the measurement of growth is not at all easy. Linear dimensions, such as height or head circumference (**fig. 3.1.25**), can be very deceptive – cake mixture will increase in both height and circumference as it cooks, but it has not grown! A balloon increases in circumference as it is blown up but it is not growing. Measuring mass also has its problems – the water content of the cells may vary greatly, particularly in plants, and more complex animals will have varying quantities of faecal material and urine held in their bodies.

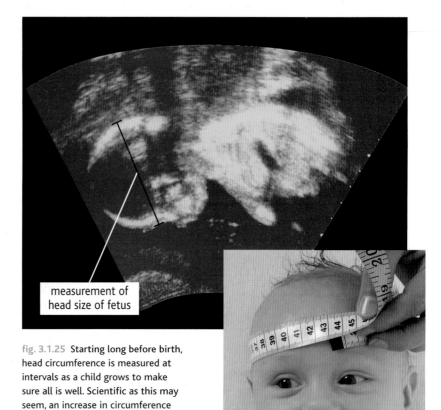

measurement of head size of fetus

fig. 3.1.25 **Starting long before birth, head circumference is measured at intervals as a child grows to make sure all is well. Scientific as this may seem, an increase in circumference does not always indicate growth.**

HSW Measuring growth in animals

Because growth involves an increase in the cell content of an organism, mass is the best and most commonly used measure of growth. However, the water content of organisms can vary greatly so **dry mass** is the most accurate way of measuring growth. The dry mass is the mass of the body of an organism with all the water removed from it. This gives an accurate picture of the amount of biological material present but has one major drawback. If you remove all the water from an organism you kill it, so that further growth cannot be measured. To get useful results from dry mass measures you need to grow large samples of genetically identical organisms under similar conditions, then take random samples and dry them to a constant dry mass. This method is very useful for plants but has obvious limitations for animals. It is not easy or ethical to maintain large colonies of genetically identical higher animals and then kill and dry them. This means that in most cases scientists use less reliable indicators such as height and wet mass to measure growth when working with animals.

Growth patterns

In spite of the difficulties in measuring growth, we have a good picture of the patterns of growth of many organisms. Growth curves show growth throughout the life of an organism, including when most growth takes place. The growth curve is very similar for most organisms (see **fig. 3.1.26**). In many animals, after an initial relatively slow start there is a rapid period of growth until maturity is reached, when growth slows down and may stop. In most land animals growth stops completely with maturity because size is limited by the weight of the animal and the ability of its muscles to move it against gravity. In plants growth often continues throughout life, and the same is true for marine animals, where the mass of the body is supported by the water. This pattern – even when it stops at maturity – is known as **continuous growth**.

Not all organisms undergo continuous growth. Insects grow in a series of **moults**. They shed one exoskeleton and then, while the new exoskeleton is soft, they expand the body by taking in air or water and 'grow'. Once the new skeleton has hardened, the air or water can be released and there is room for the tissues of the insect to increase in size and mass. This is known as **discontinuous growth**. If length is measured the insect appears to grow in a series of steps.

The development of the embryo is the time when the largest amount of growth, measured as a percentage of body mass, occurs. Different parts of the organism can grow at very different rates. For example, in the human embryo the nervous system and the head grow much faster than some other areas. Later in life – through puberty, for example – the head stops growing while the long bones and the rest of the body continue.

Right at the beginning of life, mitosis takes place at a very rapid rate. In organisms where growth slows down or stops completely at maturity, mitosis does not stop. Cells are continually becoming worn out and being replaced by mitotic divisions. This continues until the onset of **senescence** or old age, when mitosis occurs less frequently and the cells dying begin to outnumber the new cells being formed. When this process reaches a certain point, death of the whole organism will occur.

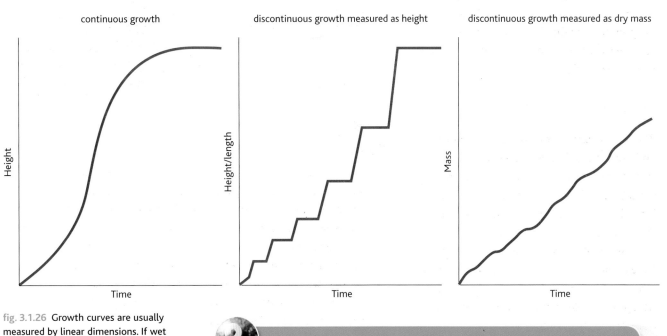

fig. 3.1.26 Growth curves are usually measured by linear dimensions. If wet mass is used the discontinuous growth curve becomes even more pronounced.

Questions

1 What are the difficulties in measuring growth in living organisms?

2 What is the role of mitosis in growth?

3 Why are the curves for discontinuous growth so different when measured as height and as dry mass?

3.2 Sexual reproduction and cell specialisation

Why sexual reproduction?

As you have seen, **asexual reproduction** can be very successful at producing new individuals but leaves the population vulnerable to changes in the environment. Relatively few organisms rely solely on asexual reproduction. Most have at the very least a back-up system of sexual reproduction which is used when conditions are tough. In the more complex organisms, particularly animals, sexual reproduction is the main way of producing offspring.

Sexual reproduction is the production of a new individual resulting from the joining of two specialised cells known as **gametes**. The individuals that result from sexual reproduction are *not* genetically the same as either of their parents, but contain genetic information from both of them. Sexual reproduction is more risky than asexual reproduction because it relies on two gametes meeting. It isn't always easy to find a mate, particularly if you are a solitary predator! It is also more expensive in terms of bodily resources because it usually involves special sexual organs. But the great advantage of sexual reproduction is that it increases genetic variety. This variety is the result of the fusing of gametes from two different individuals. In a changing environment, this gives a greater chance that one or more of the offspring will have a combination of genes that improves their chance of surviving and going on to reproduce.

fig. 3.2.1 The genetic variety in offspring produced by sexual reproduction is sometimes very easy to see!

What are gametes?

As you know from chapter 2.2, the nucleus of a cell contains the chromosomes. In the vast majority of the cells of an organism these occur in pairs. A cell containing two full sets of chromosomes is called **diploid (2n)** and the number of chromosomes in a diploid cell is characteristic for that species. However, if two diploid cells combined to form a new individual in sexual reproduction, the offspring would have four sets of chromosomes, losing the characteristic number for the species. Each new generation would become more heavily loaded with genetic material until eventually the cells would break down and fail to function. To avoid this, **haploid (n)** nuclei are formed with one set of chromosomes (half of the full chromosome number), usually within the specialised cells called gametes. Sexual reproduction occurs when two haploid nuclei fuse to form a new diploid cell called a **zygote** (see **fig. 3.2.3**), a process called **fertilisation**.

The formation of gametes

Gametes are formed in special sex organs. In simpler animals and plants the sex organs are often temporary, formed only when they are needed. In more complex animals the sex organs are usually more permanent structures that are sometimes called the **gonads**.

In flowering plants the sex organs are the **anthers** (male) and the **ovaries** (female). The male gametes, **pollen**, are produced in the anthers and the female gametes, **ovules**, are formed in the ovaries. In animals the male gonads are the **testes**, which produce **spermatozoa** (commonly known as **sperm**), the male gametes. The female gonads are the ovaries and they produce **ova** (eggs), the female gametes. The male gametes are often much smaller than the female ones, but they are usually produced in much larger quantities. This can be summarised as:

- male: many, mini, motile
- female: few, fat, fixed!

HSW Sexual reproduction in bacteria

For many years it was assumed that bacteria reproduced only asexually by fission. Then in 1946 Joshua Lederberg, a student at Yale University, and his professor, Edward Tatum, performed an experiment which showed that some sort of genetic exchange must occasionally take place in these organisms.

They grew two strains of bacteria. Strain A could grow on a minimal medium as long as the amino acid methionine and the vitamin biotin were added. These bacteria could synthesise everything else they needed. Strain B could grow on minimal medium as long as the amino acids threonine and leucine were added. Neither strain could grow on the minimal medium alone. But after mixing the two strains together, about 1 in every 10 million bacterial cells could grow on the minimal medium without any additions. The only way this was possible was if they had received some genes from strain A and some from strain B.

The electron microscope has since shown bacteria 'mating' and exchanging genetic information through a strand of cytoplasm known as the **sex pilus**. This knowledge of **bacterial conjugation** (**fig. 3.2.2**) has been vital in the development of **genetic engineering**, which will be discussed in more detail later in this chapter.

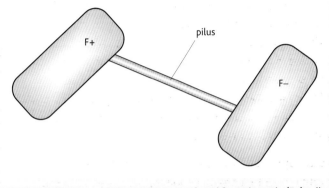

fig. 3.2.2 Bacterial conjugation: DNA is transferred from the male (F+) cell to the female (F−) cell through the pilus which joins them.

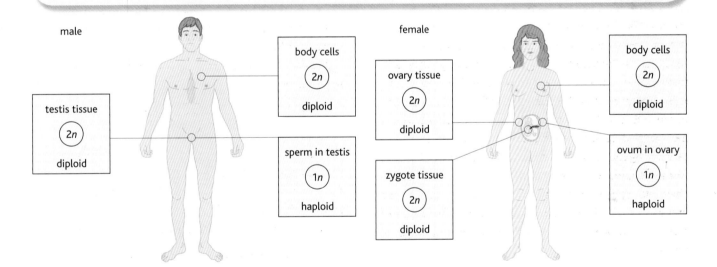

fig. 3.2.3 The only cells in the body that are haploid are the gametes.

Questions

1 Give examples of conditions when sexual reproduction would be more advantageous in the production of offspring than asexual reproduction. Explain your choices.

2 Suggest why most organisms have a stage of sexual reproduction in their life cycle.

3 Explain how the work of Tatum and Lederberg changed scientific thinking about reproduction in bacteria.

Meiosis

In chapter 3.1 you saw that when cells divide by mitosis the number of chromosomes in both the daughter cells is the same as in the original parent cell. However, in the cell divisions that form gametes the chromosome number needs to be halved to give the necessary haploid nuclei. To bring about this reduction in the chromosome number, gametes are formed by a different process of cell division known as **meiosis**. Meiosis is a reduction division and it occurs only in the sex organs. In animals the gametes are formed directly from meiosis. In flowering plants meiosis forms special cells called **microspores** (male) and **megaspores** (female) which then develop into the gametes. Meiosis is of great biological significance – it is the basis of the variation that allows species to evolve.

What happens to the chromosomes?

In meiosis two nuclear divisions give rise to four haploid daughter cells, each with its own unique combination of genetic material (**fig. 3.2.4**). These four genetically individual cells are a result of a process known as **crossing over**, as you will see below. The events of meiosis are continuous although we describe the stages as separate phases.

As in mitosis, the contents of the cell, and in particular the DNA, are replicated while the cell is in interphase. Once the cell has all the materials it needs it can enter meiosis.

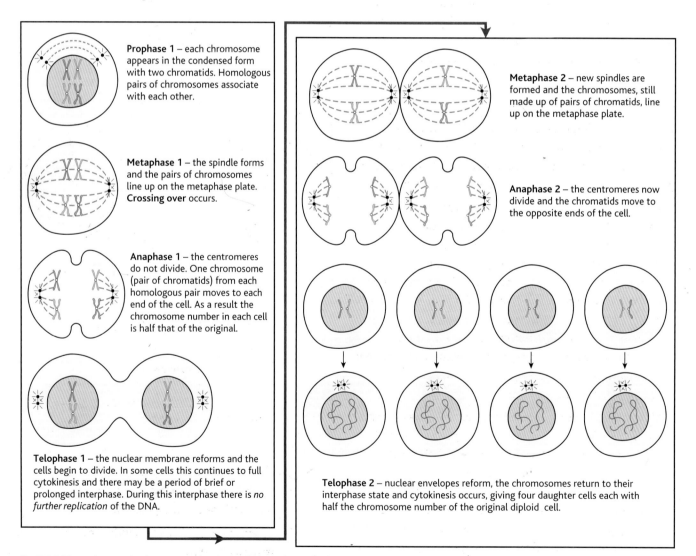

Prophase 1 – each chromosome appears in the condensed form with two chromatids. Homologous pairs of chromosomes associate with each other.

Metaphase 1 – the spindle forms and the pairs of chromosomes line up on the metaphase plate. **Crossing over** occurs.

Anaphase 1 – the centromeres do not divide. One chromosome (pair of chromatids) from each homologous pair moves to each end of the cell. As a result the chromosome number in each cell is half that of the original.

Telophase 1 – the nuclear membrane reforms and the cells begin to divide. In some cells this continues to full cytokinesis and there may be a period of brief or prolonged interphase. During this interphase there is *no further replication* of the DNA.

Metaphase 2 – new spindles are formed and the chromosomes, still made up of pairs of chromatids, line up on the metaphase plate.

Anaphase 2 – the centromeres now divide and the chromatids move to the opposite ends of the cell.

Telophase 2 – nuclear envelopes reform, the chromosomes return to their interphase state and cytokinesis occurs, giving four daughter cells each with half the chromosome number of the original diploid cell.

fig. 3.2.4 The main steps in the process of meiosis which results in the formation of haploid gametes. Meiosis is much more complex than this diagram shows.

Many of the stages of meiosis are very similar to those of mitosis, with just a couple of crucial variations.

The chromosomes replicate to form chromatids joined by a centromere as in mitosis. However, in meiosis the two chromosomes of each pair stay close together. In the first stages of meiosis all four chromatids become joined. At this stage crossing over takes place (see **fig. 3.2.5**). Just as in mitosis, the nuclear membrane and nucleolus break down and the centrioles pull apart to form the spindle.

The centromeres do not split in the first division of meiosis, so *pairs of chromatids* move to the opposite ends of the cell. The cell then immediately goes into a second division without any further replication of the chromosomes. This division is just like mitosis. The centromeres divide and chromatids move to opposite poles of the cell.

Finally the nuclear membranes re-form as the chromosomes decondense and become invisible again. Cytokinesis takes place which gives four haploid daughter cells, each with half the chromosome number of the original parent cell. These daughter cells later develop into gametes.

The importance of meiosis

Why is meiosis important? It reduces the chromosome number in gametes from diploid to haploid. It is also the main way in which genetic variety is introduced to a species. This variety is introduced in two main ways.

* **Independent** or random **assortment**: the chromosomes that came from the individual's two parents are distributed into the gametes (and so into their offspring) completely at random. For example, each gamete you produce receives 23 chromosomes. In each new gamete any number from none to all 23 could come from either your maternal or your paternal chromosomes. It has been calculated that there are more than eight million potential genetic combinations within the sperm or the egg. This alone guarantees great variety in the gametes.

* **Crossing over** or **recombination**: this process takes place when large, multienzyme complexes 'cut and stitch' bits of the maternal and paternal chromatids together. The points where the chromatids break are called **chiasmata**. These are important in two ways. First, the exchange of genetic material leads to added genetic variation in its own right. Second, errors in the process lead to **mutation** (see chapter 2.1) and this is a further way of introducing new combinations into the genetic make-up of a species.

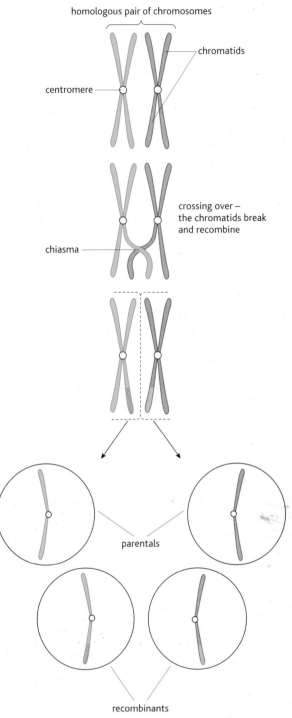

fig. 3.2.5 Chromosomes crossing over in meiosis – this process introduces more variety into the gametes.

Questions

1   Explain the importance of meiosis in a life cycle involving sexual reproduction.

2   Explain how meiosis leads to variation between offspring.

Gametogenesis

The gametes that make sexual reproduction possible are formed in a process called **gametogenesis**. Meiosis is just one stage in gamete formation, which produces different male and female cells. You are going to consider the way in which sperm and ova are made in the sex organs of mammals, using humans as an example, and also how gametes are formed in flowering plants. You will then look at how the mammalian gametes are adapted to their functions.

Gamete production in mammals

Sperm cells in the male mammal are produced in the testes. Ova in the female are produced in the ovaries. Many millions of sperm are released every time a male mammal ejaculates. The eggs in a sexually mature female are usually numbered in thousands and will eventually run out. Special cells (the primordial germ cells) in the gonads divide, grow, divide again and then differentiate into the gametes.

Spermatogenesis is the formation of spermatozoa. Each primordial germ cell in the testes results in large numbers of spermatozoa (see **fig. 3.2.6**). There are enormous numbers of primordial germ cells in the testes producing millions of spermatozoa on a regular basis.

Oogenesis is the formation of ova. Each primordial germ cell in an ovary results in only one ovum (see **fig. 3.2.6**). As a result the number of female ova (or eggs) is always substantially smaller than the number of spermatozoa. Ova contain a much higher proportion of material than sperm so there is a much greater investment of resources in each one. It would not make sense biologically to waste resources by producing too many of them.

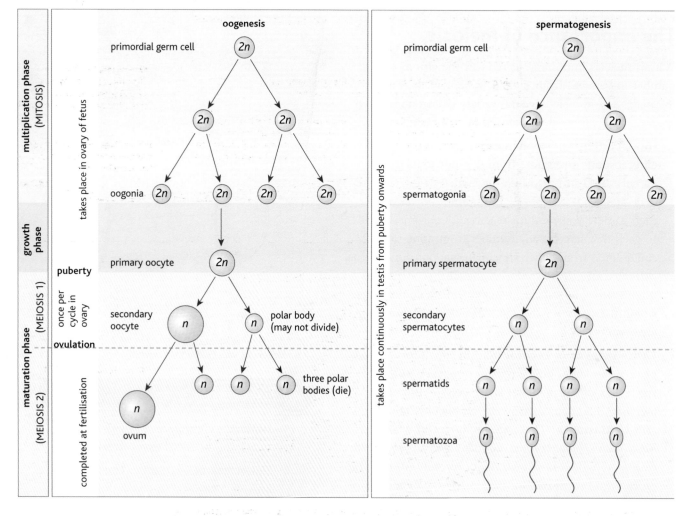

fig. 3.2.6 The formation of sperm in the testes and ova in the ovaries.

Both mitosis and meiosis play a role in gametogenesis. Mitosis provides the precursor cells. Meiosis brings about the reduction divisions that result in gametes. In human males, the process of gametogenesis continues constantly from puberty. In females, the mitotic divisions take place before birth. The meiotic divisions take place in a few oocytes each monthly cycle from puberty to menopause and are only completed if the oocyte is fertilised.

Spermatozoa: many, mini, motile

The male gametes or spermatozoa (sperm cells) of most mammalian species, including humans, are around 50 μm long. They have several tasks to fulfil. They must remain in suspension in the semen so they can be transported through the female reproductive tract, and they must be able to penetrate the protective barrier around the ovum and deliver the male haploid genome safely inside. The close relationship between the human spermatozoan's structure and its functions are shown in **fig. 3.2.7**. Millions upon millions of these motile gametes are

produced in the lifetime of a human male. The average size of a family, with only one spermatozoan needed to fertilise each ovum, gives an idea of the scale of biological wastage.

Ova: few, fat, fixed

Although spermatozoa of most animals are very similar in size, the same cannot be said for ova (eggs). These vary tremendously in both their diameter and their mass. The human ovum is about 0.1 mm across, whilst the ovum contained in the ostrich egg is around 6 mm in diameter. Eggs do not move on their own, so they do not need contractile proteins, but they usually contain food for the developing embryo. The main difference between eggs of various species is the quantity of stored food they contain. In birds and reptiles a lot of development takes place before the animal hatches, so the egg contains a large food store. In mammals, once the developing fetus has implanted in the uterus it is supplied with nutrients from the blood supply of the mother and so large food stores in the egg are unnecessary.

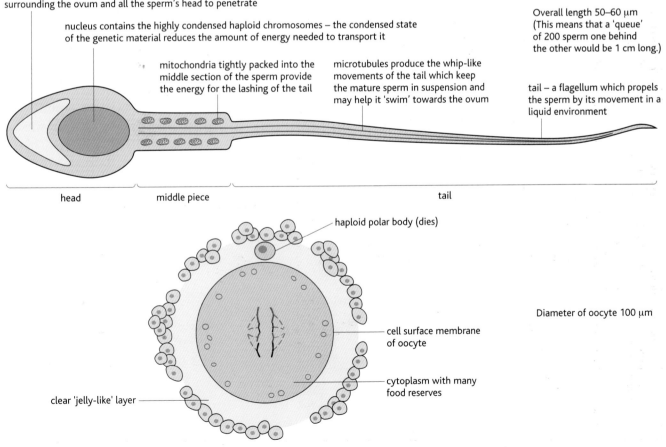

fig. 3.2.7 Human gametes – the sperm and the ova show clear specialisations which fit them for their functions.

Gametogenesis in plants

The formation of gametes in flowering plants is more complex because plants have two phases to their life cycles. The **sporophyte** generation is diploid and produces spores by meiosis. The **gametophyte** generation which results is haploid and gives rise to the gametes by mitosis.

In plants such as mosses and ferns these two phases exist as separate plants. In flowering plants, the two phases have been combined into one plant. The main body of the plant which we see is the diploid sporophyte. The haploid gametophytes are reduced to parts of the contents of the anther and the ovary. They are produced by meiosis from spore mother cells.

The formation of pollen, the microgametes

The anthers of flowering plants are analogous to the testes of animals. Meiosis occurs here, resulting in vast numbers of the pollen grains that carry the male gametes. Each anther contains four **pollen sacs** where the pollen grains develop. In each pollen sac there are large numbers of microspore mother cells. These are diploid. They divide by meiosis (see **fig. 3.2.8**) to form haploid **microspores** which are the gametophyte generation. The gametes themselves are formed from the microspores by mitosis, with one cell enveloping the other to form a pollen grain containing two haploid nuclei, the **tube nucleus** and the **generative nucleus**. The tube nucleus has the function of producing a pollen tube which penetrates through the ovary and down into the ovule. The generative nucleus then fuses with the nucleus of the ovule to form a new individual.

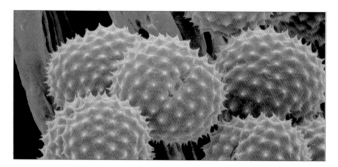

fig. 3.2.9 **These fearsome-looking pollen grains are from ragweed. The surface patterns are unique and specific to the species. Pollen grains are extremely tough and resistant to decay and can remain in the soil for thousands of years. An expert examining an ancient sample of soil or dust can identify what plant species were growing at the time, and even get an idea of how abundant the different species were, from the pollen grains that survive.**

The formation of egg cells, the megagametes

The ovary of the plant is analogous to the animal ovary. Meiosis results in the formation of a relatively small number of ova contained within ovules inside the ovary. Some plants – an example is the nectarine – produce only one ovule (egg chamber) whilst others such as peas produce several.

The ovule is attached to the wall of the ovary by a pad of special tissue called the **placenta**. A complex structure of integuments (coverings) forms around tissue known as nucellus. In the centre the **embryo sac** forms the gametophyte generation (see **fig. 3.2.10**). Diploid megaspore mother cells divide by meiosis to give rise to four haploid **megaspores**, three of which degenerate leaving one to continue to develop. The megaspore undergoes three mitotic divisions which result in an embryo sac containing an egg cell (the **megagamete**), two polar bodies and various other small cells, some of which degenerate.

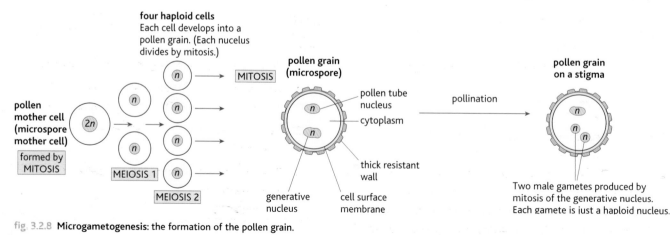

fig. 3.2.8 **Microgametogenesis: the formation of the pollen grain.**

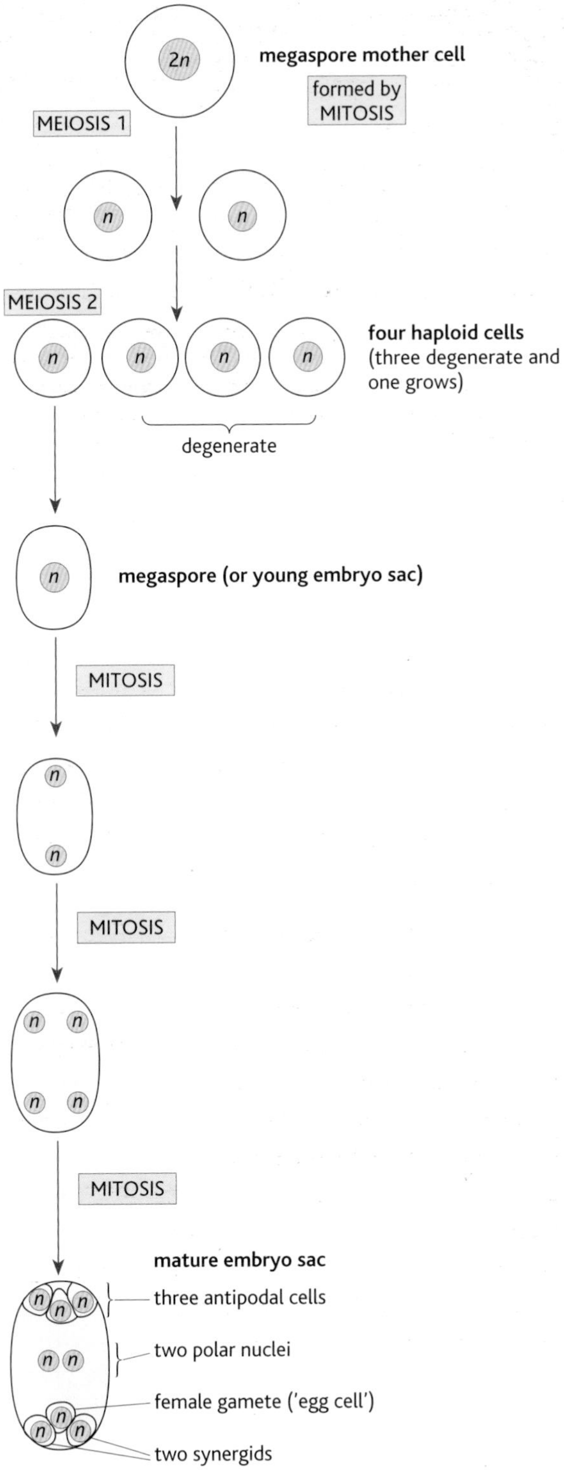
2n
megaspore mother cell
formed by MITOSIS
MEIOSIS 1
n
n
MEIOSIS 2
n
n
n
n
four haploid cells (three degenerate and one grows)
degenerate
n
megaspore (or young embryo sac)
MITOSIS
MITOSIS
MITOSIS
mature embryo sac
three antipodal cells
two polar nuclei
female gamete ('egg cell')
two synergids

fig. 3.2.10 Megagametogenesis: the formation of the egg cell.

Getting together

Asexual reproduction is a guaranteed method of passing on the genes from one individual into the next generation. For sexual reproduction to be successful, the gametes must meet. If these gametes come from two different individuals, the male gamete needs to be transferred to the female gamete. In plants, some flowers attract other organisms such as insects, birds or mammals to transfer the pollen from one plant to another (**pollination**). Others rely on the wind to carry their pollen from plant to plant.

In animals a wide variety of strategies are used to make sure the gametes meet. They fall into two main categories:

- **External fertilisation** occurs outside the body, with the gametes shed directly into the environment. This is common only in aquatic species, because spermatozoa and ova are very vulnerable to drying and are rapidly destroyed in the air. Simpler animals such as jellyfish release copious amounts of male and female gametes into the sea. It is largely a matter of chance whether fertilisation takes place. More complex animals such as fish and amphibians have evolved rituals which increase the likelihood of fertilisation by ensuring that the ova and sperm are released at the same time close to each other. In spite of these strategies many of the gametes do not meet. External fertilisation is very wasteful, and is not an option for organisms that live on land.
- **Internal fertilisation** involves the transfer of the male gametes directly to the female. This does not guarantee fertilisation but makes it much more likely. The way in which the sperm are transferred varies greatly. In many species the male produces packages of sperm for the female to pick up and transfer to her body. More complex animals such as insects and vertebrates have evolved a system whereby the male gametes are released directly into the body of the female during **mating**. This makes sure that the ova and sperm are kept in a moist environment and are placed as close together as possible, which maximises the chances of successful fertilisation.

Questions

1. What part do gametes play in sexual reproduction?
2. Explain the role of meiosis in the production of gametes. How does this vary in animals and plants?
3. How are the human male gametes adapted to fit their role?
4. Compare the adaptations of female gametes in plants and mammals.

Fertilisation

Once the male and female gametes have been brought close together, they meet and fuse in a process known as **fertilisation**.

Fertilisation in humans

For sexual reproduction to be successful in humans, as in any other species, the gametes must meet and fuse. The ovum is fully viable for only a few hours. The sperm will survive a day or two in the female reproductive tract. There is little evidence to suggest that the sperm are attracted to the egg in any way – their meeting seems to be entirely a matter of chance. In spite of this, they frequently do meet and fuse.

As sperm move through the female reproductive tract the acrosome region matures so it is able to release enzymes and penetrate the ovum.

The ovum released at ovulation has not fully completed meiosis and is called a **secondary oocyte**. It is surrounded by a protective jelly-like layer known as the **zona pellucida** and also by some of the follicle cells. Many sperm cluster around the ovum, and as soon as the heads of the sperm touch the surface of the ovum the **acrosome reaction** is triggered (see **fig. 3.2.11**). Enzymes are released from the acrosome which digest the follicle cells and the zona pellucida. One sperm alone does not produce sufficient enzyme to penetrate the protective layers around the ovum. This seems to be one reason for the very large number of sperm released in ejaculation, providing enough in the oviduct to surround the ovum and digest its defences.

Eventually one sperm will wriggle its way through the weakened protective barriers and touch the surface membrane of the oocyte. This has several almost instantaneous effects. The oocyte undergoes its second meiotic division providing a haploid egg nucleus to fuse with the haploid male nucleus. It is vital that no other sperm enter the egg, as this would result in its being fertilised by too many sperm (**polyspermy**) and would produce a nucleus containing too many sets of chromosomes. The events that follow fertilisation prevent this happening. Ion channels in the cell membrane of the ovum open and close so that the inside of the cell, instead of being electrically negative with respect to the outside, becomes positive. This alteration in charge blocks the entry of any further sperm. It is a temporary measure until a tough **fertilisation membrane** forms around the fertilised ovum. This takes over the job of repelling other sperm as the electrical charge returns to normal.

The head of the sperm enters the oocyte, but the tail region is left outside. Once the head is inside the ovum it absorbs water and swells, releasing its chromosomes to fuse with those of the ovum and forming a diploid zygote. At this point fertilisation has occurred and a new genetic individual has been formed. Fertilisation is also referred to as **conception** in the case of humans.

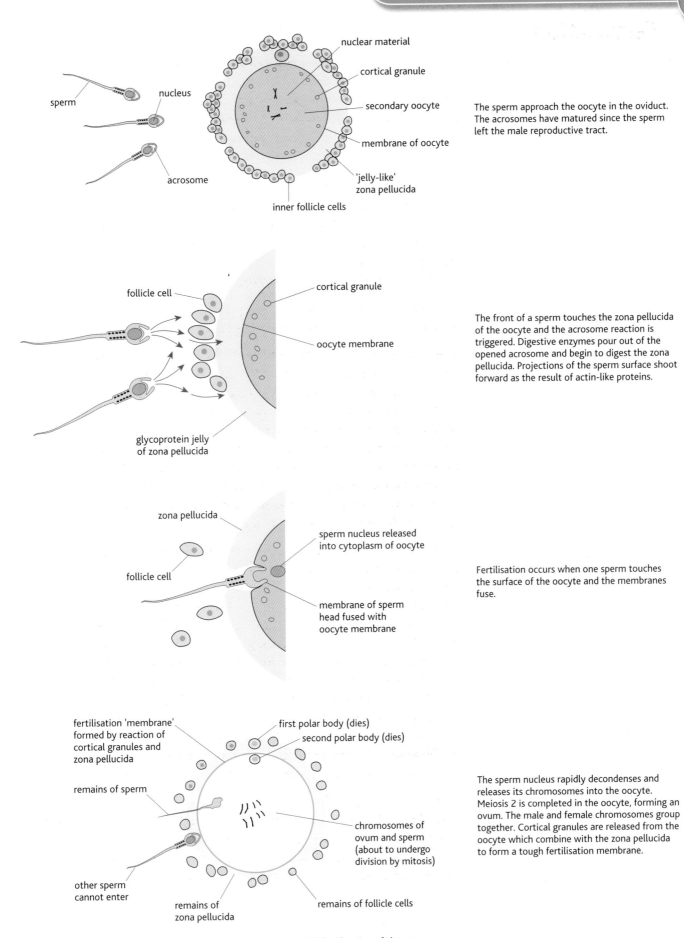

sperm

nucleus

acrosome

nuclear material

cortical granule

secondary oocyte

membrane of oocyte

'jelly-like' zona pellucida

inner follicle cells

The sperm approach the oocyte in the oviduct. The acrosomes have matured since the sperm left the male reproductive tract.

follicle cell

cortical granule

oocyte membrane

glycoprotein jelly of zona pellucida

The front of a sperm touches the zona pellucida of the oocyte and the acrosome reaction is triggered. Digestive enzymes pour out of the opened acrosome and begin to digest the zona pellucida. Projections of the sperm surface shoot forward as the result of actin-like proteins.

zona pellucida

follicle cell

sperm nucleus released into cytoplasm of oocyte

membrane of sperm head fused with oocyte membrane

Fertilisation occurs when one sperm touches the surface of the oocyte and the membranes fuse.

fertilisation 'membrane' formed by reaction of cortical granules and zona pellucida

remains of sperm

other sperm cannot enter

remains of zona pellucida

first polar body (dies)

second polar body (dies)

chromosomes of ovum and sperm (about to undergo division by mitosis)

remains of follicle cells

The sperm nucleus rapidly decondenses and releases its chromosomes into the oocyte. Meiosis 2 is completed in the oocyte, forming an ovum. The male and female chromosomes group together. Cortical granules are released from the oocyte which combine with the zona pellucida to form a tough fertilisation membrane.

fig. 3.2.11 **The acrosome reaction plays a vital role in the successful fertilisation of the egg.**

Fertilisation in plants

The male gamete is contained within the pollen grain. The female gamete is embedded deep in the tissue of the ovary. The pollen grain lands on the surface of the stigma of the flower during pollination. The molecules on the surface of the pollen grain and the stigma interact. If they 'recognise' each other as being from the same species the pollen grain begins to grow or **germinate**. Often the pollen grain will only germinate if it is from the same species but a different plant. This helps to prevent self-fertilisation, which would reduce variety. Alternatively, pollen grains from the same plant may start to germinate but be unable to penetrate the carpel.

A **pollen tube** begins to grow out from the tube cell of the pollen grain through the stigma into the hollow style.

It continues to grow down towards the ovary, and the generative cell containing the generative nucleus travels down it. The nucleus of this cell divides by mitosis as it moves down the tube to form two male nuclei. Eventually the tip of the pollen tube passes through the micropyle of the ovule. The growth of the pollen tube is very fast due to the rapid elongation of the cell.

Once the tube has entered the micropyle, the two male nuclei are passed into the ovule so that fertilisation can occur. Flowering plants undergo what is known as **double fertilisation**. One male nucleus fuses with the nuclei of the two polar bodies to form the **endosperm nucleus**, which is **triploid**. The other male nucleus fuses with the egg cell to form the diploid zygote (see **fig. 3.2.12**). At this point fertilisation is complete and the development of the seed and the embryo within can begin.

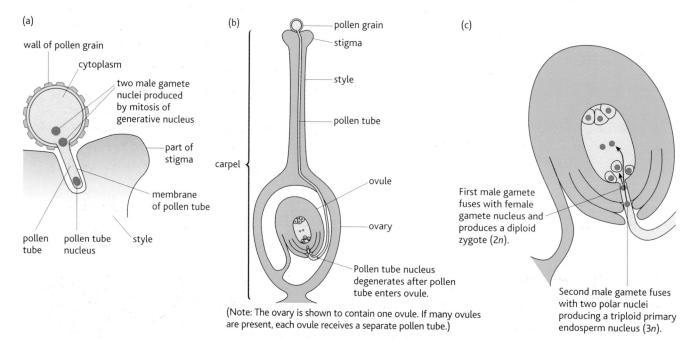

fig. 3.2.12 A summary of the events that follow pollination in a flowering plant and lead to the fertilisation of the ovule.

Questions

1 What characteristics would you look for in an organism with:

   a external fertilisation of the gametes

   b internal fertilisation of the gametes?

   Explain why these adaptations are important to the method of fertilisation.

2 What is the importance of the reaction between pollen grains and the surface of the stigma in plants?

3 What part is played by enzymes in mammalian fertilisation?

4 How is polyspermy prevented, and why is it important to stop it happening?

HSW Overcoming male infertility

As scientists have understood more about the process of fertilisation, they have been able to do more to help people who have problems conceiving. Infertility is an increasing problem – as many as one UK couple in six will have problems conceiving when they want to start a family. There are many possible causes of this infertility.

Male infertility usually results from problems with the production of sperm in the process of spermatogenesis. If no sperm are produced, then fertilisation is impossible. Even if sperm are produced, the numbers needed for fertility are huge – the lowest sperm count regarded as normal is 20 million sperm per 1 cm^3 of semen. If numbers fall below this level fertility is affected – only 1 in every 2000 sperm will make it from the cervix to the oviduct and the ovum.

For the best chance of fertilisation the sperm need to be mobile and swim strongly in straight lines with actively lashing tails. If more than 50% of them swim round in circles or have tails which only lash weakly, fertility levels drop. Another common cause of male infertility is the presence of a high proportion of abnormal sperm (fig. 3.2.13). In normal semen about 15% of the sperm are abnormal – with double heads, broken necks and other problems. If this proportion is higher, the chances of fertilisation become less.

Overcoming male infertility is not easy, but recent developments in reproductive technology have helped. Eggs can be harvested from the woman and injected with a single healthy sperm from the male partner. This overcomes all the problems of low sperm numbers, weak sperm or abnormal sperm. It avoids the need for lots of acrosome reactions to dissolve away the zona pellucida. Once cell division is established in the zygote, the embryo can be replaced inside the mother's uterus to establish a normal pregnancy. Even more recently doctors have developed a technique to harvest immature sperm cells from men who do not produce mature sperm. After careful treatment these can also be used to fertilise ova, although the success rate is not yet very high. These methods have enabled a few men affected by the genetic condition cystic fibrosis to father children overcoming the problems of missing tubes and sticky mucus.

The application of science in this way to help infertile couples brings great happiness to individuals, but there are some ethical issues to consider as well. These are questions to which science cannot supply the answers.

- Does everyone have the right to have a child?

- Research into infertility treatments such as these is expensive and takes a lot of time to help a relatively small number of people. When NHS funding is finite, the world population is growing and millions of people around the world still die of preventable diseases, is this an acceptable use of money?

- What are the issues with developing a treatment to enable people affected by genetic conditions such as cystic fibrosis to have a child?

fig. 3.2.13 If too many sperm have double heads, broken necks or miniature heads, their chances of remaining suspended in the semen, reaching the ovum and achieving fertilisation are substantially reduced.

Embryo development and cell differentiation

What happens following fertilisation?

Fertilisation starts a complex series of events that will eventually lead to the birth of a fully formed new individual. In humans, the fertilised egg cell or zygote has the potential to form all of the 216 different cell types needed for an entire new person. It is said to be **totipotent**. The future roles of individual cells are decided relatively early in the development of an embryo.

The first stage of the process is known as **cleavage**. Cleavage involves a special kind of mitosis, where cells divide repeatedly without the normal interphase for growth between the divisions. The result of cleavage is a mass of small, identical and undifferentiated cells forming a hollow sphere known as a **blastocyst** (fig. 3.2.14). In humans this process takes about a week, occurring as the zygote is moved along the oviduct towards the uterus. One large zygote cell forms a large number of small cells in the early embryo.

The tiny cells of the early human embryo are known as **stem cells**. Stem cells are undifferentiated cells, but have the potential to develop into many different types of specialised cells from the instructions in their DNA. The very earliest cells in an embryo are totipotent like the zygote. In the blastocyst the outer layer of cells goes on to form the placenta, and the inner layer of cells have already lost some of their ability to differentiate. They can form almost all of the cell types needed in future, but not tissue such as the placenta – we say they are **pluripotent**. These cells are known as pluripotent embryonic stem cells.

The formation of different cell types

Almost every cell in an organism contains the DNA instructions to make any other type of cell and, in the earliest stages of an embryo, each cell can produce any tissue. However, only days after conception, cells are already predestined or determined to become one type of tissue or another. This **cell determination** is closely linked to the position of the cells in the embryo. If the cells are surgically removed from the embryo and grown on, they will still produce the predetermined cell type – even if it is entirely inappropriate in the new setting (see the box below). No one is yet entirely sure of the mechanism of cell determination. Following this stage the cells can **differentiate** and develop into organs and tissues.

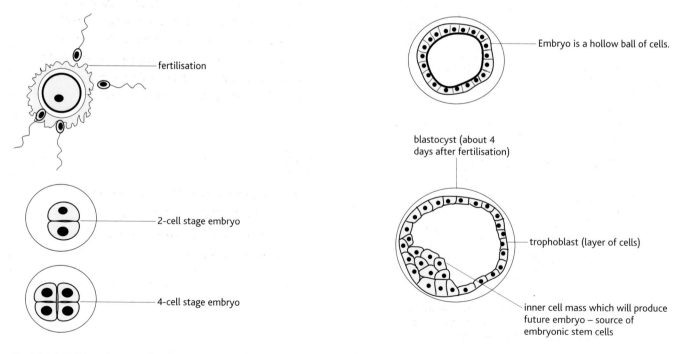

fertilisation

2-cell stage embryo

4-cell stage embryo

Embryo is a hollow ball of cells.

blastocyst (about 4 days after fertilisation)

trophoblast (layer of cells)

inner cell mass which will produce future embryo – source of embryonic stem cells

Fig. 3.2.14 At this early stage of pregnancy any one of these cells has the potential to form another entire human being.

HSW Evidence for cell determination

Scientists working on cell determination and specialisation use a variety of organisms. Obviously such research on humans would be ethically unacceptable. However, amphibians such as frogs and the fruit fly *Drosophila* have both proved very useful.

In a normal early amphibian embryo, cells undergo the first stages of differentiation and go on to become particular tissues as development continues. The hypothesis that this differentiation occurs irreversibly at an early stage was disproved by transplanting tiny patches of tissue from one area of an early embryo to another. The transplanted cells differentiated to form the tissue linked to their new position, not their original position (see **fig. 3.2.15**). When the same experiment was carried out with cells from a slightly older embryo, scientists saw that determination was complete and the cells differentiated to form the tissue determined by their original position rather than their new position.

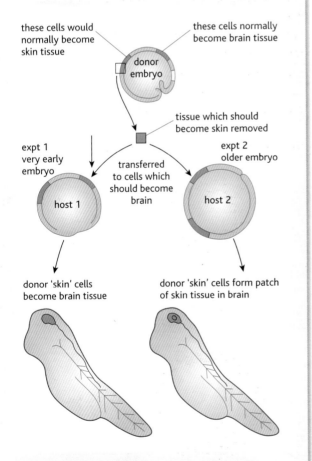

fig. 3.2.15 Scientists use evidence from experiments like this to build up a model of how cell determination and differentiation works and when it takes place.

As cell differentiation takes place to form tissues and organs, different types of cells produce more and more proteins specific to their cell type. The shape of the cell and the arrangement of the organelles will differ. In the nuclei of your cells about 2 metres of DNA is coiled and folded into a space only a few micrometres in diameter. The Human Genome Project found that the human genome consists of fewer than 30 000 individual genes. In a differentiated cell, between 10 000 and 20 000 of those genes are actively expressed. A different combination will be expressed in different cells, creating the variety of structure and function seen in cells of different tissues.

HSW Demonstrating differentiation

What happens as cells differentiate? This might seem an easy question to answer, because different types of cells often look very different. But many cells that carry out different functions in the body look superficially very similar. One way in which scientists can measure the degree of differentiation between cells is to compare the proteins they contain using gel electrophoresis (see chapter 1.3). The scientists have discovered that almost all cells have a number of 'housekeeping' proteins in common. These are the proteins involved in the structures common to almost all cells, eg the structural proteins of the membranes and the enzymes involved in cellular respiration. But each different cell type also produces specific proteins which relate to the particular function of the cell and these can be used to measure the level of differentiation that has taken place – for example, the enzymes needed to produce insulin are found only in islet of Langerhans cells in the pancreas.

Questions

1 What are the main differences between totipotency and pluripotency?

2 Explain how cell determination and cell differentiation differ.

Using totipotent cells

Totipotent plant cells

A clone is a group of cells or organisms that are genetically identical and have been produced from cells with the same genetic material. In chapter 3.1 you saw how cloning can happen naturally in animals and plants. Plant breeders have been using the process of vegetative propagation for hundreds or possibly thousands of years. New complete plants can be grown from fragments of leaves, stems or roots of many different plant species, producing clones of the original plant. However, work in the last century showed that you could start with just one cell. In the 1950s a carrot was produced by Frederick Steward from a single carrot phloem cell, grown in a rich nutritive medium. For this to happen the original cell needs to have kept the capacity, present in the cells of an embryo, to develop into any type of cell. It needs to be totipotent – with none of its genes permanently switched off.

Cloning has an ever-increasing number of uses. In horticulture, plants such as orchids are now almost always produced commercially by tissue culture to guarantee the quality of the new plants, and because they are very difficult to grow from seed. In research, it is of great value to have plants and animals that are known to be genetically identical because any differences in their development or behaviour can then more reliably be put down to the experimental variable. The cloning of plants is now commonplace and relatively easy. Plant cells remain totipotent throughout life. The cloning of animals is less easy because animal cells are only totipotent for a short time. They then become pluripotent and finally, once they are fully differentiated, they lose the ability to form different cells completely. But in spite of this animal cloning is used increasingly in the production of top-quality farm stock and in medicine.

Cloning plants

Plant **tissue culture** can be used to demonstrate totipotency in plant cells. This involves taking a small piece of plant tissue and growing it on a gel medium impregnated with plant hormones. A mass of undifferentiated, genetically identical plant cells is produced by mitosis. Transferring the cells to more gel impregnated with different hormones encourages the development of roots and shoots and hundreds or thousands of new, identical cloned plants result.

fig. 3.2.16 All of these undifferentiated plant cells have the potential to become identical adult clones.

Totipotent animal cells

There is one type of animal cloning which is relatively easy to carry out and is now in regular use in animal husbandry. This is **embryo splitting** and it relies on the fact that every cell in a very early embryo is totipotent. The technique is used particularly in cattle. Embryos are created, either in the mother (in which case the embryos are then washed out of the uterus) or in the laboratory, using gametes from the best stock animals. These individual early embryos are then cloned by separating them into individual cells to produce large numbers of genetically identical embryos. Each of these is replaced in the uterus of an ordinary cow which then carries the calf to term and delivers offspring that is not biologically hers (**fig. 3.2.17**).

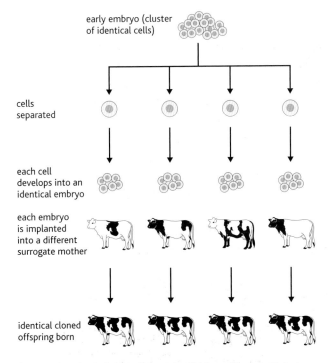

fig. 3.2.17 Embryo cloning allows many high-quality animals to be produced using ordinary cows.

Using this technique makes it possible to transport prize stock around the world very cheaply. Cloned early cattle embryos are placed in the uterus of rabbit 'surrogate mothers', where they begin to develop. The rabbits are flown cheaply to another country, where the embryos are removed and implanted into cows to continue their development. The value of this over simply transporting frozen sperm for artificial insemination is that both the cattle parents can be carefully selected for the best, disease-free features. However, the rabbits are killed to remove the embryos, so some people have ethical objections to the technique.

HSW Cloning animals

In the 1930s scientists recognised that the cells of an embryo were totipotent, but that the ability to form different types of cells was lost as the cells specialised. In 1938 Hans Spemann, a well-known German embryologist, suggested a method of testing whether the nucleus from a specialised cell still contained all the information needed to create an entire animal. His idea was to remove the nucleus from an unfertilised egg and replace it with the nucleus from a fully differentiated cell, to see if an entire animal developed.

It was many years before practical techniques were available to try this ambitious experiment. In the early 1950s scientists working on frogs managed to transfer nuclei from tadpoles into unfertilised eggs. They saw development begin, but the frogs died at the larval stages. John Gurdon in Britain was the first person to clone an adult animal successfully using the nucleus of a differentiated cell from the gut of a toad tadpole.

Since then people have tried to clone a wide range of animals and succeeded. But the technique still raises many difficulties and has a low success rate.

fig. 3.2.18 A *Xenopus* toad – the first animal species to be cloned successfully and so to show that the nucleus of a differentiated cell still contains all the information to make a whole animal.

Questions

1 Why is plant cloning so much easier than animal cloning?

2 Why is embryo splitting so useful, and what are the related ethical issues?

What causes cell differentiation?

Cell differentiation takes place as cells switch different genes on and off as needed. This switching may be in response to a stimulus from inside the cell (an **internal stimulus**) or in response to changes outside the cell which in turn affect the inside of the cell (an **external stimulus**).

In the early stages of development the different positions of cells in the embryo seem to result in different chemical gradients in their cytoplasm. These chemical differences appear to trigger the start of differentiation. Once differentiation starts and particular proteins are made, the chemical differences in the cells increase and so differentiation continues.

The most common way of controlling gene expression is by switching on and off the transcription of certain genes. Transcription describes the process by which the genetic code of the DNA is copied to a complementary strand of RNA before protein synthesis can take place (see chapter 2.1). One way in which this is brought about is by supercoiling parts of chromosomes, preventing the genes there from being transcribed, and uncoiling other areas, opening them up for transcription so that new proteins are made.

HSW Gene probes

Much of our knowledge about genes and gene expression depends on the use of **gene probes**. Gene probes allow scientists to identify particular pieces of DNA in a cell. They might find the genes for haemoglobin, or for a particular inherited disease, or for a piece of recombinant DNA in a bacterium used in genetic engineering.

Whatever their use, all gene probes use similar technology. To find a particular gene out of the hundreds of thousands present in the DNA of a human cell you need a very specific probe. The gene probe finds the unique sequence of nucleotides on the DNA that make up a gene. This can be recognised by a stretch of RNA with the complementary sequence, in a process known as DNA–RNA hybridisation.

The DNA from some cells is isolated and heated gently. This breaks the weak hydrogen bonds holding together the two strands of DNA. Radioactively labelled mRNA for the required gene is added – this is the probe. Any DNA–RNA hybridisation that takes place shows that the required gene is present. This hybridisation is pinpointed using the radioactive label on the mRNA, which shows up on X-ray film (autoradiography).

So, for example, using gene probes, you can show that both the red blood cells forming in the bone marrow (called reticulocytes) and the neurones in your brain have the gene for haemoglobin as part of their DNA. However, if you use the probe on mRNA from the cells rather than DNA from the nuclei, the haemoglobin gene shows up in only the red

blood cells. In other words, the gene for the production of haemoglobin is present in both types of cell but it is only expressed in the red blood cells.

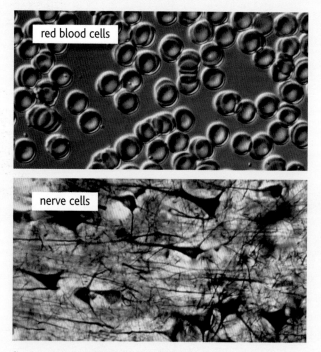

fig. 3.2.19 Using gene probes scientists have shown that both of these kinds of cells have the same genes in their nuclei – but different combinations are expressed in the cytoplasm.

Scientists are finding more and more uses for gene probes – to highlight active genes, to pick out genetic traits and diseases and as a vital feature of genetic engineering.

HSW Evidence for external stimuli controlling cell differentiation

Moulting, the shedding of the exoskeleton in insects, is controlled by two hormones. **Ecdysone**, the 'moulting and metamorphosis' hormone, controls the events of the moult itself. It is a steroid hormone that was first extracted from the pupae of silkworms. **Juvenile hormone** controls the kind of moult that occurs. As juvenile hormone levels decrease, more adult characteristics occur. When there is no juvenile hormone the pupa becomes an adult.

The way in which ecdysone has its effect has been studied using the larvae of *Drosophila* (fruit fly) and *Chironomus* (midge). In the cells of the salivary glands of these insects there are giant chromosomes, 100 times thicker and 10 times longer than normal chromosomes and easily visible with the light microscope. Banding is visible on these chromosomes due to supercoiled areas of DNA. When an insect is undergoing a moult, or when ecdysone is injected artificially into an insect, 'puffs' appear on the chromosomes. These puffs appear to be pieces of genetic material from supercoiled areas which have been opened up and made available for transcription. It is thought that they carry information about the new proteins needed in a more adult stage of the life cycle (see fig. 3.2.20). This supports the view that steroid hormones have a direct effect on the DNA of a cell. Scientists think that many steroid hormones act in a similar way. Unfortunately not many organisms possess giant chromosomes so the effect is not always easy to observe.

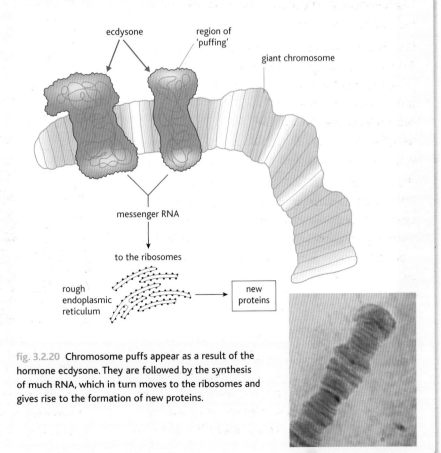

fig. 3.2.20 Chromosome puffs appear as a result of the hormone ecdysone. They are followed by the synthesis of much RNA, which in turn moves to the ribosomes and gives rise to the formation of new proteins.

The control of transcription during development

In the production of haemoglobin during the development of a human fetus, different versions of the globin genes can be seen to be switched on and off. Adult haemoglobin contains two alpha and two beta globin chains. Fetal haemoglobin, which has a stronger affinity for oxygen than adult haemoglobin, contains two alpha and two gamma globin chains. The levels of the different types of globin change through the development of a human fetus into young baby. During this time not only do genes for the different proteins get switched on and off (**fig. 3.2.21**), the genes are also activated in different tissues as development progresses. Globin production moves from the yolk sac to the liver and then the spleen in the fetus, with the genes in the bone marrow taking over almost completely by the time of birth.

In most cases a signal molecule controls the formation of a single protein. But scientists have discovered that in some cases a single signal molecule can set up a cascade of other signals which result in the formation of an entire organ. One of the most dramatic examples is in the formation of the eye. In *Drosophila* (fruit flies), mice and even human beings, a single protein from a single gene controls eye development. The power of this signal molecule has been clearly demonstrated by scientists in the fruit fly. The protein produced as a result of the activation of the Ey gene in the cells triggers the formation of a complete eye. When cells expressing the Ey gene were moved in an early fly embryo to a region of tissue that forms the legs, the flies that eventually hatched had an eye on their legs (see **fig. 3.2.22**). Scientists are still working out the details of how a single regulatory protein can control the formation of an entire organ with all its specialised cells.

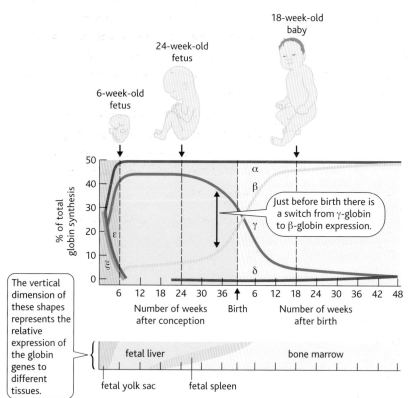

fig. 3.2.21 Different genes for the production of globin molecules are switched on and off in different tissues during the development of the human fetus.

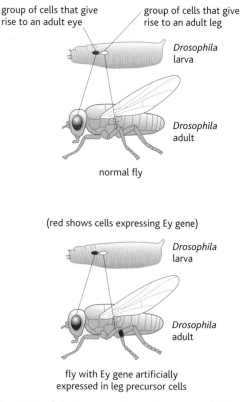

fig. 3.2.22 If the Ey gene in *Drosophila* is moved in the early embryo, the signal protein it produces triggers the formation of a complete eye in another part of the body.

Restoring totipotency

Cloning adult animals is difficult because the differentiated cells are no longer totipotent or even pluripotent. But in 1997 a team from Edinburgh produced Dolly the sheep, cloned from an adult mammary tissue cell which had totipotency restored. There are many difficulties with this technique but many different types of animals have now been cloned in this way, from cats and horses to rare breeds on the verge of extinction. With time, scientists have learned more about the conditions needed to make successful cloning more likely. This has opened the way for cloning genetically modified animals that produce human proteins and medicines in their milk. The gene for the human protein is inserted into the nucleus of the egg of a sheep or cow, which is then fertilised. As the embryo starts to develop it is cloned, and the embryos are implanted into ordinary mothers. This results in a number of **transgenic** animals which make human proteins in their milk. More animals can be produced by further adult cell cloning, avoiding a repeat of the engineering process. In this way the biological production levels of some therapeutic molecules, eg factor VII and factor IX of the blood-clotting cascade (see chapter 1.1), have been increased dramatically, improving the quality of life for many people.

HSW Human cloning?

The technology currently available means that human cloning is not as close as it might seem, but inevitably it will become possible. This prospect raises many concerns. Science can provide the tools to carry out such cloning, but society must decide if it is acceptable. Cloning human cells for medical purposes is becoming increasingly acceptable. However, the idea of cloning a whole person is not acceptable, at least in countries such as the UK and the US. The media often paint pictures of evil dictators cloning themselves to rule forever, but we should bear in mind that each of us is much more than simply the sum of our genes. Our environment before and after birth has a major effect on us too, so no one could ever have an exact duplicate of themselves in every character trait. If society worldwide decides that human cloning should not be done, the likelihood is that it will still happen somewhere – but whether it will give people what they want is another matter entirely.

Restoring totipotency or even pluripotency to differentiated cells and cloning whole organisms has proved a real challenge. But perhaps surprisingly, cloning certain types of cells has been even tougher. The pluripotent stem cells of an embryo have been very difficult to culture in the laboratory. When the breakthrough came it led to some very exciting developments, as you will see in the next chapter.

Questions

1 Why are fruit flies often used as experimental organisms for scientists looking at cell differentiation?

2 What is the value of gene probes in investigating cell differentiation?

3 Some people have ethical objections to the cloning of animals. What arguments could you put for and against the development of this technology? What factors do you think should be taken into consideration when deciding whether the technology should be allowed to go ahead?

4 The cloning of early human embryos is allowed in the UK. However, the development of any cloned human beyond 14 days is not permitted. Give some of the arguments for and against human cloning.

3.3 Stem cells and beyond

Stem cells and where we find them

In chapter 3.2 you considered the way in which the early cells of an embryo, **embryonic stem cells**, have the potential to become any cell in the body. These cells become increasingly specialised to carry out a particular function within a multicellular organism as the organism develops. There are other sources of stem cells, some within the adult body, that retain the ability to develop into a range of cells. In recent years scientists have begun to recognise the potential of stem cells in medical treatments for conditions where body cells fail to work normally, such as after an accident, in a condition such as Parkinson's disease or as a result of an inherited disease such as cystic fibrosis. This is a fast-moving area of biology which is fraught with ethical issues and controversies.

Embryonic stem cells

The earliest embryonic cells are totipotent. By the blastocyst stage, when the embryo would implant in its mother's uterus, the inner cells of this ball are pluripotent – they will eventually form most but not all of the cells of the baby. Pluripotent stem cells become more specialised as the embryo develops, eg forming blood stem cells which give rise to blood cells and skin stem cells which give rise to skin cells. By around three months of pregnancy the cells have become sufficiently specialised that when they divide they only form more of the same type of cell.

In 1998, in a breakthrough which caused ripples of excitement through the scientific and medical world, two American scientists managed to culture human embryonic stem cells that were still pluripotent in the laboratory. James Thomson and his research team at the University of Wisconsin maintained a culture of human embryonic stem cells for several months. They originally obtained the cells from spare embryos which had been produced during *in vitro* fertility (IVF) treatments. Couples had donated their spare embryos for scientific research rather than having them destroyed.

At the same time, John Gearhart and his group at the Johns Hopkins University were also culturing human embryonic stem cells. They used cells from fetuses which had been aborted at 5–9 weeks of gestation.

fig. 3.3.1 John Gearhart and his research team were the one of the first groups to publish their findings on culturing human embryonic stem cells.

The culturing of embryonic stem cells has caused a major stir in scientific, medical and political circles. In theory at least, the pluripotent cells could be encouraged to grow into almost any type of cell needed in the body.

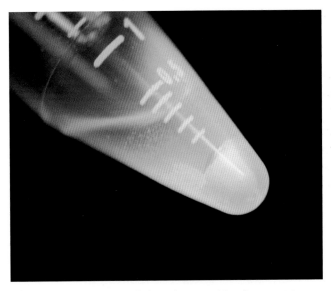

fig. 3.3.2 Embryonic stem cells have the potential to form most types of human cells. Many scientists regard them as having tremendous potential in treating human disease.

Umbilical cord stem cells

The blood that drains from the placenta and umbilical cord after birth is a rich source of pluripotent stem cells. If this blood is frozen and stored, those stem cells will be available throughout the life of the child should they – or their family – need them for treatment later in their lives. It may become possible to store stem cells from every newborn baby, ready for when they might need them. However, there are several problems – it would take a lot of storage space to do this for everyone, and would cost money. Also there is some evidence that the precursor cells of conditions like leukaemia are already present in the blood at birth. Because the benefits are as yet largely unproven, the only way parents in the UK can store umbilical cord blood for their children at the moment is to pay around £500 to do it privately.

Adult stem cells

An adult human is made up of many different types of highly specialised cells. However, adult stem cells do exist in the form of undifferentiated cells found among the normal differentiated cells in a tissue or organ. They can differentiate when needed to produce any one of the major cell types found in that particular tissue or organ, eg white bone marrow contains stem cells that can form white blood cells. Another term for adult stem cells is **somatic stem cells**.

There are only a very small number of adult stem cells in each different tissue. They are also very difficult to extract and most of them form a very limited range of differentiated cells – they are said to be **multipotent**. What is more, they are extremely difficult to grow in the laboratory. Scientists are working hard to develop better ways of culturing adult stem cells because huge cell cultures are needed for stem cell therapy. They also need to find the triggers that persuade the stem cells to differentiate into specific cell types.

We have known about the stem cells in bone marrow, capable of forming all the different kinds of blood cells, for about 30 years, and bone marrow transplants are used regularly in the treatment of certain cancers and immune system diseases. These transplants need to be taken from close relatives or from strangers who have matching immune systems, otherwise the body will reject and destroy the transplant – as you will see later.

Gradually over the years scientists have discovered stem cells in the bone marrow that can generate bone, fat, cartilage and fibrous tissue, and in the 1990s stem cells were discovered in the brain that can form the three main types of brain cells. Adult stem cells have been found in many different organs and tissues. These cells could be extracted from a patient and treated so that they develop into cells that the patient needs. Some examples of this kind of treatment already exist (see page 181) although there is still much more research to be done. Using the patient's own adult stem cells avoids the risk of rejection of new tissue.

Stem cell cloning

Stem cell cloning or **therapeutic cloning** is an experimental technique to produce healthy tissue using a method similar to the one that produced Dolly the sheep. The aim is to treat someone with a disease caused by faulty cells, such as type 1 diabetes or Alzheimer's disease. The first step is to produce healthy cloned cells from the patient. This is done by removing the nucleus from one of their normal body cells, and transferring it to a human ovum which has had its original nucleus removed. After a mild electric shock the new pre-embryo cell starts to develop, producing a collection of embryonic cells with the same genetic information as the patient. This is a cloned human embryo, but it has been formed with no intention to clone a whole human being. This embryo is simply a source of stem cells with DNA that matches the patient perfectly. Stem cells will be harvested from the embryo, which is destroyed in the process. The embryonic stem cells will be cultured in a suitable environment so that they differentiate into the required tissue. These tissue cells will then be transferred to the patient, where they can do their job without the risk of the immune system rejecting them. However, scientists are still trying to determine the triggers that control cell differentiation, so there is a long way to go before particular cell types or organs can be produced on demand. What is more, in the treatment of genetic diseases, the adult stem cell nucleus would need to be genetically modified before being added to the empty ovum, otherwise the cultured stem cells would carry the genetic mutation that caused the problem in the first place.

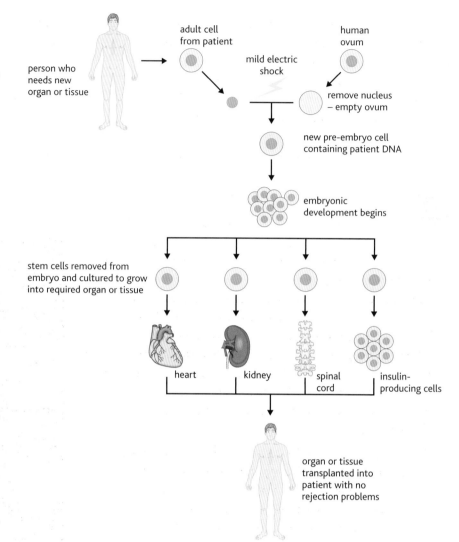

fig. 3.3.3 The process of therapeutic stem cell cloning.

Pitfalls and potential benefits of stem cell therapy

There are problems that relate to the uses of all kinds of stem cells. For example, at the moment no one is quite sure just how the genes in cells are switched on or off to form particular types of tissue – kidney rather than liver, or liver rather than heart. Scientists will need to be able to control this properly in stem cells in order to produce the cells they need for a patient.

There are also risks. There are concerns that stem cells could cause the development of cancers in the body. There is some evidence from people who have been given bone marrow transplants (stem cell treatment) to help them overcome leukaemia are at higher risk of developing other cancers later.

However, there are also great potential advantages. At the moment there are no cures for many of the conditions that stem cell therapy might solve. The ability to produce tailor-made cells to take over the function of damaged ones would revolutionise medicine. Organ transplants to treat people with damaged tissue caused through disease or accidents create their own problems. Glycoproteins on the surface of your cell membranes act as part of your cell recognition system. Your immune system recognises your own cells (self) and different cells (non-self) – and destroys the non-self cells. This is great if you pick up an infection, but potentially lethal if your immune system attacks a transplanted organ. After a transplant people have to take immunosuppressant drugs for the rest of their lives, and this puts them at higher risk of

infectious diseases. One advantage of using embryonic stem cell therapy is that it could avoid this risk of rejection. The immune system does not attack and destroy a developing embryo even though it has different markers on its cells from those of its mother. Maybe new cells or organs created from embryo stem cells would enjoy this same protection.

HSW The human therapeutic cloning scandal

In 2004 and 2005 the scientific community became very excited by the results of the work of an international team of scientists in South Korea, led by Woo Suk Hwang. The research, published in two papers in the journal *Science*, used 242 eggs donated by 16 healthy women. The team took a nucleus from a somatic cell of each woman and placed it in her own empty ovum. They created 30 cloned embryos which grew to around 100 cells each. Hwang managed to harvest embryonic stem cells from one of the embryos. He showed that the stem cells could differentiate into other tissues, and claimed to have developed sets of cells (stem cell lines) for individual patients.

There was great praise for the work, particularly because cloning primates has been regarded as almost impossible. Therapeutic possibilities suddenly became a reality. Some people began campaigning for a complete ban on reproductive human cloning – the technology to make cloned human babies appeared to have arrived.

However, early in 2006 came the news that Hwang's work was faked. There had been no human stem cell lines from cloned embryos. DNA fingerprints, photos and all the data had been faked. The journal *Science* withdrew the papers, and Hwang was disgraced and dismissed from his post. In a further twist to the story, it looks possible that Hwang did achieve a real and different success by getting the egg cells to develop into embryos. But this work needs to be repeated by other teams before it is accepted.

fig. 3.3.4 Woo Suk Hwang publicly apologised for the faking of his team's stem cell results.

Questions

1 Why is there excitement at the ability to clone and maintain cultures of embryonic stem cells?

2 Suggest why scientists accepted the work of Wuk Soo Hwang and his team, even though many people were struggling to produce human stem cell clones. What do you think will be the impact of the work turning out to be faked?

3 There has been considerable opposition to work based on the findings of Gearhart and Thomson. Some countries, including the US, have put severe restrictions on funding. Why do you think there are these reservations? What role do you think strongly held ethical views should have in determining the progress of scientific and medical research? Justify your position.

Who could benefit from stem cell therapy?

Stem cell therapy could be used for a wide range of diseases that are caused by faulty cells. Here are some examples.

Parkinson's disease

Parkinson's disease is a brain disorder affecting about 2% of people over 65 years old. Nerve cells in the brain that produce dopamine (dopamine neurones) stop working and are lost. As dopamine levels fall, people develop uncontrollable tremors in their hands and body. Their body becomes rigid and eventually they cannot move normally at all. Although drug treatments have improved a lot in recent years, there is still no long-term cure. Scientists hope that stem cell transplants will allow them to replace the lost brain cells and restore dopamine production, letting people return to a normal life.

Recently research scientists led by Lars Björkland in the US managed to get mouse embryonic stem cells to form dopamine neurones. These cells were transplanted into the brains of rats that had the symptoms of Parkinson's disease. The cells grew and released dopamine – and the ability of the rats to control their movement improved. This showed that the mouse cells would develop and produce effective dopamine, encouraging further research.

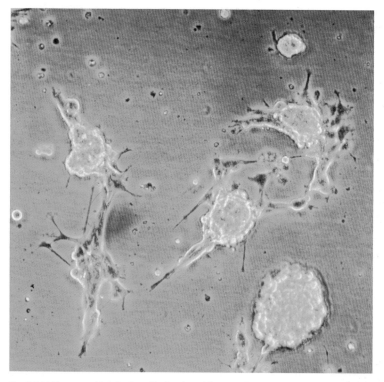

fig. 3.3.5 **These working brain cells developed from embryonic stem cells that were persuaded to become immature nerve cells and then transplanted into the brains of young mice.**

Diabetes

Type 1 diabetes mellitus usually develops when people are young. The insulin-secreting islet of Langerhans cells in the pancreas are destroyed or stop making insulin, so the blood glucose concentration is uncontrolled. This can be very serious or even fatal. Although insulin injections work well enough, people affected by diabetes have to monitor their food intake and blood glucose concentration and inject insulin regularly. Stem cell therapy could give them working pancreas cells again, restoring insulin production and so blood glucose control.

Scientists have succeeded in getting some mouse embryonic stem cells to form a group of cells that looked and worked just like insulin-producing tissue. Some of these cells were transplanted into mice with diabetes where they produced a rise in the blood concentration of insulin and improved control of blood glucose.

Scientists have also successfully transplanted insulin-producing islet of Langerhans cells from one person to another and cured diabetes.

Damaged nerves

So far there is no medical cure for damaged and destroyed nervous tissue in the brain and spine. These nerves do not usually regrow, and so someone who suffers a major injury to their spine may be permanently paralysed below the level of the damage. Embryonic stem cells have been transplanted into mice and rats with damaged spines and the animals regained a certain amount of control and movement of limbs that had been paralysed. Examining the spinal cords of these rats and mice showed that the embryonic stem cells had grown into working adult nerve cells and the damaged spinal cords had at least partly been rejoined, offering hope for future human treatments.

HSW New heart cells – an untried treatment

In 2003 Dimitri Bonnville, a 16-year-old American boy, was shot in the heart with a nail gun in a bizarre accident. He then suffered a massive heart attack which destroyed large areas of heart muscle and meant that each heartbeat was pumping only 25% of the blood out of his heart instead of the normal 55–65%. He needed a heart transplant, but no heart was immediately available. The hospital was about to start a programme of research looking at using adult stem cells to help repair damaged hearts, so they rushed through a special treatment regime just for Dimitri. He is the first patient in the world to undergo an experimental adult stem cell treatment to help repair the damaged tissue.

Doctors gave Dimitri drugs to stimulate the production of stem cells in his blood. They then harvested the stem cells and transplanted them directly into his coronary artery, which carried them to his damaged heart muscle. Within a short time the capacity of his heart increased to 35% with each beat – enough to allow Dimitri to return home and begin to live a more normal life. And because Dimitri received his own cells, there were no problems of rejection. Trials of similar treatments are now taking place in several hospitals.

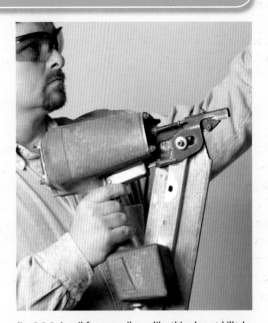

fig. 3.3.6 A nail from a nail gun like this almost killed Dimitri Bonnville before doctors used his own blood stem cells to mend his badly damaged heart with a revolutionary new treatment.

Organs for transplants

Many people die because they have damaged organs that no longer function properly. Hearts, kidneys, livers and many other organs can be replaced by transplant, but only if there is a suitable donor organ available. There is a desperate need for new organs – and preferably ones that will not cause rejection problems when they have been transplanted. A team of researchers at the Weizmann Institute of Science in Israel has taken human and pig kidney stem cells and transplanted them into mice. The cells formed miniature functioning kidneys that produced urine. What is more, the kidneys produced seemed to avoid rejection by the immune system of the recipients. Could embryo stem cells provide the huge supply of organs we need?

Questions

1 What is the biggest obstacle scientists need to overcome to enable them to develop useful therapies with embryonic stem cells?

2 What are the main advantages and disadvantages of adult stem cells compared with embryonic stem cells?

3 Explain the importance of the use of experimental animals in developing new stem cell treatments.

4 The treatment used on Dimitri Bonnville was untested – was it ethical to try it on him?

HSW The ethics of using stem cells

There have been some very powerful reactions to this new area of biotechnology. As well as the many practical problems to be overcome before stem cell therapy becomes a standard treatment, society has many ethical issues to deal with. Science can provide only some of the answers when such radical new technologies appear.

Embryonic stem cells

Perhaps the greatest problems with research into pluripotent cells are ethical ones. The cells used come either from aborted embryos or from spare embryos in fertility treatment.

Many people think the new work is a major breakthrough with the same potential to change healthcare as the discovery of antibiotics more than 70 years ago, offering hope of a cure to millions of people for whom there is, at the moment, no hope. Also, the vast majority of human embryos never make it beyond the early stages of development to form living babies, so the argument for using a small number of early embryos is acceptable in this context. People who are in favour of stem cell research suggest that once tissue lines from a relatively small number of willingly donated embryos are established, the need to use further embryos will be reduced.

What is more, many supporters of stem cell research feel that adult stem cells do not offer a good alternative because they are much more limited in their scope for forming new and different tissues. They want research funding to be directed mainly at embryonic stem cell work.

Other people feel that the use of embryonic tissue is wrong and an abuse of human rights. Many people, including many religious groups, think it is wrong to use a potential human in this way. Some objectors feel that every early human embryo has the potential to become a living human being and so should be afforded the same human rights as a fully grown adult. Others have strong religious convictions that using embryos is killing, and therefore wrong. They think that no medical advances are worth the moral evil of using embryonic tissue as a source of stem cells.

Many of these people feel that the use of adult stem cells offers an exciting and acceptable possible alternative to embryonic cells and they campaign for research funding to be directed to projects using these ethically less sensitive cells. The use of embryonic stem cells from the umbilical cord of newborn babies may help to overcome many of these reservations.

fig. 3.3.7 **Stem cell research is an area of science that raises many ethical questions. For some people the science can provide the answers – for others, it never will.**

In therapeutic stem cell cloning the embryos are not created to develop into a new human being – they are simply there to provide embryo stem cells. This is why many people are very optimistic about the future potential of this new technology and see no major ethical stumbling blocks to overcome. However, other people fear that if the cloning is allowed for therapeutic purposes it could easily be taken further, with the cloned embryos implanted into a uterus to produce a cloned baby. What is more, even if the embryo used as a source of stem cells is produced in an unorthodox way, it is still an embryo. Many people have the same ethical and/or religious objections to using these for research as they do to the use of any other embryos. So therapeutic stem cell cloning can raise even more ethical problems than embryonic stem cell research.

Beyond embryonic stem cells?

In 2006 a team of researchers in Japan led by Dr Shinya Yamanaka made an astonishing breakthrough. They took adult mouse cells and, using genetic engineering techniques, they reprogrammed them to become pluripotent again. Effectively they produced stem cells without an embryo.

The Japanese team and James Thomson and his team at Wisconsin – one of the groups who first cultured human embryonic stem cells – then worked furiously to do the same thing in human cells. Towards the end of 2007, both groups announced that they had succeeded. The Japanese work, published in the journal *Cell*, described how they used harmless, genetically modified viruses to carry a new group of four genes into skin cells from a 36-year-old woman and connective tissue from a 69-year-old man.

In the US, Thomson used a very similar technique (two genes the same, two different) to genetically modify immature skin cells from a fetus and from the foreskin of a newborn baby boy. Again the cells they produced seemed to act like pluripotent stem cells, and the work was published in the on-line version of *Science*. They have since used their technique on cells from an adult.

These induced pluripotent stem cells (iPS) appear to be very similar, although not identical, to human embryonic stem cells in their behaviour. The Japanese team even managed to make their iPS cells develop into brain cells and heart muscle cells.

One of the greatest potential benefits of this new technology is that it overcomes the ethical objections to using embryonic tissue, even embryos specifically created as a source of stem cells. There is also no risk of rejection if cells from an individual are used to provide their own stem cells.

Persuading the cells to become pluripotent isn't easy. Making them differentiate is much more difficult. How well, and for how long, they will behave as pluripotent stem cells remains to be seen. Perhaps most worryingly, the new stem cells show a tendency to become cancerous very quickly.

In spite of the difficulties many scientists think that iPS cells may well be the way forward in stem cell medicine. As there are still major problems to be overcome, there is a strong feeling in the scientific community that research into both embryonic stem cells and iPS cells should continue until it is clear which will produce the best treatments for patients in the long term.

Questions

1   What are the main ethical objections to using embryonic stem cells, and do you think they are issues that can be answered by scientific argument?

2   What are the advantages and disadvantages of umbilical blood and adult tissue as alternative sources of stem cells?

3   Choose one of the methods for producing embryonic stem cells and the new method of producing iPS cells from skin cells. Prepare arguments for a debate on which technique should receive major funding in the next 10 years or whether both should be funded.

3.4 Expressing the genome

How genes interact

As you saw in chapter 2.2, some of your characteristics are very clearly the result of the genes you have inherited. A very few depend on information carried on a single gene that has just two alternative alleles. Many depend on a single gene that has **multiple alleles**, or are **polygenic** traits in which phenotypic features result from the interaction of a number of different genes. The expression of these genes can get very complicated.

Epistasis

In some polygenic traits one gene affects or alters the expression of another. This is known as **epistasis** (meaning 'to stand over') and one example is coat colour in the domestic cat. Cats have a number of genes affecting coat colour, two of which code for coat pattern. The agouti gene causes banding and is associated with tabby coat patterns of any colour. Unless the agouti allele is present, a cat will have a solid coat colour. The agouti alleles are **A** and **a**. **AA** or **Aa** gives an **agouti** coat – grey hairs with a yellow band, resulting in a wide range of brownish colours. depending on the width and position of the bands. The **aa** combination gives a pure black or grey coat.

The tabby coat pattern gene has three alleles: **T**, **T^B** and **t^b**. **TT** or **Tt^b** gives a mackerel-tabby coat pattern with vertical curving black stripes, whilst **t^{b}t^b** is recessive and gives the classic blotched-tabby coat with swirls of black. **T^{B}T**, **T^{B}T^B** or **T^{B}t^b** gives a ticked or freckled appearance. However, without at least one dominant agouti allele on the agouti locus none of the tabby patterns will show up in the phenotype because the cat will have a solid coat colour.

Fig. 3.4.1 shows just one possible outcome of a cross between an agouti blotched-tabby and a pure black cat. Because the **aa** coat colour gene is epistatic to the coat-marking genes it was impossible to tell by looking at the black cat which coat-marking alleles it might have. All of the kittens have a mackerel-tabby pattern which they inherited from their black parent, revealing the effect of the coat-marking genes.

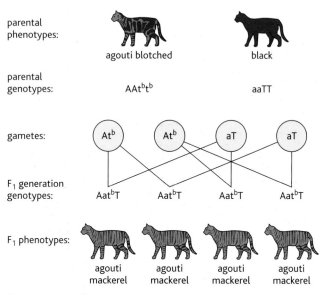

fig. 3.4.1 **Coat colour in cats – an epistatic gene in action.**

Don't confuse epistasis with dominance. Dominance occurs between two alleles of the same gene. Epistasis happens between two genes coding for quite separate phenotypic features, when one locus masks the expression of another.

X chromosome inactivation

Another example where the phenotype cannot be predicted precisely from the genotype alone occurs with X chromosome inactivation. Early in the development of the embryonic cells of a female mammal, one of the X chromosomes is inactivated by supercoiling, to form a mass known as the **Barr body** (see **fig. 3.4.2**).

It seems to be random which X chromosome is inactivated in each embryonic cell, but all the cells later descended from that cell will have the same inactivated X chromosome. So females are made up of mosaics of cells, some with one X chromosome inactivated and some with the other. For example, a female tortoiseshell cat is heterozygous for the coat colours black and orange. Half of the X chromosomes carry an allele for black fur and half for orange. The X chromosome that is inactivated is random, so each tortoiseshell cat has a unique combination of orange and black patterns in the fur. The mixture of colours reflects the mosaics of cells with different X chromosome inactivation.

An example of X chromosome inactivation in humans is the mutation on the X chromosome that causes **anhidrotic ectodermal dysplasia**. If a man inherits the faulty X chromosome he will have no teeth, body hair or sweat glands. If a woman inherits one faulty X chromosome she will have a random pattern of missing teeth and patches of the body with no sweat glands, but some normal areas.

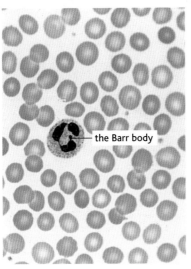

the Barr body

fig. 3.4.2 **A human cell stained to show the Barr body.**

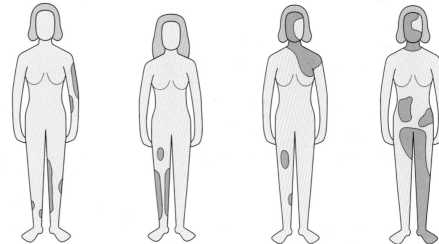

first generation ⟶ second generation ⟶ third generation

fig. 3.4.3 **The random nature of X chromosome inactivation can be clearly seen in this unusual human condition – the last two women are identical twins.**

HSW Barr bodies and cheating in athletes

It is not only by taking banned drugs that some athletes cheat to win. In 1964 the women's world track records in the 400 m and 800 m races were broken by an athlete who turned out to be a man. The 1966 winner of the women's downhill ski title later married and fathered a child!

A simple scientific test was needed to show definitively whether someone is male or female. By looking for Barr bodies in the cells taken from the inside of the cheek, the International Olympic Committee felt that cheats could be identified. 'Female' athletes without a Barr body in their cells would be given the opportunity of further testing, or a quiet withdrawal from their event.

The test was used for about 30 years. Several athletes quietly retired rather than face further testing.

Unfortunately, most of those athletes were not men – or were men with biological issues which made them effectively women. For example, people with XO have only one X chromosome, so no Barr bodies, but no Y chromosome either. Some were men who tested as female, eg with XXY chromosomes. As a result the test is no longer used in sport. It has been replaced with DNA analysis and karyotypes when needed.

Questions

1 Even if you know which alleles a person has for one particular gene, you may not be able to predict which phenotypic feature they will show. Explain why.

2 Look at fig. 3.4.1. Draw at least one more possible cross from parents with the same phenotype.

3 Explain why you never see a male tortoiseshell cat. From the box above, develop a hypothesis suggesting how a male tortoiseshell cat might in rare circumstances be possible.

Interactions between genes and the environment

Your genetic make-up obviously plays a very large part in determining your phenotype or appearance. However, the genotype is not the end of the story. Genetically identical plants (clones) grow very differently when exposed to varying amounts of light and soil nutrients. They can be used to demonstrate very clearly that the environment in which an organism finds itself also has a big impact on its appearance. The ability of animals to achieve their full genetic potential in terms of growth also depends heavily on environmental factors such as the amount of food available. But it is more difficult to investigate the impact of environment on the phenotype with animals because of the difficulty of producing large numbers of cloned, genetically identical organisms, and the ethical aspects of experimentation.

The impact of environment is seen in Siamese cats and certain rabbit breeds that have dark 'points' on the ears, the muzzle and the paws. The genotype of these animals suggests that they should have dark fur all over the body, as a result of melanin produced in a process involving the enzyme tyrosinase (see chapter 2.2). However, a mutation in Siamese cats results in a version of tyrosinase that is inactive at normal body temperatures and only works at lower temperatures. So in Siamese cats, the fur over the majority of the body is pale, but at the extremities – the ears, paws and nose – where the temperature is lower, the enzyme is not denatured and as a result the fur is dark!

fig. 3.4.4 The environment of the animal's own body interacts with the genotype to give the unique markings of a Siamese cat.

HSW *E. coli* and the lac operon

Environmental factors that affect the phenotype of an organism are often described in fairly loose and general terms, such as nutritional levels or temperature. But by looking at the phenotypes of bacteria scientists have shown that it is possible in some cases to see the environmental effect take place at a biochemical level. In the 1950s and 1960s two French geneticists, François Jacob and Jacques Monod, were working with the common gut bacterium *E. coli* (*Escherichia coli*).

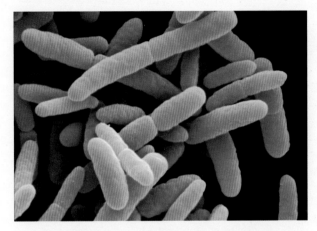

fig. 3.4.5 *E. coli* have been used to demonstrate that the environment can have a very direct effect on the phenotype.

Lactose is a disaccharide found in milk that is broken down into galactose and glucose by the enzyme β-galactosidase. If there is no lactose present in the environment of the *E. coli* bacterium, then fewer than 11 molecules of this enzyme are found in each bacterial cell. But if lactose becomes available to the bacterium (which happens in your gut when you eat ice cream or drink a milkshake) each bacterial cell will soon contain thousands of molecules of β-galactosidase and two other related enzymes. The presence of lactose induces the bacterial cells to make the enzyme.

Jacob and Monod showed that this effect takes place at the level of the genes. They found a cluster of genes which function together to make the enzymes for breaking down lactose, and they called the cluster the *lac* **operon**. They developed a model of a regulatory gene which produces a repressor substance when there is no lactose around. This repressor binds with the genetic material in a part of the gene which blocks the binding of RNA polymerase, preventing the code for the lactose-digesting enzymes from being transcribed. This means mRNA is not made, and so neither is the lactose-digesting enzyme

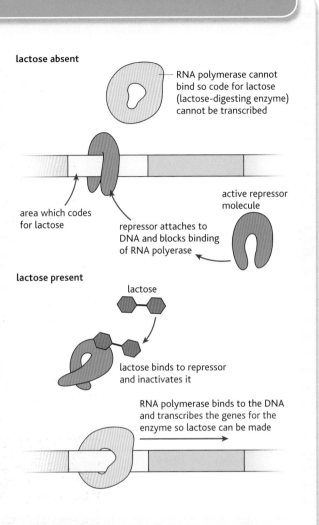

fig. 3.4.6 **The presence of lactose in the diet changes the way the DNA is transcribed and translated – lactose switches on the production of the enzyme that digests it.**

When lactose is present, it binds to the repressor substance, making it change shape. This new complex can't bind to genetic material so RNA polymerase has free access to the DNA and transcription takes place, closely followed by translation at the ribosomes to produce the enzymes. The bacterial cell can now use lactose as a food source. The change in the phenotype from not producing β-galactosidase to producing active enzyme is brought about purely in response to changes in the environment.

Operon systems like this give very sensitive control of the metabolism of the cell, allowing it to change in response to changes in the environment. They also provide an excellent research tool because, by changing the nutrients given to microorganisms growing in cultures, scientists can change their phenotypes and investigate further the mechanisms by which such changes come about.

Studying variation in humans

In mammals such as humans, the environment in the uterus affects the development of the fetus even before birth. If the mother is malnourished, or the placenta is not very effective, the fetus may be deprived of vital nutrients and not fulfil its full growth potential. If a mother smokes, her fetus will be deprived of oxygen and this in turn can affect the growth of both the body and the brain. Heavy use of alcohol or some drugs, or illness in a mother, can all have serious consequences for the phenotype of the fetus.

There are several difficulties in studying the interaction of genotype and environment in human beings. It is very important during any experiment on this interaction that all the organisms are subjected to the same conditions. Then as far as possible any differences between them can be seen as the results of genetic differences alone. But in human beings, imposing conditions like these is impossible. Scientists need other ways of answering questions about the interaction of nature and nurture.

HSW Twin studies

One strategy is to consider genetically identical individuals and try to disentangle the effect of genes and environment. In humans this involves twin studies (as you saw in chapter 1.4). Identical twins are human clones – they have the same genetic material. Non-identical twins are normal siblings, with closely related but not identical DNA, but being the same age they are more likely to have a similar environment than ordinary siblings. Ordinary siblings are useful as a control group.

In twin studies the differences between identical (monozygotic) twins are studied and compared with the differences between non-identical twins and normal siblings.

Occasionally identical twins are separated at birth to be adopted by different families. This provides a rare and very useful resource. Scientists have traced many pairs of separated twins and compared them with non-separated identical twins, non-identical twins and ordinary siblings. Very often the twins will still show remarkable similarities even after being separated as babies. If twins who have been reared apart show strong similarities in a trait, such

as height or weight, this suggests that the influence of the genotype on the characteristic is very strong. On the other hand, if twins reared together are quite similar for a trait, but twins reared apart show a greater difference, it suggests that the environment has a stronger influence on that characteristic.

From the study summarised in **table 3.4.1**, height appears to have a strong genetic component and is influenced relatively little by environmental factors. On the other hand, body mass also seems to be affected by external factors such as the family eating habits, and IQ – one measure of intelligence – seems to be a combination of both, with environment playing a distinct and important role.

A team at University College, London studied 5000 pairs of twins aged between 8 and 11 years that were brought up together. Their results, published in 2008, showed the 77% of the variation in BMI and waist circumference (chapter 1.3) of the children was caused by their genes and 23% to their home environment.

Trait	Identical twins reared apart	Identical twins reared together	Non-identical twins	Non-twin siblings
Height difference	1.8 cm	1.7 cm	4.4 cm	4.5 cm
Mass difference	4.5 kg	1.9 kg	4.6 kg	4.7 kg
IQ score difference	8.2	5.9	9.9	9.8

table 3.4.1 These data show the results from a US study based on 19 pairs of identical twins reared apart, along with 50 pairs each of identical twins reared together, non-identical twins and non-twin siblings, by Newman, Freeman and Holzinger at the University of Chicago in 1937.

Our knowledge of the genetic basis of human traits is still far from complete, but work like these twin studies helps us move our understanding forward.

Fig. 3.4.7 shows some conclusions about relative genetic and environmental influences drawn from a number of similar studies.

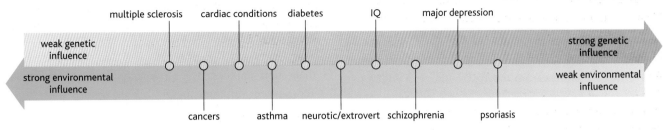

fig. 3.4.7 Nature versus nurture – the evidence suggests that both play a part in many of our characteristics.

HSW Epidemiological studies

One way of separating environmental and genetic influences in humans is to look at huge numbers of individuals and consider a genetic feature along with information about lifestyle. For example, the Health Statistics Center of the West Virginia Bureau for Public Health in the US tracked the effect of a mother smoking during pregnancy by looking at data on all the live babies born in the region for the 10-year period 1989–98. These data – an enormous sample – were collected from West Virginia certificates of live birth, which includes a question regarding the mother's smoking habits during pregnancy. The difference in birth weights and premature births between the babies born to mothers who are smokers compared to mothers who are non-smokers is quite striking and gives a clear picture of the impact of smoking on the phenotype of the baby. Because the study was so large and spanned a 10-year period, these data paint a reliable picture of the impact of smoking on the phenotype of a baby.

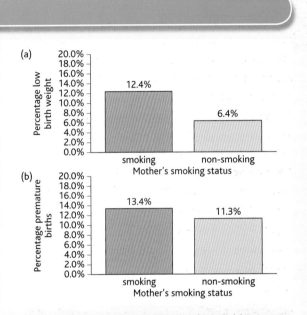

fig. 3.4.8 Smoking has a clear impact on phenotype. (a) babies do not grow to their genetic potential if their mother smokes during pregnancy, and (b) they are also likely to be born earlier, which in turn can affect other aspects of their growth and development.

Questions

1 Explain the following observations in terms of the interaction of the genotype and the environment.

   a A patch of white fur is removed from the back of a rabbit which has a chocolate point pattern to its fur. A cool pack is kept on the area of skin as the fur grows back. The new fur which grows is dark.

   b If a Siamese cat needs surgery, an area of fur will be shaved off the body to leave a patch of bare skin. The fur that grows back over this skin is dark. It is not until after the next moult that the fur returns to its normal cream colour.

2 Why are data from identical twins reared apart even more useful than data from twins brought up together?

3 How reliable are the results from the study shown in table 3.4.1? Explain your answer.

4 What is the importance of the work on gut bacteria carried out by Jacob and Monod?

Variation

Different species are often obviously very different from each other, but individuals of the same species can also show remarkable variety. Some of the variety you see around you comes from the genotype of the organism, but some of it comes from the environment in which the organism has grown.

Discontinuous and continuous variation

In any population of organisms there are two types of variation – **discontinuous** and **continuous**. Discontinuous variation is shown by features that are either present or not, such as blood groups or sex (male or female). These features are generally determined by one or at most a very few genes, and the environment does not usually have an effect – you are either male or female, and your blood group is either A, B, AB or O. (There are exceptions of course – exposure to high levels of sex hormones in the uterus or rare chromosome mutations can make the sex of an individual difficult to determine, but usually discontinuous variation is very clear-cut.)

Characteristics that show continuous variation include weight and height in an animal species, or the number of leaves on a plant. Factors such as these are often determined by a number of genes (they are polygenic) but they are also very much affected by the environment.

Studying continuous variation

Height in humans has a very strong genetic component – tall parents tend to have taller children than short parents. It is a polygenic feature – different genes affect different factors related to size, such as the length of the bones in the legs or the size of the vertebrae. But height also has an environmental element. If an individual has a balanced diet throughout their growing years, then they are more likely to fulfil their genetic height potential than if they were malnourished during one or more of their major growth periods. Because of the variety of factors that influence height, it shows continuous variation.

When studying continuous variation in a population you need to take large samples, because sheer chance could make a difference to the results. If you take only a small sample from one particular area you might, quite by chance, pick on a group of the shortest individuals in the population, or those with the most vivid coloured plumage. You also need to collect your large sample randomly from as much of the organism's habitat as possible. If you collect from only one area you might find a regional effect due to climate or diet, in which case you might not get an accurate picture of the whole species. Data like these can be displayed using a graph or histogram, to show the **frequency distribution** of the characteristic clearly.

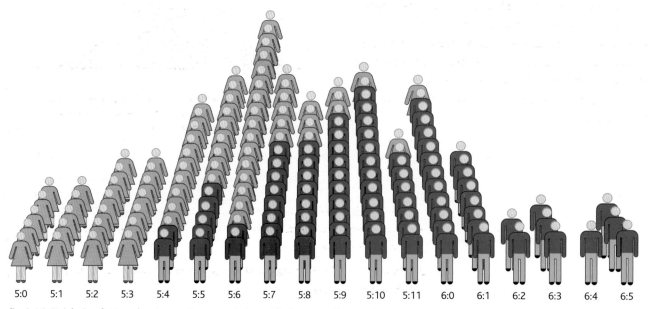

fig. 3.4.9 **Height is a feature showing continuous variation, with clear sex differences but with overlap between taller women and shorter men.**

HSW Normal distributions and standard errors

If a feature of an animal or plant shows continuous variation, then the frequency distribution will be a **normal distribution curve**. This is the typical 'bell-shaped curve' which you can see in **fig. 3.4.10**. This shape of curve will be found when studying many kinds of variation, such as height in the human female population. The curve shows you that most women in the population are somewhere in the middle of the height range. At the extremes there will only be a few individuals who are very, very short or tall. The mean height (the average height of all the women measured) is exactly at the centre of the normal distribution curve, and all the heights are distributed equally about the mean.

A useful concept in helping you to judge the amount of variation in a sample is the idea of **standard deviation**. The standard deviation is a statistical measure of the amount of difference from the mean in the sample. It depends on sample size as well as the distance of the measurements from the mean. About 68% of the sample will be within one standard deviation of the mean, 95% will be within two standard deviations and almost all will be within three standard deviations of the mean. If the standard deviation is small, then there is very little variety in your population – all the sampled individuals are very close to the mean. On the other hand, if the standard deviation is large this indicates lots of variation – there is a wide range of measurements away from the mean.

When looking at data that have been collected on a particular feature, eg height or intelligence, it is important to draw out the normal distribution curve and then check the standard deviation. If the standard deviation is small, with all the results clustered together, then the results are probably reliable. If the standard deviation is large, then the results are more likely to be random rather than the result of a linking factor, and it is best not to use the data to draw clear conclusions.

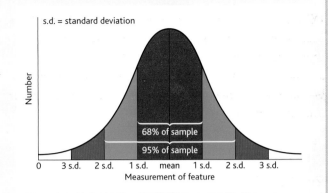

fig. 3.4.10 **Most of the measurements in a normal distribution are within three standard deviations from the mean of the sample.**

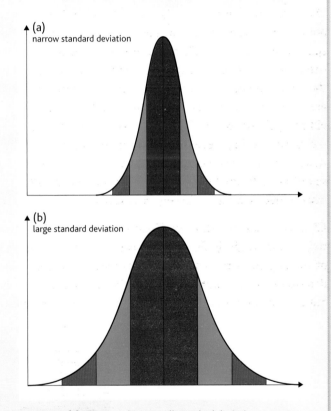

fig. 3.4.11 **(a) This curve has a small standard deviation – most of the individuals are close to the mean. (b) This curve shows a population with much greater variation.**

Questions

1 Explain why characteristics affected by both genotype and environment show continuous variation.

2 What is the value of using a normal distribution to analyse data on a phenotypic feature?

Genes and environment in human features

Monoamine oxidase A

Monoamine oxidase A (**MAOA**) is an enzyme found in the nervous system, the liver and the gut, where it breaks down monoamines. Monoamines are eaten as part of the diet. However, many of the chemicals involved in transmitting nerve impulses across the synapses (**neurotransmitters**) are also monoamines, such as **serotonin**, **noradrenaline** and **dopamine**. Once these transmitters have been released into the synapse, they are broken down by monoamine oxidase.

If too much or too little of the enzyme is formed, this can lead to a number of different illnesses and problems such as Parkinson's disease and some forms of depression. If there is too much of the enzyme present, it breaks down too many of the neurotransmitter molecules and this can have a profound effect on your mood – hence the link with depression – and on the functioning of different areas of your brain. People with this form of depression may be given medication to inhibit the action of the MAOA enzyme, allowing higher levels of neurotransmitters such as serotonin to build up.

Levels of monoamine oxidase are determined genetically and a number of different mutations can cause raised or lowered levels of this enzyme. Low levels of MAOA seem to be linked not just to depression but also to addictive behaviour (particularly a tendency to become dependent on alcohol) and in some cases criminal behaviour. High levels of the enzyme have been linked to risk-taking and aggressive behaviour. Controversial research has linked some of these genetic tendencies to certain racial groups such as the Maoris in New Zealand. However, environmental factors seem to combine with genetics to determine the phenotype – particularly the stress levels the person experiences. There is still much work to be done and some of the evidence from different research teams is conflicting.

Fig. 3.4.12 **No-one is forced into crime, but it looks as if genetic factors such as the levels of MAOA in our brain may affect the choices some of us make.**

HSW Evidence for the MAOA level/behaviour link

fig. 3.4.13 Some exciting research on the interaction between genotype, environment and phenotype has come from a longitudinal study in New Zealand following 1037 people through childhood to adulthood.

In the early years of the twenty-first century the results of a longitudinal study were published in the journal *Science*. You saw in chapter 1.4 how a longitudinal study follows a group of individuals of the same age over a period of years. The data came from 1037 children born at the same time in New Zealand. The children were first assessed when they were three years old; by the time they were 26, scientists had assessed them nine times. Researchers measured many factors, including levels of MAOA. They also calculated a factor for maltreatment between the ages of 3 and 11. Maltreatment included lack of continuity in the people looking after the children, rejection by their mother, or physical or sexual abuse. When the children had grown into young adults the team assessed them using four different measures of antisocial behaviour, including convictions for violent crime.

They looked for any links between the behaviour of the young adults and their genes or upbringing. They discovered that there appeared to be a clear correlation between low MAOA levels, maltreatment and subsequent antisocial behaviour. People who had suffered severe maltreatment but had high MAOA levels were no more likely to be antisocial than anyone else. Only 12% of the group had both low MAOA levels and experienced maltreatment, but they accounted for 44% of the convictions for violent crime. In fact, 85% of the males with both of these risk factors went on to show some form of antisocial behaviour.

Males were more likely to be involved in antisocial behaviour than females. The MAOA gene is carried on the X chromosome, so girls have double the chance of inheriting the dominant allele for high MAOA levels. This raises a possible explanation for the fact that, worldwide, males are more likely to be involved in antisocial and criminal behaviour than females.

This evidence seems quite clear-cut, and backs up the findings of an earlier study on a Dutch family group. However, it is important to make sure that an apparent link such as this is not just an example of correlation. A lot more work is needed before it is safe to say that a particular combination of genes and environmental conditions causes a young person to become a criminal, or even to suggest an increased likelihood that they will be involved in criminal behaviour.

Cancer

Cancer cells do not respond to the normal mechanisms that control cell growth and the cell cycle (see chapter 3.1). They divide rapidly to form a mass of abnormally growing cells (**a tumour**) which invades the surrounding tissues. A tumour may split up, releasing small clumps of cells into the blood or lymph systems, where they circulate and then lodge in a different area of the body, continuing their uncontrolled division and forming secondary tumours. This splitting is called **metastasis**. A tumour that invades surrounding tissues and metastasises is known as a **malignant** tumour. Not only do cancer cells divide more rapidly than normal cells, they also live longer. The enlarging tumour completely disrupts normal tissues, often to the point of killing the organism.

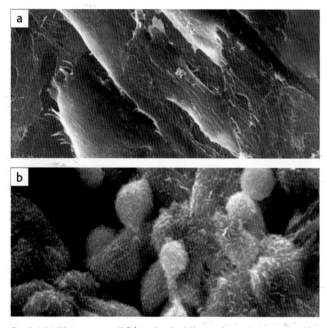

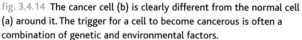

fig. 3.4.14 The cancer cell (b) is clearly different from the normal cell (a) around it. The trigger for a cell to become cancerous is often a combination of genetic and environmental factors.

The causes of cancer are many, but they include both genetic and environmental factors. About 15% of human cancers are the result of a viral infection of the cells, where the viral genetic material triggers uncontrolled cell growth. For example, cervical cancer is almost always the result of infection by the human papillomavirus. However, many cancers arise from mutations in the DNA of normal body cells as they reproduce and divide. It usually takes several mutations for a cell to become cancerous. This is why cancers become more common as you get older, because the cells accumulate mutations with age.

Mutations that cause cancer

The mutations that result in cancer affect the control of the cell cycle, usually by disrupting the chemical systems that control the stages of the cycle. The multiple mutations leading to cancer often interfere with the accurate replication of the DNA, decrease the efficiency of DNA repair in the cell and/or increase the likelihood of chromosomes breaking and rearranging. However, some cancers are caused by a single gene mutation. Genes known as **proto-oncogenes** code for the proteins that stimulate the cell cycle. If one of these genes mutates, it forms an **oncogene**. The oncogene produces uncontrolled amounts of these proteins so the cell cycle is constantly stimulated, causing cancer. The oncogene allele is dominant.

Another family of genes involved in cancer are the **tumour suppressor genes**. These normally produce chemicals that suppress the cell cycle, acting as a brake on cell division. If a mutation occurs in these genes, the brake on the cell cycle is removed and the cell goes into uncontrolled growth – again causing cancer.

The role of the environment

Environmental factors play an important part in cancer by increasing the likelihood of these cancer-causing mutations occurring. So factors such as the tar in cigarette smoke, the chemicals in alcoholic drinks, asbestos and ionising radiation all increase the likelihood of a mutation in your DNA which in turn causes cancer.

A clear example is the ultraviolet light in sunlight. For many years people regarded getting a suntan a very healthy thing to do. They would often consider that suffering sunburn for a few days was a price worth paying for a golden tan. And as more people started to go abroad for holidays, the tans got deeper – and redder! We may think a suntan looks good, but it is a sign that the cells of your skin have already been damaged. They are making more melanin to try and protect themselves from any more burning.

Doctors noticed a long time ago that people who spent much of their working day out in the sun, such as farm workers, were more likely to get skin cancers than office workers. Then a few years ago it became increasingly clear that it was the ultraviolet light in the Sun's rays that caused mutations in the skin cells. These mutations lead to very aggressive malignant tumours. **Melanoma**, a cancer of the cells that produce the pigment melanin, tends to metastasise very rapidly and can be fatal. As more people took holidays abroad, the cases of melanoma began to rise steadily (**fig. 3.4.15**). It became clear that getting sunburnt, particularly in childhood, was very dangerous indeed, as it greatly increased the risk of mutations that give rise to melanomas. This is not only a cancer of old age – relatively young adults can and do die from skin cancer.

Damage of this type is a particular problem for pale-skinned people – anyone with a brown or black skin has plenty of melanin, which acts as a natural sunscreen and protects the skin from damage, although extra sunscreen does no harm. So around the world, in countries like Australia, the US and the UK, there has been a strong move to persuade white-skinned people in particular to wear sunscreens when they go out in the sun. These either absorb the ultraviolet radiation or reflect it back away from the skin, preventing damage. The risk from ultraviolet radiation is increasing all the time as the ozone layer thins, so it is hoped that wearing sunscreen and covering up in the heat of the day will become a life-saving habit for everyone.

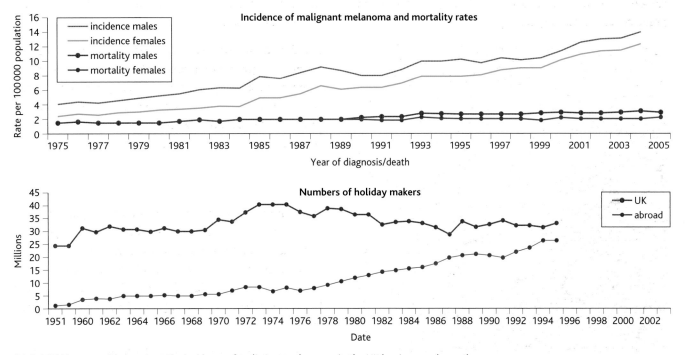

fig. 3.4.15 These graphs show how the incidence of malignant melanomas in the UK has increased over the last 30 years, along with the change in the numbers of people holidaying abroad over the last 50 years.

Questions

1 How is monoamine oxidase A linked to disease?

2 What is the value of a longitudinal study in looking for interactions between genotype, environment and phenotype?

3 Explain how the genotype and environment may interact at a cellular level to cause a cell to become cancerous.

4 What do the graphs in fig. 3.4.15 tell you about the incidence of malignant melanoma in the UK over the last 30 years? Can you suggest any reasons for the apparent differences between females and males in both incidence and mortality?

5 How might the data in the second graph suggest a possible cause for the increase in melanomas? What type of evidence would you need to decide whether any apparent link was correlation or causation?

Examzone: Topic 3 practice questions

1 Copy the table below which refers to some features of prokaryotic and eukaryotic cells.

If the feature is present, place a tick (✔) in the appropriate box and if the feature is absent, place a cross (✗) in the appropriate box.

	Prokaryotic cell	Eukaryotic cell
Nuclear envelope		
Cell surface membrane		
Ribosomes		
Microtubules		
Mitochondria		

(Total 5 marks)

2 Insulin is a protein.

a i State precisely where proteins are made in a cell. (1)

   ii Explain how proteins are transported from their site of production to the outside of the cell. (2)

b A person who has Type 1 diabetes cannot make enough of the hormone insulin. This is because the beta cells in the pancreas have been destroyed by the immune system. It is possible that, in the future, we will be able to replace beta cells with ones produced from embryonic stem cells.

   i Explain what is meant by embryonic stem cells. (1)

   ii Explain how stem cells might provide a way of obtaining beta cells. (1)

c Some people object to the use of embryonic stem cells and consider it to be ethically wrong. State whether you are for or against stem cell research. Give reasons for your choice. (3)

d The occurrence of Type 1 diabetes is more common in some families than others, through many generations even if the members of the family now live in different parts of the world. In pairs of identical twins one twin may develop the condition and the other may not.

Explain what the information above tells you about the causes of Type 1 diabetes. (2)

(Total 10 marks)

3 The flow diagram below shows the sequence of cells formed during spermatogenesis in the mammalian testis.

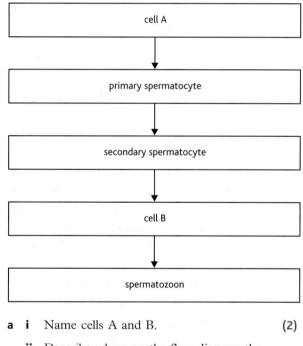

a i Name cells A and B. (2)

   ii Describe where on the flow diagram the second division of meiosis occurs. (1)

b Give two features of a spermatozoon that enable it to carry out fertilisation. (2)

(Total 5 marks)

4 a The diagrams below represent the chromosomes during stages in the process of mitosis.

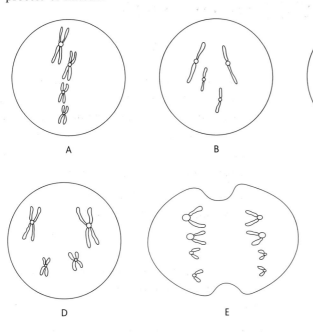

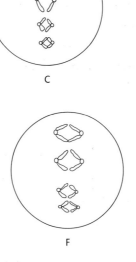

Write the letters in the order that represents the sequence in which these stages occur. **(1)**

b State *two* ways in which meiosis differs from mitosis. **(2)**

c Explain the significance of mitosis in living organisms. **(3)**

(Total 6 marks)

5 A Himalayan rabbit has a coat colour which is white with black tips to the ears, nose, feet and tail. Albino rabbits are entirely white. A cross was made between two Himalayan rabbits. The offspring were initially kept in a warm environment and all had entirely white coats. They were then moved to an environment with lower temperatures and 75% of them developed the Himalayan markings.

a i Using the symbols A^h for the Himalayan allele and a for the albino allele, state the genotypes of the parent rabbits in this cross. **(1)**

ii Draw a genetic diagram to show this cross. **(2)**

b i Comment on the expression of the allele for the Himalayan coat colour. **(2)**

ii Suggest a reason for the distribution of black fur in Himalayan rabbits. **(1)**

(Total 6 marks)

6 The diagram below shows a vertical section through part of a flower.

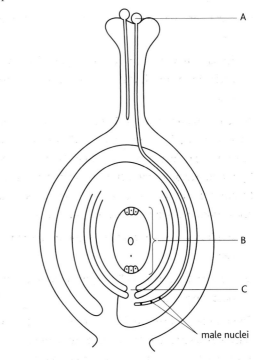

a Name the parts labelled A, B and C. **(3)**

b Name the type of cell division which gives rise to the male nuclei. **(1)**

c Describe the process of fertilisation in a flowering plant. **(3)**

(Total 7 marks)

Topic 4 Biodiversity and natural resources

This topic deals with the use of plants as natural resources and how the structures in the plant are related to the way we use them. It also looks at biodiversity, how it is created through evolution and how it is being reduced by human activity.

What are the theories?

We use plants in many ways, such as for food, to make clothing, and for building materials. We also use the chemicals they contain to make drugs to treat illness. In the future we may be able to use plants to replace some products that we make from oil, including plastics, so that we can become more sustainable and reduce pollution.

Biodiversity, in terms of species-richness and genetic diversity, is increased through speciation which is driven by evolution due to natural selection. As a result of humans needs and use of resources, many other species have become extinct or are endangered. Maintaining biodiversity is important to the health of the planet, and conservation of species aims to reduce our impact on biodiversity.

What is the evidence?

You will learn how the usefulness of plant chemicals developed as drugs comes from a rigorous testing process, aimed at reducing the risks to humans as much as possible. You will also find out how new methods of analysis are changing our view of the interrelationships between species, and discover how science is attempting to assess biodiversity and its importance to the health of the planet. You will have the opportunity to carry out your own investigations, for example determining the tensile strength of plant fibres and investigating plant mineral deficiencies and the antimicrobial properties of plants.

What are the implications?

Science can provide us with tools to answer questions about how things work, and can provide the data on which we can make choices. However there are questions it cannot answer. Such as should we test drugs in a way that risks the health of animals and humans in the trials. How do we choose which areas of biodiversity to protect? The answers to questions like these depend on our point of view and on the society that asks them. And the answers will change our future and that of the planet.

The map opposite shows you all the knowledge and skills you need to have by the end of this topic. The colour in each box shows which chapter they are covered in and the numbers refer to the Edexcel specification.

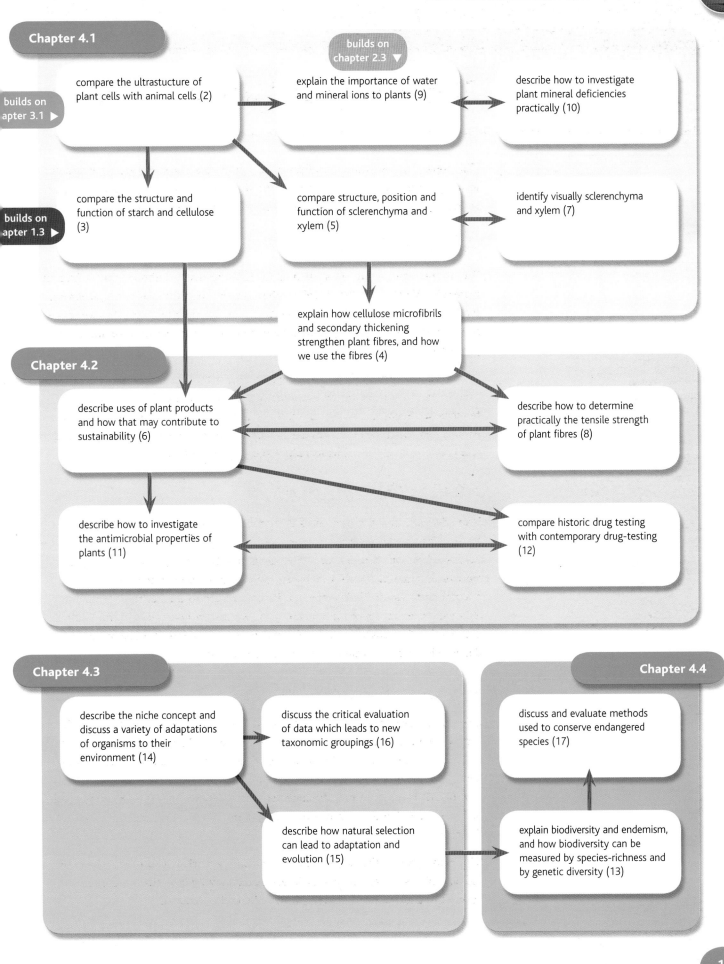

Chapter 4.1

builds on apter 3.1 ▶

compare the ultrastucture of plant cells with animal cells (2)

builds on chapter 2.3 ▼

explain the importance of water and mineral ions to plants (9)

describe how to investigate plant mineral deficiencies practically (10)

builds on apter 1.3 ▶

compare the structure and function of starch and cellulose (3)

compare structure, position and function of sclerenchyma and xylem (5)

identify visually sclerenchyma and xylem (7)

explain how cellulose microfibrils and secondary thickening strengthen plant fibres, and how we use the fibres (4)

Chapter 4.2

describe uses of plant products and how that may contribute to sustainability (6)

describe how to determine practically the tensile strength of plant fibres (8)

describe how to investigate the antimicrobial properties of plants (11)

compare historic drug testing with contemporary drug-testing (12)

Chapter 4.3

describe the niche concept and discuss a variety of adaptations of organisms to their environment (14)

discuss the critical evaluation of data which leads to new taxonomic groupings (16)

Chapter 4.4

discuss and evaluate methods used to conserve endangered species (17)

describe how natural selection can lead to adaptation and evolution (15)

explain biodiversity and endemism, and how biodiversity can be measured by species-richness and by genetic diversity (13)

4.1 Plant structure

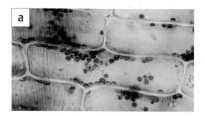

numbered chloroplasts

numerous chloroplasts

tonoplast (membrane around vacuole)

starch grain

cell wall (cell surface membrane underneath wall)

sap vacuole

nucleus

cytoplasm

10 μm

Plant cell structure

It is easy in our modern technological world to lose sight of the importance of plants. Plants harness the energy of the Sun in the process of photosynthesis, and that energy becomes available to us as food. Plants and their products have shaped our practical and economic world. In this chapter you are going to look at some of the features of plants that make them so important to us, and focus on some of the ways in which we use them.

The structure of a typical plant cell

A typical plant cell has many features in common with a typical animal cell (see chapter 3.1). Plants cells have many membranes with the same basic chemical make-up as those of animal cells – lipoproteins with protein pores and some carbohydrate surface markers. They have similar properties and control what moves through them in a similar way.

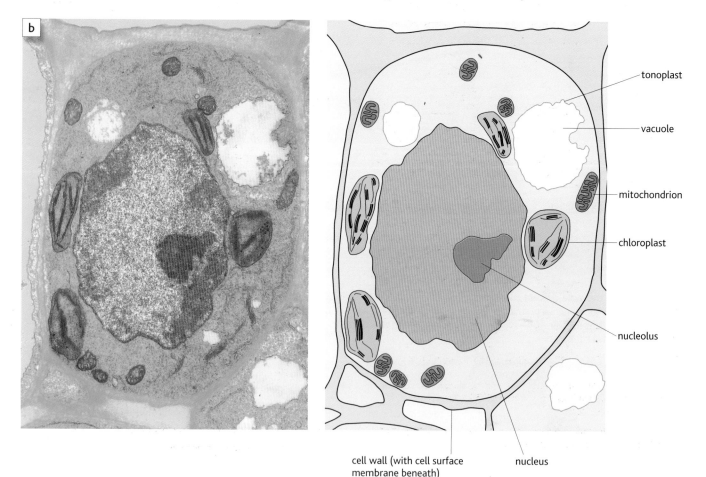

tonoplast

vacuole

mitochondrion

chloroplast

nucleolus

cell wall (with cell surface membrane beneath)

nucleus

fig. 4.1.1 (a) The light microscope shows the major differences between plant and animal cells.
(b) The electron micrograph reveals similarities and differences in more detail.

Plant cells contain cytoplasm and a nucleus. You will find rough and smooth endoplasmic reticulum spreading throughout the cytoplasm, along with active Golgi bodies. Mitochondria produce ATP which is as vital to the working of the plant cell as it is to the animal cell. However, there are several quite fundamental differences.

The plant cell wall

Animal cells can be almost any shape. Plant cells tend to be more regular and uniform in their appearance. This is largely because each cell is bounded by a **cell wall**. You can visualise a plant cell as a jelly-filled balloon inside a shoe box. The cell wall (shoe box) is an important feature which gives plants their strength and support. It is made up largely of insoluble **cellulose**. Cellulose has much in common with the complex carbohydrates starch and glycogen which you studied in chapter 1.3. It is made up of long chains of glucose joined by glycosidic bonds. However, glucose comes in two different forms (**isomers**), α-glucose and β-glucose. The two isomers result from different arrangements of the atoms on the side chains of the molecule (see **fig. 4.1.2**).

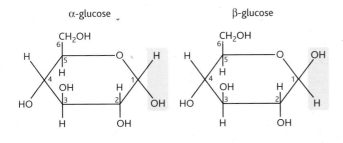

or, even more simply:

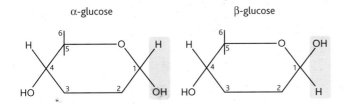

In these diagrams, the positions of carbon atoms are represented by their numbers only.

Note carefully the different arrangement of atoms around the carbon I atoms in α-glucose and β-glucose. This small difference can give the molecules different properites.

fig. 4.1.2 The difference between α-glucose and β-glucose may seem small but it has a large biological impact.

The different isomers form different bonds between neighbouring glucose molecules, and this affects the polymers they form. In starch the monomer units are α-glucose. In cellulose they are β-glucose and they are held together by 1,4-glycosidic bonds (see chapter 1.3), where one of the monomer units has to be turned round (inverted) so the bonding can take place. This linking of β-glucose molecules means that the hydroxyl (–OH) groups stick out on both sides of the molecule (see **fig. 4.1.3**). This means **hydrogen bonds** can form between the partially positively charged hydrogen atoms of the hydroxyl groups and the partially negatively charged oxygen atoms in other areas of the glucose molecules. This is known as **cross-linking** and it holds neighbouring chains firmly together.

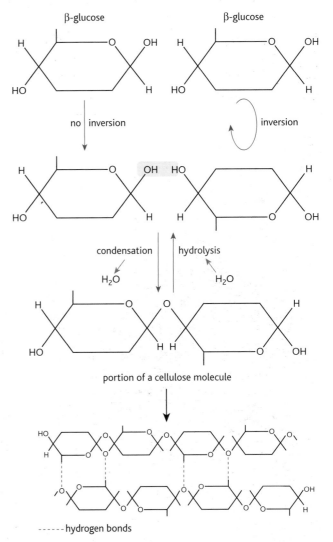

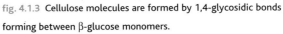

------ hydrogen bonds

fig. 4.1.3 Cellulose molecules are formed by 1,4-glycosidic bonds forming between β-glucose monomers.

Many of these hydrogen bonds form, making cellulose a material with considerable strength. Cellulose molecules do not coil or spiral – they remain as very long, straight chains. In contrast, starch molecules, with 1,4- and 1,6-glycosidic bonds between α-glucose monomers, form compact globular molecules that are useful for storage. This difference in structure between starch and cellulose gives them very different properties.

Starch is an important source of energy in the diet for many animals. However, most animals do not possess the enzymes needed to break the 1,4-glycosidic bonds between the molecules of β-glucose and so they cannot digest cellulose. Animals such as ruminants use the cellulose-digesting enzymes from bacteria living in their gut to digest their food, while termites use the enzymes in various protoctista. It is the cellulose in plant food that acts as roughage or fibre in the human diet – an important part of a healthy diet even though you can't digest it.

In the cell wall, groups of 10–100 000 cellulose molecules form **microfibrils** which can be seen under the electron microscope (see **fig. 4.1.4**). These cellulose fibrils are laid down in layers held together by a **matrix** of hemicelluloses and other short-chain carbohydrates which act as a kind of glue, binding to each other and

fig. 4.1.4 These cellulose microfibrils consist of thousands of cellulose chains held together by hydrogen bonds. The orientation and packing of the microfibrils affect the strength of the cell wall.

to the cellulose molecules. Examples of the sugars involved include mannose, xylose and arabinose. The combination of the cellulose microfibrils in the flexible matrix makes a **composite material**, combining the properties of both these materials in the plant cell wall. The cells are firm (**turgid**) most of the time, giving the strength to support the plant in a vertical position, yet the plant can wilt when water is in short supply and the cells become floppy (**flaccid**).

It isn't always easy to remember that a plant cell is a three-dimensional structure. Individual cellulose molecules do not spiral, but the microfibrils do. They are arranged in spirals around the 'box' which is the cell wall. The more vertical the spiral and the closer together the turns, the stronger is the structure of the cell in the vertical direction. This of course is the direction in which the weight of the plant acts downwards, so this is where the strength needs to be.

In normal circumstances the cell wall is freely permeable to everything that is dissolved in water – it does not act as a barrier to substances getting into the cell. However, the cell wall can become impregnated with **suberin** in cork tissues, or with **lignin** to produce wood. These compounds affect the permeability of the cell wall so that water and dissolved substances cannot pass through it.

The plant cell wall consists of several layers. The **middle lamella** is the first layer to form when a plant cell divides into two new cells. It is made largely of **pectin**, a polysaccharide which acts like glue and holds the cell walls of neighbouring plant cells together. Pectin has lots of negatively charged carboxyl ($-COO^-$) groups and these combine with positive calcium ions to form calcium pectate. This binds to the cellulose that forms on either side.

The cellulose microfibrils and the matrix build up on either side of the middle lamella. To begin with these walls are very flexible, with the cellulose microfibrils all orientated in a similar direction. They are known as **primary cell walls**. As the plant ages, secondary thickening may take place. A **secondary cell wall** builds up, with the cellulose microfibrils laid densely at different angles to each other. This makes the composite material much more rigid. Hemicelluloses harden it further. In some plants, particularly woody perennials, lignin is then added to the cell walls to produce wood which makes the structure even more

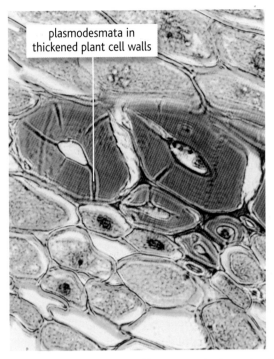

plasmodesmata in thickened plant cell walls

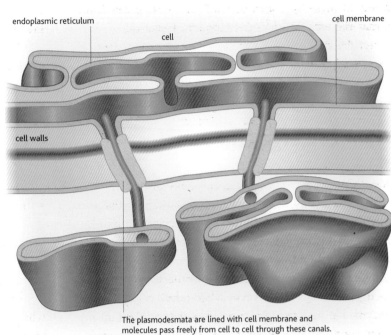

endoplasmic reticulum   cell   cell membrane

cell walls

The plasmodesmata are lined with cell membrane and molecules pass freely from cell to cell through these canals.

fig. 4.1.5 Plasmodesmata provide a communication system between plant cells – but scientists are still working out exactly what goes on.

rigid.

Within the structure of a plant there are many long cells with cellulose cell walls which have been heavily lignified. These are known as **plant fibres** and people use them in many different ways, as you will see in chapter 4.2.

Plasmodesmata

In spite of being encased in cellulose cell walls, plant cells seem to be in close communication with each other. Intercellular exchanges seem to take place through special cytoplasmic bridges between the cells known as **plasmodesmata** (**fig. 4.1.5**). The plasmodesmata appear to be produced as the cells divide – the two cells don't separate completely, and threads of cytoplasm remain between them. These threads pass through gaps in the newly formed cell walls and signalling substances can pass from one cell to another through the cytoplasm. The interconnected cytoplasm of the cells is known as the **symplast**

HSW Evidence for the importance of plasmodesmata

Scientists are still working hard to discover exactly how plant cells communicate through plasmodesmata. One clear piece of evidence showing that these intercellular junctions are vital in the life of plants comes from work with plant grafts. If a rose is grafted onto a hardy root stock, the graft tissue only starts healthy cell division and growth once plasmodesmata bridges are established between the host tissue and the graft tissue.

Questions

1 Explain how the chemical structure of cellulose differs from that of starch, and how this affects their functions in a plant cell.

2 What role do cell walls play in the structure of a plant, and how does their structure fit them for their function?

3 How does the plant cell wall change as the cell grows and develops, and how does this affect the cell?

4 Explain why plasmodesmata are an important feature of plant cell structure.

Plant cell organelles

Plant cells contain several kinds of organelle that are not found in animal cells. These include vacuoles and chloroplasts.

Plant cell organelles

Vacuole

A **vacuole** is any fluid-filled space inside the cytoplasm surrounded by a membrane. Vacuoles occur quite frequently in animal cells, but they are only temporary, being formed and destroyed when needed. In most plant cells the vacuole is a permanent structure with an important role. The vacuole of a plant cell is surrounded by a membrane called the **tonoplast**. It is filled with **cell sap**, a solution of various substances in water. This solution causes water to move into the cell by **osmosis** (see chapter 2.3) and the result is that the cytoplasm is kept pressed against the cell wall. This in turn keeps the cells turgid and the whole plant upright.

As well as fulfilling the important role of maintaining the plant cell shape, the many different types of vacuoles in plants carry out a range of different functions. Vacuoles are used for the storage of a number of different substances. Many vacuoles store pigments – eg the betacyanin pigment of beetroot is normally stored in the vacuoles of the cells and does not leak out into the cytoplasm unless the root is cut. If the tissue is heated, the characteristics of the membrane around the vacuole will change and so pigment will leak out more rapidly. Vacuoles can be used to store proteins in the cells of seeds and fruits, and in some plant cells they contain lytic enzymes and have a function rather like lysosomes in animal cells. Vacuoles are often used to store waste products and other chemicals. For example, digitalis which is the active drug/poison found in foxgloves (see chapter 4.2) is stored in the vacuoles of the cells.

Chloroplasts

Of all the differences between plant and animal cells, the presence of **chloroplasts** in plant cells is probably the most important because they enable plants to make their own food. Not all plant cells contain chloroplasts – only those cells from the green parts of the plant. However, almost all plant cells contain the genetic information to make chloroplasts and so in some circumstances different areas of a plant will become green and start to photosynthesise. The exceptions are parasitic plants such as broomrape. Cells in flowers, seeds and roots contain no chloroplasts and neither do the internal cells of stems or the transport tissues. In fact the majority of plant cells do not have chloroplasts – but these organelles are very special and unique to plants.

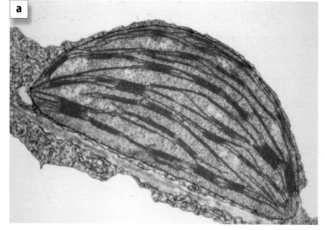

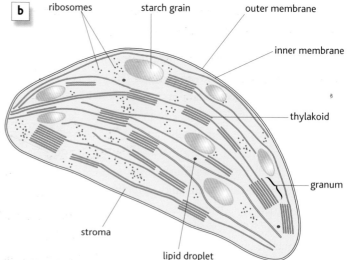

fig. 4.1.6 (a) Micrograph of a chloroplast. (b) Labelled diagram to show structures in a chloroplast.

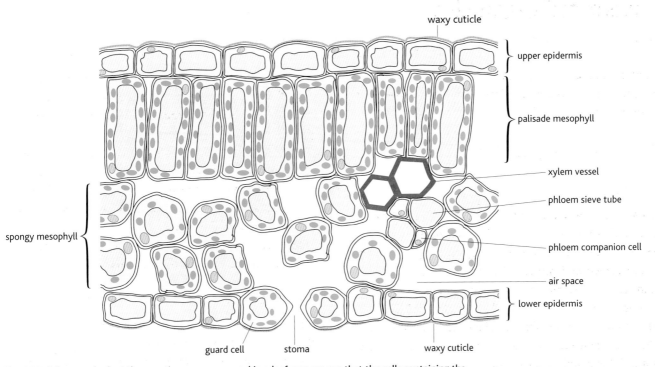

fig. 4.1.7 **When you look at the way tissues are arranged in a leaf you can see that the cells containing the majority of chloroplasts are found at the top, right under the colourless protective layer of the epidermis.**

There are some structural similarities between chloroplasts and mitochondria. Like mitochondria, chloroplasts are large organelles. They have a biconvex shape with a diameter of 4–10 mm and are 2–3 mm thick. They contain their own DNA and are surrounded by an outer membrane. Also like mitochondria, chloroplasts have an enormously folded inner membrane which gives a greatly increased surface area on which enzyme-controlled reactions take place. Scientists think that chloroplasts, like mitochondria, were once free-living prokaryotic organisms which were engulfed by and became part of other cells at least 2000 million years ago.

Chloroplasts are the site of photosynthesis. They contain **chlorophyll**, the green pigment that is largely responsible for trapping the energy from light, making it available for the plant to use. Chloroplasts are formed from a type of relatively unspecialised plant 'stem cell' known as a **leucoplast**. There is more detail about the structure of chloroplasts in relation to their function in A2 Biology.

Amyloplasts

Amyloplasts are another specialised plant organelle and, like chloroplasts, they develop from leucoplasts. They are colourless and are used to store **amylopectin**, a form of starch. This can then be converted to glucose and used to provide energy when the cell needs it. Amyloplasts are found in large numbers in areas of a plant that store starch, eg potato tubers.

Questions

1 Amyloplasts and chloroplasts come from the same type of unspecialised cell. How do the two structures differ?

2 Compare and contrast the structure of a typical plant cell with the structure of a typical animal cell.

3 Explain why chloroplasts are found only in particular parts of a plant. What do you think happens to make part of a plant, eg a potato tuber, turn green when exposed to light?

The structure of plant stems

Many plant cells are specialised and adapted for a particular role in the plant. As a result they no longer look like the 'typical' plant cell. They may be organised into tissues and organs that carry out a specific function in the plant. One example of a plant organ is the stem.

Providing support and transport

The primary function of a stem is support, to hold the leaves in the best position for obtaining sunlight for photosynthesis. Stems also support the flowers in a way that maximises the likelihood of pollination occurring. The stem has to provide flexible support, because plants are frequently buffeted by wind and rain. Stems need to bend to withstand the forces the elements exert upon them, and yet have the strength to stay upright.

The other major function of stems is the movement of materials about the plant. They provide the route along which the products of photosynthesis are carried from the leaves where they are formed to other parts of the plant where they are needed. Water moves through the stems in a steady stream from the roots up to the leaves, carrying mineral ions needed for the synthesis of more complex chemicals.

Most stems are green – that is, they contain chlorophyll. They carry out a small amount of photosynthesis, but this is not a major function.

Not all plants have stems. The liverworts have a simple flat structure and the mosses have leaves which arise directly from a pad of rhizoids. Both of these groups have no specialised transport tissues and grow close to the ground. However, the majority of the more complex plants do possess stems.

The tissues that make up the stem

Stems are organs and they contain many different tissues as you can see in **fig. 4.1.8**.

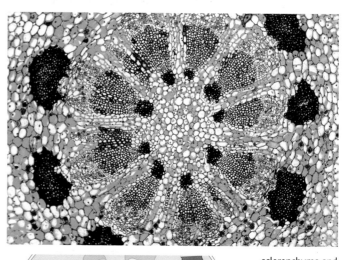

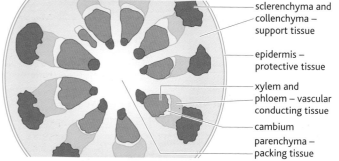

sclerenchyma and collenchyma – support tissue

epidermis – protective tissue

xylem and phloem – vascular conducting tissue

cambium

parenchyma – packing tissue

fig. 4.1.8 **The distribution of the different tissues in the stem of a plant.**

Epidermis

The outer layer of the stem is the **epidermis**, which plays no role in support but protects the cells beneath it. Epidermal cells secrete **cutin**, a waxy substance which helps to prevent water loss from the stem surface and protects against the entry of pathogens. The epidermal cells may also form hairs, either as an extension of a single cell (like root hairs) or from several modified epidermal cells. These hairs may act as an insulating layer, trapping moist air to reduce water loss. Some are hooked and help climbing plants to grip their supports. Other hairs are protective (see **fig 4.1.9**), stiff and bristly or may be loaded with irritant chemicals.

fig. 4.1.9 **These hairs on the stem of a stinging nettle give protection against attack by animals.**

Parenchyma and collenchyma

Much of the stem is made up of packing tissue which consists of the most common type of plant cells, known as **parenchyma** cells. These are unspecialised cells, but they can be modified in a variety of ways to make them suitable for storage and photosynthesis. For example, the outer layers of parenchyma cells in the stem may contain some chloroplasts. Some of the parenchyma in the stem is modified into **collenchyma** and **sclerenchyma**.

Collenchyma cells have thick primary cell walls which are even thicker at their corners. This gives the tissue its strength. These cells are found around the outside of the stem, just inside the epidermis, and they give plenty of support but remain living, so they stretch as the plant grows. For example, it is collenchyma that makes up the 'strings' in celery and gives it the crunch when you bite into it!

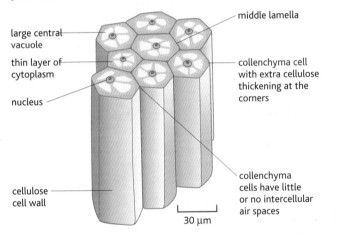

large central vacuole

thin layer of cytoplasm

nucleus

cellulose cell wall

middle lamella

collenchyma cell with extra cellulose thickening at the corners

collenchyma cells have little or no intercellular air spaces

30 μm

fig. 4.1.10 Collenchyma cells.

Sclerenchyma

Sclerenchyma is another type of modified parenchyma tissue found in stems. It develops as the plant gets bigger, to support the increasing weight of the upper part of the plant. Sclerenchyma tissue is found around the vascular bundles in older stems and in leaves. All sclerenchyma cells have strong secondary walls made of cellulose microfibrils laid down at right angles. Some sclerenchyma forms fibres, very long cells often found in bundles or cylinders around the outside of a stem or root. Lignin is deposited on the cell walls of these fibres in either a spiral or a ring pattern, and this makes the fibres strong yet flexible. The strength of the fibres depends on their length and how much they are lignified. Once the fibre is lignified the cell contents

die because lignin is impermeable to water, and so the fibres are hollow tubes. Once the cells have died they can no longer grow, so growth has to take place higher up the stem.

Sclerenchyma cells can also become completely impregnated with lignin, when they form **sclereids**. These very tough cells may be found in groups throughout the cortex of the stem or individually in plant tissue. For example, the gritty texture of pears is a result of individual sclereids in the flesh.

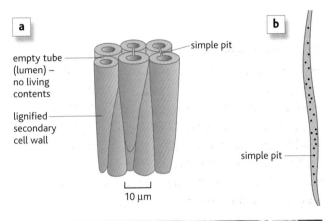

a

empty tube (lumen) – no living contents

lignified secondary cell wall

simple pit

10 μm

b

simple pit

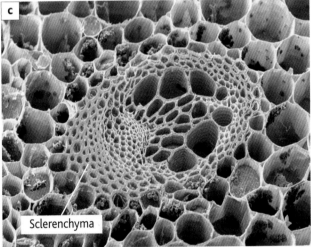

c

Sclerenchyma

fig. 4.1.11 The structure of (a) sclerenchyma cells and (b) sclerenchyma fibre. (c) This cross-section of a buttercup stem shows the thickened sclerenchyma cell walls.

Questions

1 Explain how the different tissues in a stem enable it to carry out its function of support. In addition, explain how the support tissues change as a plant grows and what effect this has on the plant.

2 Explain how the epidermis of a plant stem is adapted to its functions.

Transport tissues in plants

Vascular bundles are found through the plant, including the stem (**fig. 4.1.12**), and they contain the transport tissues xylem and phloem.

- **Xylem** tissue carries water and dissolved minerals from the roots to the photosynthetic parts of the plant. The movement in the xylem is always upwards. Xylem is made up of several different types of cells, most of which are dead. Long tubular structures called xylem vessels are the main functional units of the xylem.

- **Phloem** is living tissue made of phloem cells which transport the dissolved product of photosynthesis (sucrose) from the leaves to where it is needed for growth or storage as starch. The flow through phloem can go both up and down the plant.

- Cambium is a layer of unspecialised cells which divide, giving rise to more specialised cells that form both the xylem and the phloem.

Xylem

Xylem starts off as living tissue. The first xylem to form is called **protoxylem**. It is capable of stretching and growing because the walls are not fully lignified. The cellulose microfibrils in the walls of the xylem vessels are laid down more or less vertically in the stem, which increases the strength of the tube and allows it to withstand the compression forces from the weight of the plant pressing down on it. As the stem ages and the cells stop growing, increasing amounts of lignin are laid down in the cell walls. As a result the cells become impermeable to water and other substances. The tissue becomes stronger and more supportive but the contents of the cells die. This lignified tissue is known as **metaxylem**. The end walls between the cells largely break down so the xylem forms hollow tubes running from the roots to the tip of the stems and leaves. Water and minerals are transported from one end to the other in the transpiration stream (see page 210). Water moves out of the xylem into the surrounding cells either through unlignified areas or through specialised **pits** (holes) in the walls of the xylem vessels.

The lignified xylem vessels are very strong and play a very important supportive role in the stems of plants, particularly larger plants. In smaller, non-woody plants support comes mainly from the turgid parenchyma cells in the centre, and the sclerenchyma and collenchyma. This is why young plants wilt if too much water is lost. As woody plants grow older, more xylem tissue is lignified to increase support. In trees this is taken to the limit and lignified xylem makes up the bulk of the trunk of the tree (the wood). The living cells around the cambium are on the outside of the trunk of the tree, just under the bark. A new ring of vascular tissue is formed each year, so the growth rings of the tree are a record of the xylem produced in each growing season.

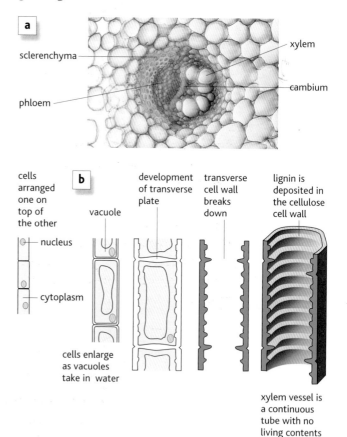

fig. 4.1.12 (a) Xylem forms part of each vascular bundle. (b) As the xylem vessels develop, they change from living vessels to non-living tubes of lignin.

We will look at how water and dissolved solutes move through xylem tissue in more detail in the following pages.

HSW Evidence for the movement of water through the xylem

Evidence for the movement of water through the xylem can be obtained in several ways.

- If the cut end of a shoot is placed in a solution of eosin dye, the dye is carried into the transport system and through to the vascular tissue of the leaves (see **fig. 4.1.13**).

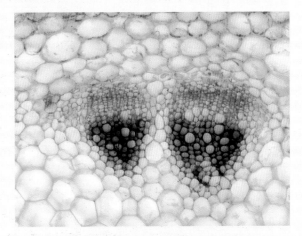

fig. 4.1.13 Eosin dye is transported in the xylem. When sections of the stem and leaves are examined under the light microscope, the dye can be seen in the xylem vessels only.

- Ringing experiments involve removing, or killing with a steam jet, a complete ring of bark. This destroys the living phloem cells but not the xylem cells. Eosin dye placed in the water shows that the upward movement of water through the plant is unaffected.

- If the plant is provided with water containing radioactive isotopes, these can be traced by autoradiography as they move through the plant. Water is seen to travel up the xylem. The movement of minerals in the xylem can be followed in the same way.

The technique of **autoradiography** is very useful for following the transport of substances around plants. It involves several steps.

1 The plant is given a radioactively labelled version of the substance being studied. For example, water containing deuterium (^{2}H, a radioactive isotope of hydrogen) instead of normal hydrogen can be used to investigate the movement of water through the xylem.

2 The radioactive substance is taken up in the same way by the plant as the normal isotope.

3 The substance can then be tracked by placing the plant against photographic film for a while to produce an autoradiograph. The labelled substance causes the photographic film to shadow, revealing the areas where it has accumulated (see **fig. 4.1.14**). The radioactive label can also be traced by examining each area of the plant separately using a scintillation counter. This shows which parts of the plant, or even which organelles, have incorporated the radioactive substance.

fig. 4.1.14 An autoradiograph of a plant to show the movement of water up from the roots to the leaves.

Questions

1 When young trees are planted by forestry workers or gardeners they usually have plastic tubes around the lower part of their trunk to protect them. What do you think they are being protected from and why is this necessary?

2 Which of the methods of demonstrating the movement of water through the xylem would be best for the following situations? Explain your choice in each case:
a in a sixth-form science investigation
b in a university laboratory
c in a year 7 science investigation.

Translocation of water

The movement of substances around plants is usually called **translocation**. Plants do not have mechanical systems (like a heart) to force materials along the narrow tubes of the xylem and phloem – they use a variety of physical processes instead.

Plants have to move water up from the roots where it is absorbed to the aerial parts. The xylem vessels are dead tubes with an inner diameter of only 0.01–0.2 mm, so there is a great resistance to movement through them. Yet water has been shown to move up through the xylem vessels at speeds from 1 to 8 m h^{-1}, and to heights of up to 100 m above the ground in the tallest trees.

Transpiration

The movement of water in the xylem of plants depends on **transpiration**, which is the loss of water vapour from the surface of the plant, mainly from the leaves.

Once in the leaves, water moves by osmosis from the xylem in the veins of the leaves into the mesophyll cells (see **fig. 4.1.16**). Water then evaporates from the cellulose walls of the spongy mesophyll cells into the air spaces. The water vapour moves through open stomata into the external air along a diffusion gradient. Even on a windy day, each leaf has a layer of still air around it. The thickness of this layer varies with the wind speed. The water vapour diffuses through this still layer before it is swept away by the mass of moving air. The amount of water lost by a plant due to transpiration can be surprisingly large. A sunflower may transpire 1–2 dm³ in a day, whilst a large oak tree can lose up to 600 dm³ in the same period.

HSW Demonstrating water loss from a plant

You can demonstrate the loss of water from the surfaces of a plant very easily. First seal the pot of a potted plant in a plastic bag to prevent evaporation of water from the soil surface interfering with the experiment. Then seal the plant in a bell jar. As water is lost a colourless liquid collects on the glass of the bell jar. You can show that this contains water by using cobalt chloride or copper sulfate paper.

It is not as easy to measure the amount of transpiration taking place. However, you can measure the uptake of water by a plant. As most of the water taken up by a plant is used for transpiration, this can be considered a close estimate. Uptake of water is demonstrated using a **potometer** (see **fig. 4.1.15**).

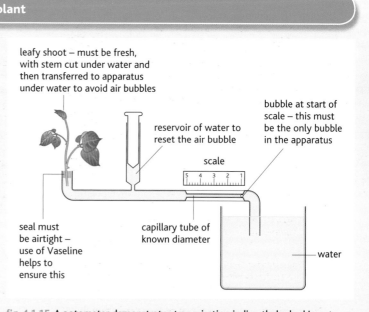

fig. 4.1.15 A potometer demonstrates transpiration indirectly by looking at water uptake. It is important to prevent air entering the apparatus.

The transpiration stream

A giant redwood tree regularly raises a column of water more than 30 m in its xylem. Water is moved up to 100 m in the tallest trees. What makes this possible?

When water is lost by transpiration from the leaves, it moves by osmosis across the leaf from cell to cell, all the way from the xylem. When molecules of water leave the xylem to enter a cell by osmosis, this creates tension in the column of water in the xylem, and this tension is transmitted all the way down to the roots. This is due to the **cohesion** of the water molecules. Because of their dipolar nature and the hydrogen bonds that form between them, water molecules 'stick together' which gives the column of water a high **tensile strength** – it is less likely to break.

The molecules also adhere strongly to the walls of the narrow xylem vessel and (probably more importantly) to the millions of tiny channels and pores within the cellulose cell walls of the leaf. **Adhesion** is the attraction between unlike molecules and it is sufficient to support the entire column of water in the xylem. The combination of adhesive and cohesive forces pulls the whole column of water in the xylem upwards. More water is continuously moved into the roots by osmosis from the soil to replace that lost from the leaves by transpiration (see **fig. 4.1.16**).

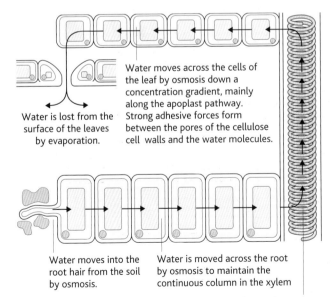

Water moves across the cells of the leaf by osmosis down a concentration gradient, mainly along the apoplast pathway. Strong adhesive forces form between the pores of the cellulose cell walls and the water molecules.

Water is lost from the surface of the leaves by evaporation.

Water moves into the root hair from the soil by osmosis.

Water is moved across the root by osmosis to maintain the continuous column in the xylem.

As water molecules are lost by evaporation and moved out of the xylem, cohesion between the water molecules means that the whole column of water in the xylem is pulled upward.

fig. 4.1.16 The transpiration stream is set up as a result of physical processes – and a pressure of around 4000 kPa can result, moving the water upwards. This is enough to supply water to the tops of the tallest trees.

 HSW Artificial transpiration – a model solution

In 1893 Josef Bohm used a model to demonstrate neatly the effect of evaporation on a column of water. Using a porous pot, he showed that adhesive forces between the water molecules and the pores of the porous pot are strong enough to support an enormous column of water, and cohesive forces between the water molecules stop the column breaking under the strain. This gives us our best model so far of the transpiration stream.

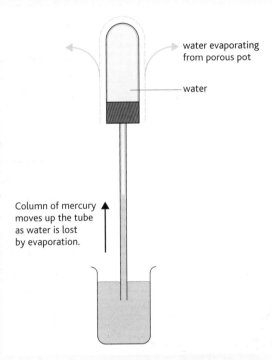

water evaporating from porous pot

water

Column of mercury moves up the tube as water is lost by evaporation.

fig. 4.1.17 Drawn by the evaporation of water in the experiment, the column of mercury rises to over 1000 mm. It is calculated that if there was only water in the system, the column could be pulled to a height of more than 1 km – far greater than the height of any living plant!

Questions

1 What are the differences between transpiration and translocation?

2 Why is transpiration an important process in a plant?

3 Design an experiment which you could use to investigate the effect of one environmental condition on the rate of transpiration in a plant.

The uptake of water by plants

An important component of soil is soil water. Even soil that is fairly dry can have a thin film of water around the soil particles which plants can take in through their roots. Water is absorbed mainly by the younger parts of the roots where the majority of the **root hairs** are found. These microscopic hairs are extensions of the membranes of the outer cells of the root, and they greatly increase the surface area for absorption. The root hairs allow close contact with the soil particles.

At its simplest, uptake of water by the roots depends on the concentration gradient across the root from the soil water to the xylem. Water moves from the soil into a root hair cell down a concentration gradient by osmosis. This makes the root hair cell more dilute than its neighbour, so water moves from cell to cell by osmosis across the root to the xylem. However, this model is very simple – the detailed mechanism is more complicated.

There is a concentration gradient across the root from the root hair cells to the cells closest to the xylem. This is the result of two effects. Water is continually moved up the xylem by transpiration, and the solute concentration increases in the cells across the root towards the xylem. But the water does not simply flow from one cell to another – there appear to be three alternative routes into the xylem vessels (see **fig. 4.1.18**).

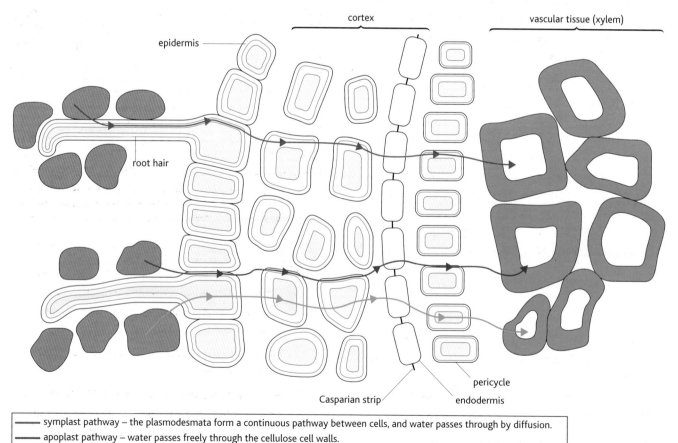

symplast pathway – the plasmodesmata form a continuous pathway between cells, and water passes through by diffusion.
apoplast pathway – water passes freely through the cellulose cell walls.
vacuolar pathway – water moves through the vacuoles by osmosis from the cytoplasm.

fig. 4.1.18 **The cells of the roots are organised into a system which is very efficient at taking up water from the soil.**

- In the **vacuolar pathway** water moves by osmosis across the vacuoles of the cells of the root system. The water moves down a concentration gradient from the soil solution to the xylem.

- In the **symplast pathway** water moves down the same concentration gradient from the root hair cells to the xylem. However, in this pathway the water moves through the interconnected cytoplasm (**symplast**) of the cells of the root system through the plasmodesmata, strands of cytoplasm which go through pores in the cellulose cell walls.

- In the **apoplast pathway** water is pulled by the attraction between water molecules across adjacent cell walls (the **apoplast**) from the root hair cell to the xylem. Because of the loose, open-network structure of cellulose, up to half of the volume of the cell wall can be filled with water. As water is drawn into the xylem, attraction between the molecules ensures that more water is pulled across from the adjacent cell wall and so on. Water entering the root hair from the soil has mineral ions dissolved in it, and they are drawn through the apoplast pathway too. The water moves across the cells of the root in the cell walls until it reaches the endodermis, which contains a waterproof layer called the **Casparian strip**.

Whichever route the water and minerals have taken across the root, once they reach the Casparian strip they enter the cytoplasm of the cell temporarily. Minerals may need to enter the cytoplasm up a concentration gradient, so this often involves active transport. This seems to be a way in which the cells control how much water and minerals move from the soil into the xylem. In spite of the barrier of the Casparian strip, the end result of all the pathways is a continuous stream of water across the root to the xylem.

Root pressure

Transpiration is not the only method by which water is moved through the xylem. Transpiration seems to be a passive process, but aspects of water transport are affected by metabolic inhibitors and lack of oxygen, both of which suggest a more active mechanism.

During the night, when transpiration rates are extremely low, drops of water may be forced out of the leaves in a process known as **guttation**. In some plants, if the plant is cut off from the root, root sap will continue to ooze from the root xylem. This is a result of **root pressure**.

In tomatoes and some other plants, quite a strong root pressure can be measured when the top of the plant is cut off. This pressure disappears if the root cells are killed by steam or poisoned. This suggests that root pressure is based on active transport. The current model is that root pressure is produced by the active secretion of salts from the root cells into the xylem sap, increasing the concentration gradient across the root. This increases the movement of water into the cells by osmosis. The root pressure generated is about 100–200 kPa. This is not enough to explain all of the water movement in the xylem of many plants, but it certainly contributes, particularly in situations when the transpiration rate is low.

fig. 4.1.19 Sometimes water appears out of the edge of leaves when transpiration rates are low. This guttation is part of the evidence for the role of root pressure in water movements through the xylem in plants.

Questions

1 Summarise the main similarities and differences between the three routes by which water appears to move from the soil into the xylem.

2 Explain how the structure of root hairs is adapted to their function.

3 How might root pressure be measured? Why is the presence of root pressure alone not enough to explain the movement of water up from the roots to the leaves of a plant?

The uptake of minerals by plants

Why do plants need minerals?

Although plants can synthesise their own carbohydrates by photosynthesis, they also need other molecules such as proteins and fats. Certain minerals are needed to synthesise these and other substances essential for healthy growth. Plants must extract these minerals from the soil.

Nitrogen

Nitrate ions are used to make amino acids and therefore proteins. These proteins include plant enzymes without which the cells could not function. Nitrates are also needed for the plant to make DNA and many hormones, as well as a range of other compounds in plant cells. When plants lack nitrates the older leaves turn yellow and die, and growth is stunted.

Calcium

Calcium ions in the middle lamella of plant cells combine with pectin to form the calcium pectate which holds plant cells together. Calcium ions also play a role in the permeability of membranes. When plants lack calcium the growing points die back and the young leaves are yellow and crinkly.

Magnesium

Magnesium ions are needed to produce the green pigment chlorophyll. Magnesium is also needed for the activation of some plant enzymes and the synthesis of nucleic acids. Without magnesium yellow areas develop on the older leaves and growth slows down.

Phosphates

Phosphate ions are needed for the phosphate groups in ADP and ATP, which are involved in energy transfers in cells. They are also integral to some of the structural molecules that offer support in plant cells and to the nucleic acids. Plants lacking phosphates have very dark green leaves with purple veins and their growth is stunted.

Studying mineral deficiencies

You can use nutrient solutions, each lacking a different mineral, to investigate the effect of mineral deficiencies on the growth of a plant.

lacking nitrogen

lacking phosphate

lacking magnesium

fig. 4.1.20 These plants are each deficient in one mineral and show classic deficiency symptoms, particularly in the fastest-growing tissues such as the new leaves.

Moving minerals around the plant

The minerals in the soil water are often present only in very low concentrations – often lower than the concentrations in the cytoplasm of the root cells. So to get minerals into the cells and into the xylem for transport around the plant, they often have to be moved up a concentration gradient by active transport.

Any minerals dissolved in water absorbed from the soil are carried in the apoplast pathway and move through the adjacent cell walls until they reach the endodermis. They may also be moved through the symplast pathway in the cytoplasm, possibly by active transport using energy. Either way the result is that the parenchyma cells of the root cortex are bathed in a very dilute solution of mineral ions.

When they reach the impermeable Casparian strip the mineral ions can no longer move through the cellulose wall. So they may enter the cytoplasm of the cells, either by diffusion down a concentration gradient or by active transport if they are being moved up a concentration gradient. The minerals reach the xylem to be transported in the water that moves upwards continuously in the transpiration stream. When the mineral ions reach tissues where they are needed, they move out of the xylem into the cells either by diffusion or by active transport. This will depend on the permeability of the cell membranes and the relative concentrations of the ions inside and outside the cells.

HSW Making the most of mineral uptake

All plants take up and transport the minerals they need. But a few plants take up, transport and store a rather bigger range of minerals than usual. In the early 1970s, a tree was discovered in New Caledonia which produced a blue sap. This turned out to contain 26% nickel in its dry mass. Other plants have been discovered that can accumulate large amounts of other metals such as nickel, cobalt, cadmium, zinc and even gold. The plants take up the minerals in an active process. In the 1980s scientists suggested plants might be used to extract certain metals from the earth. Now these special mineral-rich plants are used in several ways.

In California, *Streptanthus polygaloides* plants are grown on nickel-rich soils where they take up so much nickel that it makes up as much as 1% of their dry mass. Farmers then burn the plants and the ash is smelted to produce the metal. The energy produced by burning the plants is used to generate electricity to power the extraction process and any excess electricity is sold to the local power company. Farming nickel pays better than growing wheat if you live on the right soil!

Another exciting possibility is to use plants which take up and transport metal ions to clean up ground which has been contaminated by toxic heavy metals such as thallium and lead. Land polluted like this cannot be used for growing crops or grazing animals. However, using specialised plants, such as *Alyssum* species which take up and transport the unwanted ions removes the toxic metals from the soil in a process called phytoremediation, making the soil safe to use again.

fig. 4.1.21 **Using plants to extract metals from the earth is a much greener approach than traditional mining!**

Questions

1   Why are minerals important in a plant? What would you need to consider when investigating mineral uptake by plants in the lab?

2   Explain how mineral ions are transported from soil water to where they are needed in a plant.

3   Suggest an investigation that could prove the role of active transport in the uptake of mineral ions by plants.

4.2 Plants as natural resources

Food for thought

People have always exploited plants to provide materials for building and clothing, for medicines, food and drinks, for dyes and for fuel. The structure of plants is adapted to their functions, and these same adaptations provide ideal materials for many of the raw materials of life. This chapter looks at some of these uses.

Plants are central to the human diet, providing not only the macronutrients of carbohydrates, lipids and proteins but also many micronutrients in the form of vitamins and minerals. Plants also contain fibre in the cellulose walls of their cells, which cannot be digested but which helps the working of the gut.

Some plants are grown as **food staples** – the basic energy-supplying foods in the diet. Many of these have cells filled with amyloplasts. These starch-storage organelles may be used by the plant to survive difficult conditions, e.g. the storage tubers of the potato plant. They may also be used to reproduce the species – eg many seeds, such as the cereals wheat and rice, contain very rich stores of starch. This store of energy for the developing embryo makes cereal seeds a very good foodstuff for people, providing plenty of carbohydrate, some protein and oils as well as small amounts of valuable micronutrients.

fig. 4.2.1 **Foods such as these are a staple part of the diet around the world. The starch stored in the plant tissues provides energy for people.**

We use other seeds, such as sunflowers, linseed, oil-seed rape and many nuts, for the oils they contain. Pulses such as beans, peas, lentils, soya beans and chickpeas provide much of the protein requirement for people who eat little or no meat. Fleshy and succulent fruits are also important foods, as sources of sugars as well as vitamins.

In addition to eating many parts of plants ourselves, the animals that we raise as a source of food are fed on fodder plants such as grass.

HSW The Irish potato famine and plant breeding

A single plant can have a huge effect on society, as illustrated by the story of the Irish potato famine. For centuries Ireland remained an undeveloped country, using basic methods of ploughing and harvesting, struggling with a difficult climate and geographical isolation. When the potato reached Ireland in the eighteenth century it grew easily in the prevailing conditions. This meant almost everyone could grow enough food for their families – and the population exploded. But in 1845 potato blight, a fungal disease, struck the potato crop and wiped it out. Suddenly, all over Ireland, people were starving. From an Irish population of around 9 million, 1 million men, women and children died of famine or associated disease. By the time of the First World War (1914–18) about 5.5 million Irish people had emigrated. These events had a long-lasting effect on Irish history, the make-up of America and the relationships between both these countries and Britain.

Since that time scientists have used selective breeding to develop new varieties of potato that are largely resistant to blight. Selective breeding has also been used to improve the cereal crops rice, wheat, maize, oats and barley. These are related to wild grasses, but they have been selected and bred for generations to increase their yield, to help them thrive in very different conditions in various areas of the world and to maximise their resistance to disease. Now they are being given added traits by genetic modification, enabling crops to make their own pesticides, or to indicate in some way when disease or pest damage occurs. However, both of these techniques lead to other problems.

- Selective breeding has resulted in only a small number of varieties of crop plants being grown commercially, because they are the best varieties. Within a variety there is only limited genetic variation, so a new pest or disease can be catastrophic. In modern farming **monoculture** (a field containing only one type of genetically similar plants) is common. So scientists are now preserving the genetic resources of many wild plants so they can be used for developing future new varieties to survive changing conditions (see chapter 4.4).

- Genetic modification introduces completely new traits into plants (from other plants, or even from animals or bacteria), such as making their own pesticide. One concern is that these transferred genes could become part of wild plants, such as weed species, through cross-pollination. To avoid this risk, the modified plants are also altered to make them infertile. So farmers using disease- or drought-resistant plants cannot keep seeds from one year to the next as they always have done. They have to buy seed each year, which is much more expensive.

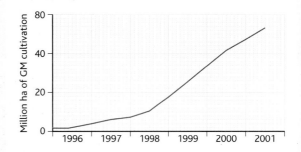

fig. 4.2.2 **In spite of concerns, genetically modified plants are being increasingly cultivated across the world.**

This is another example of a situation where the scientific answer to a specific problem raises other questions which science alone cannot answer.

Questions

1   Which features of plant cells make plants a useful source of food for people?

2   Find five plants commonly used as food sources around the world. In each case explain which part of the plant is used, how it is adapted for its role in the plant and how that adaptation makes it useful as a food source for people.

3   What are the main ethical issues raised by selective breeding and genetic modification of crop plants?

From construction to clothing

As you have seen, the structure of cellulose fibres gives them great strength. These fibres may be further toughened and strengthened by lignin, turning them into what we recognise as wood. The properties of fibres and wood make them very useful to us.

Fibres

Long and strong

We have used plant fibres such as hemp, jute, manila, flax and sisal for centuries to make ropes, paper and cloth. The fibres usually have to be extracted from the plant first. The fibres are very long sclerenchyma cells and xylem tissue and are usually very tough. Cellulose and lignified cellulose are not easily broken down either by chemicals or by enzymes. On the other hand, the matrix of pectates and other compounds around the fibres (including lignin) can usually be dissolved or removed.

Plant fibres have great **tensile strength** – they cannot easily be broken by pulling. This, along with their flexibility, makes them very useful. They usually occur in bundles of fibres which are much stronger than the individual cells.

Determining tensile strength

Different types of fibres have different strengths. Tensile strength also depends of the cross-sectional area of the fibre. You can investigate the tensile strengths of different fibres. See if you can relate your results to the way the fibres are used.

How fibres are processed to make products

Paper is usually made from fibres from wood. Wood fibres are not easy to extract because the matrix around the cellulose fibres contains a lot of lignin. So wood is soaked in very strong alkalis such as caustic soda to produce a pulp consisting of cellulose and lignified cellulose fibres in water. Thin layers of pulp are then pressed onto frames where they dry to form paper.

Many traditional methods of producing fibres such as flax (often used for ropes) simply relied on the natural actions of decomposers to break down the material around the fibres. This is known as retting. In developed countries manufacturing processes have replaced natural retting, using chemicals and enzymes to do the job much more quickly.

Probably the best known and most widely used of the natural fibres is cotton (**fig. 4.2.3**). One of the great advantages of cotton is that it is produced in the form of almost pure fibres, packed around the seeds. So there is no need for retting or other treatment.

fig. 4.2.3 Cotton bushes produce 'ready-to-use' cotton fibres around their seeds.

Even though single cotton fibre cells are very long, they are not long enough to be useful on their own. Spinning pulls out the short, single fibres and twists them together to form a long, apparently continuous thread. Spinning can be done on a small scale by individual people, but usually happens on a massive industrial scale. The resulting threads are then woven together to make a fabric. Similar processes are used with other plant fibres such as jute, sisal and hemp.

In the twentieth century synthetic fibres, eg nylon and polyester, were developed. They were new, exciting, relatively cheap, very hardwearing and did not crease. But by the twenty-first century the limitations of these artificial fibres had become apparent. Fabrics made from them do not 'breathe' and they do not absorb liquid so they do not soak up body fluids such as sweat. They are made from chemicals derived from crude oil, a non-sustainable resource which gets increasingly expensive and is rapidly being used up.

Sustainability – using materials which can be replaced – is an increasingly important idea. Plants are vital in developing sustainable resources. They soak up carbon dioxide from the atmosphere and lock it into their cell structures. People are realising that natural plant fibres are a much more sustainable alternative for a whole range of uses, from clothing fabrics to ropes and insulating materials. They can also be more comfortable to wear because they are more absorbent. The properties of many plant products mean that they will play an increasingly large role in providing what we need in the future.

Wood

Wood is a **composite material**, made of lignified cellulose fibres embedded in hemicelluloses and lignin. The great benefit of a composite is that it has the properties of both materials – reinforced concrete is an obvious example. The cellulose fibres make the wood very resistant to compression (squeezing by weight) so it is excellent for weight-bearing in buildings and can be used in supporting columns as well as in horizontal beams. Wood also retains some of the flexibility of the matrix and, because of the intermeshing cellulose fibres, doesn't tend to crack as a stiff material would do. So you can hammer a nail into it, or cut out small pieces to make joints, without damaging the strength of the wood.

Wood has a great many uses, from making baskets, fencing hurdles, boats, cricket bats or furniture to building homes (**fig. 4.2.4**). Wood is also a good insulator so homes built substantially from wood need less heating in the winter and cooling in the summer than a brick house. Wood also locks up carbon dioxide and is a sustainable resource if it is managed carefully with replanting programmes. Even if it is burned, wood can be **carbon neutral** – taking in carbon as it grows and releasing it as it is burnt – but it has the great advantage of being a renewable energy source as well.

fig. 4.2.4 Wood is strong and flexible – in fact it has a tensile strength 4–5 times greater than the equivalent mass of steel.

Questions

1   Which features of plants make them useful in the construction industry?

2   Why are cotton fibres easier to use than the fibres from jute and hemp?

3   Sustainability is becoming an increasingly important concept. What is the link between this and the increased use of plant-based materials in many areas?

HSW Bioplastics – into the future

The rise and fall of plastics

Over the last century or so, the use of natural materials in the developed world has declined with the development of new synthetic materials produced from oil-based chemicals, particularly plastics. Plastics are synthetic polymers – long-chain molecules made up of repeating units of small monomer molecules such as ethene and propene. Plastics vary from soft flexible solids with low melting points to hard brittle materials with very high melting points. They are used to make a wide range of products, from packaging to artificial joints and from cutlery to parts of cars. However, in the twenty-first century modern materials are being developed from natural products as the environmental problems caused by plastics are becoming increasingly obvious.

Most plastics, such as polyethene and PVC, are polymers made from petrochemicals, originating from oil which is a non-renewable resource. These plastics cannot be broken down by decomposers – they are non-biodegradable, which has led to plastic pollution on a grand scale. Some plastics can be melted down and recycled, but many cannot.

New horizons – biological polymers

Scientists are increasingly looking at the possibilities of **bioplastics** – plastics based on biological polymers such as starch and cellulose. These have two large potential benefits.

- They are a sustainable resource. The starch or cellulose comes from plants such as maize, wheat, potatoes and sugar beet. These plants can be grown easily to supply the needs of the bioplastics industry.

- Bioplastics are biodegradable. Because they are based on biological molecules, bacteria and fungi can usually break them down, even if the process can be very slow.

Bioplastics are increasingly being used to replace traditional plastics in roles ranging from packaging and car parts to computers and mobile phones. Bioplastics have actually been around for a long time. In 1869 the American inventor John Wesley Hyatt Jr patented a compound made from cellulose which he used to coat non-ivory billiard balls. The problem was that it caught fire if put close to a lighted cigar! But this was the beginning of celluloid, widely used in photographic film and movie film. So one of the first widely used plastics was actually a bioplastic!

In the 1920s Henry Ford, the first person to mass produce cars, experimented with plastics made from soya beans and even produced a plastic car in 1941. Then the Second World War, and the growth of the petrochemical industry, took the emphasis away from bioplastics.

fig. 4.2.5 Plastics are found almost everywhere on the planet. They are causing great environmental damage and most will not degrade and disappear.

Different types of bioplastics

Cellulose-based plastics are usually made from wood pulp (like that used in the paper industry). They are largely used to make plastic wrapping for food. Cellophane has been familiar for many years in this guise.

Thermoplastic starch is the best known and most widely used bioplastic. It is made mainly from starch extracted from potatoes and maize, which is then mixed with other compounds such as gelatine which change the properties of the starch. One of its main uses is in the pharmaceutical industry to make capsules to contain drugs. Thermoplastic starch is smooth, shiny and easy to swallow, yet the plastic absorbs water and is readily digested –perfect for the job!

fig. 4.2.6 **Bioplastics can usually do the same jobs as synthetic polymers but, as they are biodegradable, they cause fewer environmental problems when we have finished using them.**

Other bioplastics include polylactic acid (PLA) which has very similar properties to polyethene but is biodegradable. It is largely produced from maize or sugar cane grown in the US. Uses of this bioplastic include computer casings, mobile phones and drinking cups. Poly-3-hydroxybutyrate (PHB) is a stiff biopolymer rather like polypropene. It is used in ropes, bank notes and car parts and is made largely with products from the South American sugar industry.

We can burn bioplastics when their useful life is over. You might think that this would be polluting and unnecessary when they will break down anyway. However, when bioplastics are broken down by decomposers they can produce methane, a greenhouse gas which is 25 times more potent than the carbon dioxide released when they are burned. And the energy released during burning can be used to generate electricity and make more plastics.

Will bioplastics take over from oil-based plastics? The science and technology needed to produce them are becoming increasingly available. However, the plastics made from petrochemicals have extremely useful properties and it is not always easy to achieve these same properties in bioplastics. Economics and ethical considerations are also important. Bioplastics are still much more expensive than the oil-based alternatives. This is partly because the technology is still very new, and partly an effect of the economies of scale. Around 150 times more conventional plastic is made each year worldwide than bioplastic. People have to be prepared to pay more for a similar product to enable the bioplastics industry to develop. There is also tension between the use of crops such as maize, wheat, sugar cane and sugar beet for food, for biofuel and for bioplastics. At the moment there simply aren't enough crops to go around. In the face of limited crops, who decides whether they are used for food to satisfy the immediate hunger of people around the world, or for biofuels or bioplastics to try and work towards a sustainable future for everyone? These questions are more for society rather than for science and scientists.

Questions

1   Why are starch and cellulose good starting points for the manufacture of bioplastics?

2   What are the advantages and disadvantages of using bioplastics?

3   Look for scientific comparisons between the performance of petrochemical plastics and bioplastics. Which come out best? How important is this type of evidence when making decisions about using plastics, and what other factors might be considered?

Plant pharmacies

Plants produce a vast range of chemicals, some with the function of deterring animals that try to eat the plant or of destroying microorganisms that might cause disease. Over centuries of general experimentation, people have found that some of these chemicals are also of great benefit in helping the human body fight discomfort and disease. Even today, according to the World Health Organization, 75–80% of the world population relies at least partly on plant-based medicines, particularly in rural and isolated areas such as the tropical rainforests. For much of this time, nobody has really known how plant cures work. Scientists are studying some of them to find out what they do in the body, so we can develop better treatments for illnesses.

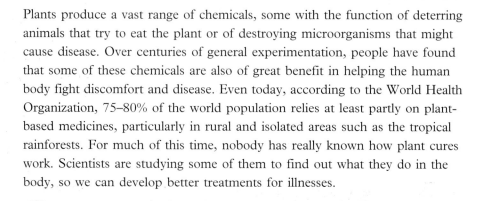

Antimicrobial plant extracts

Some plants and fungi have been shown to have antimicrobial properties – they contain chemicals that kill bacteria. You can investigate the antimicrobial properties of different plant extracts using culture plates in the lab using discs of filter paper soaked in plant extract placed on an agar plate. If the plant extract kills the bacteria, or stops them growing, a clear area of jelly can be seen around the disc.

fig. 4.2.7 Taking aspirin in tablet form is a more reliable form of pain relief than chewing willow bark or beaver anal glands.

Extracting the active ingredient

Salicylic acid – a modified version of which is known as **aspirin** – is an everyday example of a drug derived from plants, in this case a species of willow. For centuries willow bark was chewed or brewed up into a drink to relieve pain and fever. People even chewed on the anal glands of (dead!) beavers to get pain relief. Beavers eat willow bark, and the salicylic acid becomes concentrated in the anal glands. Scientists discovered the active ingredient in the bark was salicylic acid and developed a method to extract and purify it. Now we take a carefully measured dose of a closely related but safer compound, acetylsalicylic acid, in the form of a small white tablet known as an aspirin.

One of the major advantages of extracting and purifying the beneficial drugs found in plants is that it is possible to give known, repeatable doses of the active ingredient. The levels of a chemical present in any part of a plant will vary with the age of the plant, the season of the year or even the time of day. By extracting the chemicals and purifying them you can measure out an exact dose every time. However, enormous amounts of plant material are needed for this. This is why scientists work to isolate healing chemicals from plants, analyse their chemical structure and then synthesise the drug on an industrial scale. In many cases the original plant product is modified to make it even more effective.

Both for individuals and for society the impact of these drugs is far-reaching. People are ill less often, they are less severely ill and they are living longer. Not only that, but plant-based medicines have opened up parts of the world which were previously closed to development because of the prevalence of diseases. For example, malaria is a life-threatening disease that is spread by mosquitoes common in many tropical areas. The discovery of quinine which comes from the cinchona tree, and is used both to prevent and treat malaria, made it possible for loggers and developers to work in the Amazon Basin. As a result, the rich flora of the rainforests may well be destroyed before a fraction of its true potential as a source of new medicines has been tapped.

fig. 4.2.8 Foxgloves are the source of digoxin, a drug which can act as a poison but also as a cure.

Developing drugs from plants

Digitalin is a chemical found in foxgloves that has been used as a poison for centuries. However, there were also many reports about the role of foxgloves in curing the dropsy – the swelling (**oedema**) that results when the circulation is failing. Dropsy causes a long, slow death as organs like the kidneys fail, the legs swell and lungs eventually fill with fluid. It wasn't until the work of William Withering that the medical potential of foxgloves was fully realised.

HSW The work of William Withering

William Withering (1741–99) was a British doctor and a keen botanist who eventually published a book on plants. However, his real work was medicine and he was very successful in this field. In 1775 a patient came to him with the symptoms of a serious heart condition. Withering had no effective treatment to offer so his patient went off to see a local 'wise woman', who used herbs to cure a number of conditions. Withering was very impressed when his patient recovered and so he persuaded the woman to let him buy the recipe. It contained about 20 different herbs – Withering guessed that foxglove contained the active ingredient. Over the next 10 years he tested a variety of potions made from foxgloves on 163 patients whom he treated for dropsy at Birmingham General Hospital. Many of his patients got better, though some almost died of digitalin poisoning. He discovered that the side-effects of the drug included nausea, vomiting and worse. But when he got the dose right, the patient started to produce large quantities of urine and their heart beat more regularly and strongly. By the end of his period of careful observations Withering had discovered that the best treatment for dropsy and heart failure was to give a patient tea made from the dried and powdered leaves of the foxglove – but the leaves had not to be boiled as it reduced the effect of the drug. What Withering was doing was effectively extracting the chemical digitalin from the leaves of the foxgloves. Drugs based on the chemicals in foxgloves are still in regular use by GPs and in hospitals today, around 230 years later.

In the twenty-first century the development of new medicines has to go through many stages and safety checks before they get anywhere near a patient. Administering leaves from a recipe from a travelling healer would not be considered good practice today!

Questions

1 It is claimed that many plants have healing properties. Why do scientists extract the active ingredient rather than using the whole plant?

2 Which aspects of Withering's work on digitalis have a scientific basis? Which aspects would be totally unacceptable today? Explain your answers.

Modern drug development

Testing promising new medicines

Although in the past, like Withering and his use of foxgloves, herbal remedies could be used as a source of new drugs, this is not the way it is done in the twenty-first century. Every medicine that comes onto the market today is the result of years of research and development (R & D). A new medicine has to be:

- effective – it cures or prevents the disease it is designed for or relieves the symptoms

- safe – non-toxic and without unacceptable side-effects

- stable – able to be stored for some time and used under normal conditions

- easily taken into and removed from your body – able to get to its target in your body, and to be got rid of once it has done its job

- capable of being made on a large scale – able to be manufactured in a very pure form, in large quantities and relatively cheaply.

It takes around 10 years and about £550 million to develop a new medicine that achieves all of these criteria.

One way in which scientists look for new medicines is to investigate chemicals that bind to our protein receptors or to the active sites in our enzymes. Researchers often use computer models to fit new structures into the active site of enzymes or receptors that are thought to play a significant role in disease processes. This may identify a useful starting point for further work.

When scientists think they have a compound that might make a useful medicine they will patent it. A patent gives the inventor the right to be the only one to make and sell their invention for the next 20 years. It is like a 'reward' for all the work that goes into discovering a possible new drug. The only problem is that quite a few of those 20 years will be taken up with more testing.

The new compound is first tested on cell cultures, tissue cultures and whole organs in the lab. These tests are designed to see if the compound does what the scientists thought it would. Many chemicals fail at this stage because they don't work in living tissue or because they have harmful effects. But if the compound passes these tests it moves out of research and into development.

fig. 4.2.9 Testing a new drug involves years of work from many people in the research labs of the pharmaceutical companies – no wonder drugs cost so much!

Drug development and animal testing

Before a drug can be tried on people you need a way of getting it into them – in other words, a good delivery system. This might be tablets, a liquid medicine, injections or a nasal spray. You also need to make sure the drug is stable so there is no risk of it breaking down to form something toxic or inactive before it works.

At this stage the potential drug will be tested on animals to find out how it works in a whole organism. This will also show if the drug gets taken into the cells, if it is changed chemically in the body and if it is excreted safely.

Mammals are used which are as similar as possible to humans. The most widely used animals for initial tests are mice and rats. Some tests have to be carried out in two species, a rodent and a non-rodent. Animal testing is very expensive and time consuming and is the centre of much ethical debate. Animals are replaced by tissue cultures and computer models wherever possible, the numbers of animals used are kept to a minimum and the tests used are refined to cause the minimum of distress. However, at present the information from computer modelling and from tests on cell or tissue cultures is not sufficient to test drugs on people safely without animal testing. So the law states that animal testing must be carried out at this stage.

Some people have ethical objections to the use of animals in this way. But the use of mice and rats is perceived as being much less emotive than the use of dogs, cats or monkeys. The rodents provide valid models, the genetic make-up of the strains is well known and they are small and relatively easy to keep in humane conditions.

Clinical trials

If the animal testing has been successful, the very first human trials follow. Before you can try the new drug on people in the UK, you have to apply for a clinical trial authorisation with the Medicines and Healthcare products Regulatory Agency (MHRA). As part of their job, they take decisions about the testing and licensing of new medicines. Only if they are happy with all the tests carried out so far will they allow the drug to be trialled on people.

In **phase 1 trials** the new drug is given to a small number of healthy volunteers. This is to check that the drug works as expected in the human body and doesn't cause any unexpected side-effects. At the same time in animal trials scientists continue looking at the effect of longer-term use of the drug.

fig. 4.2.10 In a double-blind phase 3 trial, neither the doctor nor the patient has any idea whether the drug being given is the active substance or a placebo.

If a drug is successful in phase 1 human trials it goes into **phase 2 trials**, when the new drug is used with patients affected by the target disease. Between 100 and 500 patient volunteers are given the new drug. This is the first chance for scientists and doctors to see how the new medicine affects the disease in a real patient. The volunteer patients are closely monitored to find out more about the ideal dose, the effectiveness of the drug and any side-effects. Success at this stage means the new compound has a good chance of becoming a useful medicine.

Before a new drug is fully approved it must be used on thousands of patients with the target disease. These are the **phase 3 trials** and over 5000 volunteer patients are used.

HSW Are phase 1 trials ethical?

Computer simulations, tests on cells in laboratory conditions and even animal testing can give only limited information about a new drug and how it works. Eventually the drug has to be given to a human in a phase 1 trial. The volunteers are **altruistic** (doing something which benefits others without any direct benefit to themselves) but, because they are exposing themselves to a calculated risk, they are often paid to cover their expenses and for their time. Students are often willing volunteers in these trials, because they are keen to make some extra money.

Relatives' fury at drug test firm that had never experimented on people before

HELL OF HUMAN GUINEA PIGS

THE unbearable agony of the six drug trial victims was graphically described last night.

Within minutes of being given the first dose, they were complaining of feeling hot and suffering headaches.

They began shaking and vomiting, then shouting and writhing in intense pain. Their 'living hell' was witnessed by 23-year-old Raste

By **Michael Seamark** and **Jenny Hope**

Khan, one of two men on the trial who were given a placebo, or dummy drug. He said: 'Some screamed that their heads felt like they were about to explode. It was terrifying.'

He spoke as it emerged that the drug's German makers have never tested on people

Turn to Page Four

fig. 4.2.11 **A phase 1 drug trial went disastrously wrong in 2006. However, a later survey showed that the public were surprisingly calm about the risk because they seemed well aware of the potential benefits.**

When someone is given a new drug for the first time there are obviously risks, although everything is done to minimise them. The reality was demonstrated in the UK in 2006. Six healthy young men were given a new type of drug which had worked extremely well in all the pre-human trials. Completely unexpectedly, they reacted badly to the treatment. Five of the volunteers ended up in intensive care with severe inflammatory reactions, and some may have suffered long-term damage to their immune systems. The only volunteer not affected received a **placebo** (something that looks like the drug but has no active ingredient).

Volunteers have to give informed consent by law. However, since this disaster many questions have been asked about how 'informed' the consent might be. The reading age of the consent forms is higher than the average reading age of the UK. There were statements that appeared to deliberately play down the risk associated with the treatment. Yet the concentration used was 1/500 of the dosage used in animal trials, so the risk should have been as small as possible.

The disastrous drug trial had a number of consequences. The firm that was developing the drug was ruined financially. The ethics of testing a new drug on healthy volunteers has been questioned – some people suggest that the first human trials should be on people affected by the condition being treated. The whole process of human drug testing is now being debated and new guidelines have been produced. Science can help with some of the answers to these questions, but not all of them. Society wants the benefits of new medicines and it has to make tough decisions about the risks it is prepared to take to get them.

Phase 2 and 3 trials are normally carried out as **double-blind trials**, where neither the doctor nor the patient knows whether the patient is receiving the new medicine, a control medicine or a placebo. Patients often appear to respond to a treatment because they believe that it will do them good – there is much that is not understood about how the mind affects the body. This response is known as the **placebo effect**. The use of a double-blind trial allows this to be measured. Instead of a placebo, sometimes a control medicine, the best-performing available treatment, is used to avoid any patient being denied treatment when they take part in the trial.

Phase 3 trials are used to confirm the effectiveness and safety of the new drug. Because the numbers of patients involved are large, they also have a better chance of showing up any unexpected adverse side-effects. Patients are randomly allocated to receive the new medicine or the control/placebo. Data on effectiveness and side-effects and other information are collected and assessed to see if there are any statistically significant differences between the new medicine and the placebo, or the currently available drug.

It is difficult to achieve a complete set of results in clinical trials, because many patients stop taking the medicine for various reasons or do not take it regularly (see **fig. 4.2.12**).

In some trials the new drug or drug combination is so successful that the trial is halted early. When the evidence shows that a new treatment is particularly effective it becomes unethical to deny the new treatment to the patients who are receiving the old treatment or placebo.

If a new drug is found to be safe and effective in phase 3 trials, a marketing authorisation application will be sent to the MHRA for the UK, or the EMEA (European Medicines Evaluation Agency) to seek approval for the medicine to be sold in Europe. All the data submitted are assessed to evaluate the beneficial effects against the possible harmful effects of the medicine before a decision is made on whether to grant a licence for the product. Such a licence is needed before you can put a new drug on the market or use an existing drug to treat a different disease.

Even once a new medicine is being used to treat patients, trials still continue. The medicine will be monitored for safety and effectiveness as long as it is used. Any adverse reactions suffered by patients are reported and recorded, to make sure that the benefits of using a medicine always outweigh the risks.

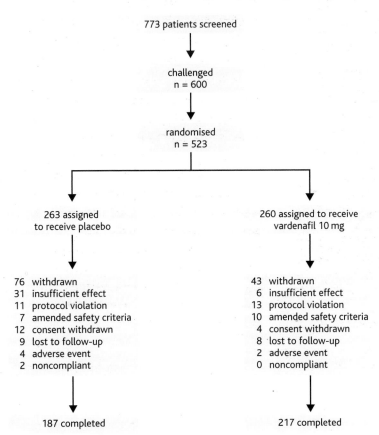

773 patients screened

↓

challenged
n = 600

↓

randomised
n = 523

263 assigned
to receive placebo

260 assigned to receive
vardenafil 10 mg

76 withdrawn
31 insufficient effect
11 protocol violation
7 amended safety criteria
12 consent withdrawn
9 lost to follow-up
4 adverse event
2 noncompliant

43 withdrawn
6 insufficient effect
13 protocol violation
10 amended safety criteria
4 consent withdrawn
8 lost to follow-up
2 adverse event
0 noncompliant

187 completed

217 completed

fig. 4.2.12 The number of patients enrolling on a trial is not necessarily the same as the number of patients who complete it.

Questions

1 Compare William Withering's methods with modern drug testing.

2 Make a flow chart to summarise the main stages in the development of a new medicine from a plant thought to have medicinal properties.

3 Discuss the ethics of carrying out phase 1 human trials on healthy volunteers. Is there an alternative?

4 Sometimes a doctor may want to prescribe a drug before it has completed full human trials. Is it ethical to use a drug before it has undergone all stages of testing? What arguments might be used for or against this happening?

4.3 Species and evolution

The background to biodiversity

Biodiversity is a buzz word of the moment because the Earth's biological diversity is reducing rapidly. Many scientists think this may affect the future health of the planet. At its simplest, biodiversity is a measure of the variety of living organisms and their genetic differences. You can find out about biodiversity in detail in chapter 4.4. In this chapter you will be looking at some of the biology you need in order to understand biodiversity.

Naming organisms

People have attempted to name and organise living organisms for thousands of years. It was the Swedish botanist Carolus Linnaeus in the eighteenth century who founded the classification system we use today. Biologists classify organisms to identify them and to show the way they are related to each other – to identify their common ancestry. In Linnaeus's system every organism is given two Latin names – the **genus** name (a genus is a group of similar species) and the species name, eg *Homo sapiens* (human), *Zea mays* (maize).

What is a species?

Classifying living organisms is central to the monitoring of changes in the species in any given area of the world. **Species** are defined in many different ways. One widely used definition is: *a group of closely related organisms that are all potentially capable of interbreeding to produce fertile offspring*. However, scientists make decisions about which organisms belong in the same species, and how they are related, in a number of ways.

Originally scientists just looked closely at the outer and sometimes inner appearance or **morphology** of the organism. The degree of difference – or similarity – of the phenotypic characters was used to group them into species, genus and so on. In many cases you can tell just by looking at an organism what it is – you would never mistake a lion for a domestic cat, for example. However, as you saw in chapter 3.4, the appearance of an organism can be affected by many different things, and there can be a huge amount of variation within a group of closely related organisms.

Today there are more sophisticated ways of comparing organisms. The fundamental chemicals of life – such as DNA, RNA and proteins – are almost universal. However, whilst these chemicals are broadly similar across all species, differences are revealed when the molecules are broken down to their constituent parts. Scientists use these differences to build up a new science of **molecular phylogeny**, as you will see later in this chapter.

fig. 4.3.1 Careful observation is useful in classifying species, but sometimes molecular phylogeny is needed. These moles are not closely related, but the plants are both species of willow.

HSW Identifying organisms

Questions about identifying different organisms are both absolute (is it species X or species Y?) and also comparative (is it a new species that has not been identified before, or just new to a particular scientist or area?). Information technology provides an ideal tool to help answer these questions.

The Natural History Museum in London is home to millions of specimens of different organisms from all over the world, collected over several hundred years. The great majority were identified by their external features, and details recorded on handwritten and typewritten index cards which are filed in the museum's vast archives. New specimens are regularly sent to the museum for identification. To reduce the time spent searching the cards, scientists at the museum and the University of Essex are developing an IT system called VIADOCS (Versatile Interactive Archive Document Conversion System) to scan and 'read' the card archives, and convert them into an internet-based database and a paper-based catalogue. Once VIADOCS is completed it will not only make searching for a particular organism much easier but it will also give scientists access to classification information while working in the field.

fig. 4.3.2 The new computer system known as VIADOCS was tested out on the 38 000 name cards held in the Natural History Museum, which describe the 27 500 species of *Pyraloidea* moths.

Another problem with identifying organisms is that a number of different classification systems have developed over the years. Prometheus is another new system which compares different classification systems. It allows scientists to use the best aspects of several systems, and also helps to avoid a false picture of biodiversity developing with the same organisms appearing several times in different systems under different names.

The reproductive definition of species

Another widely used definition of species depends on **gene flow** between individuals. If two individuals from different populations mate, they are considered the same species if fertile offspring are produced and genes are combined or 'flow' from the parents to the offspring. So, for example, horses and donkeys look similar, but the offspring produced from a horse and a donkey is a mule, which is sterile. So they are not the same species. But the offspring produced between a Shire horse and a Shetland pony is fertile – they are extreme variants of the same species. However, this definition is not foolproof. For example, lions and tigers are different species, but most of the offspring produced if a lion and tiger mate are fertile. One of the biggest difficulties with this definition is finding the evidence. This is particularly true if a new species is found which is similar to an existing species. Setting up a breeding programme is time-consuming and expensive and may not prove anything.

Other definitions of species

Other definitions of species include:

- **ecological species** which are based on the ecological niche occupied (see later in this chapter)
- **recognition species** which is a concept based on unique fertilisation systems, including behaviour
- genetic species which are based on DNA evidence.

Each of these definitions has its strengths and weaknesses, and the vast majority of the species we know today were identified by their morphology.

Questions

1 Compare the advantages and the practical difficulties of using classic morphology and gene flow to decide if an organism belongs to a particular species.

2 What important advantages will a system like VIADOCS bring to the classification of organisms and the study of biodiversity worldwide?

HSW Molecular phylogeny and the three domains

For centuries classification has been based on detailed observations of anatomy, such as counting the hairs on the foreleg of a fly or the petals of a flower, or even seeing how similar embryos are, to work out relationships between organisms. Now biochemical relationships are increasingly being used to back up or clarify the relationships based on morphology. Scientists need to analyse the structures of many different chemicals and genes to identify the inter-relationships between groups of organisms. This analysis is known as **molecular phylogeny**, and not all scientists interpret the results in the same way.

The evidence from biochemical analysis may support or conflict with relationships based on morphology. For example, all green plants have similar complex pathways for making glucose from sunlight using chlorophyll, so it seems safe to assume they all evolved from a common ancestor. In contrast, American porcupines and African porcupines occupy similar niches and look very similar, but biochemical analysis suggests that they are only very distantly related.

Here are some other examples.

- The vertebrates and the echinoderms (eg starfish and sea urchins) appear, from the evidence of comparative anatomy and embryology, to come from one line of ancestors and the annelid worms, molluscs and arthropods (including the insects) from another. This isn't obvious from looking at the organisms – starfish and sheep don't seem to have much in common. However, biochemical evidence shows that phosphagens, molecules that provide the phosphate group for the synthesis of ATP in muscles, are of two different sorts. Phosphocreatine occurs almost exclusively in the muscle tissue of vertebrates and echinoderms whilst phosphoarginine occurs in the other groups.

- Blood pigments are important in many animal groups. Analysis has shown that any one group contains only one type of blood pigment – all vertebrates and many of the invertebrates have **haemoglobin**, all polychaete worms have chlorocruorin and all molluscs and crustaceans have haemocyanin. This allows scientists to build up a more detailed picture of the relationships between the different groups.

- Analysis of the sequence of amino acids in particular proteins can help show the relationships within higher groups, such as a phylum. For example, in mammals the analysis of fibrinogen, the protein involved in the clotting of the blood, reveals how closely the different mammalian groups are related.

For many years biologists thought there were two domains (the largest groups of living organisms). These were the **eukaryotes** (with a complex cell structure) and the **prokaryotes** such as bacteria (see chapter 3.1). The theory was that eukaryotes had evolved from prokaryotes billions of years ago (fig. 4.3.3).

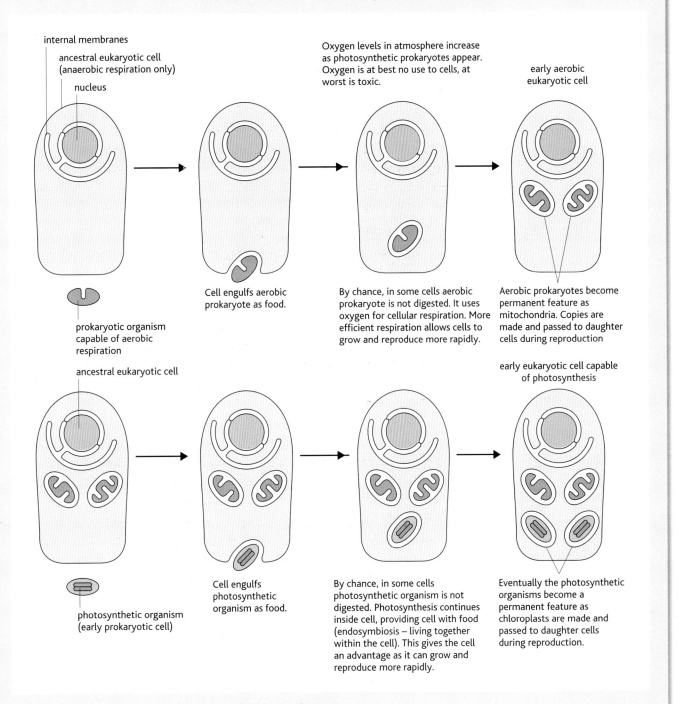

internal membranes
ancestral eukaryotic cell (anaerobic respiration only)
nucleus

Oxygen levels in atmosphere increase as photosynthetic prokaryotes appear. Oxygen is at best no use to cells, at worst is toxic.

early aerobic eukaryotic cell

prokaryotic organism capable of aerobic respiration

Cell engulfs aerobic prokaryote as food.

By chance, in some cells aerobic prokaryote is not digested. It uses oxygen for cellular respiration. More efficient respiration allows cells to grow and reproduce more rapidly.

Aerobic prokaryotes become permanent feature as mitochondria. Copies are made and passed to daughter cells during reproduction

ancestral eukaryotic cell

early eukaryotic cell capable of photosynthesis

photosynthetic organism (early prokaryotic cell)

Cell engulfs photosynthetic organism as food.

By chance, in some cells photosynthetic organism is not digested. Photosynthesis continues inside cell, providing cell with food (endosymbiosis – living together within the cell). This gives the cell an advantage as it can grow and reproduce more rapidly.

Eventually the photosynthetic organisms become a permanent feature as chloroplasts are made and passed to daughter cells during reproduction.

fig. 4.3.3 **The traditional view of how eukaryotes evolved from prokaryotes.**

Some scientists think that chloroplasts became part of 'eukaryotic ancestor' cells first, while others think that mitochondria were the first eudosymbionts. Possibly both processes were happening at the same time – evidence one way or another is almost impossible to obtain.

Some scientists used the techniques of molecular phylogeny to investigate this theory. They looked at the internal structures of the prokaryotes and eukaryotes, their ribosomes and enzymes and then analysed chemicals such as DNA and RNA. As a result a new theory developed that there are in fact three domains – two prokaryote domains, the Archaea and the Bacteria, and the eukaryotes (Eukaryota). The Archaea and the Bacteria are as different from each other as they are from the eukaryotes (see **table 4.3.1**). Genetic studies show that all three groups probably had a single common ancestor around three billion years ago. Some evidence suggests the Archaea are more closely related to eukaryotes (including us!) than the bacteria – our last common ancestor was probably around two billion years ago.

Work establishing these links has been published in peer-reviewed journals, and the experimental techniques have been archived so that other scientists can repeat the procedures to verify the results. Yet in spite of this, there is still considerable debate among some biologists about these three domains and their origins. For example, Professor Larry Moran at the University of Toronto refutes three-domain evolution. He claims to have evidence that shows that the genes of the ribosomes did not evolve together, and that the logic underpinning the whole three-domain premise is flawed. He also feels that the eukaryotes are no closer to the Archaea than to the Bacteria. He would present the evolution of organisms as a complex web of life rather than a clear simple tree.

Characteristic	Bacteria	Archaea	Eukaryota
Membrane-enclosed nucleus	Absent	Absent	Present
Membrane-enclosed organelles	Absent	Absent	Present
Peptidoglycan in cell wall	Present	Absent	Absent
Membrane lipids	Ester-linked, unbranched	Ester-linked branched	Ester-linked unbranched
Ribosomes	70S	70S	80S
Initiator tRNA	Formylmethionine	Methionine	Methionine
Operons	Yes	Yes	No
Plasmids	Yes	Yes	Rare
RNA polymerases	1	1	3
Ribosomes sensitive to chloramphenicol and streptomycin	Yes	No	No
Ribosomes sensitive to diphtheria toxin	No	Yes	Yes
Some are methanogens	No	Yes	No
Some fix nitrogen	Yes	Yes	No
Some conduct chlorophyll-based photosynthesis	Yes	No	Yes

table 4.3.1 **Some of cellular and molecular characteristics of the three domains of life on Earth.**

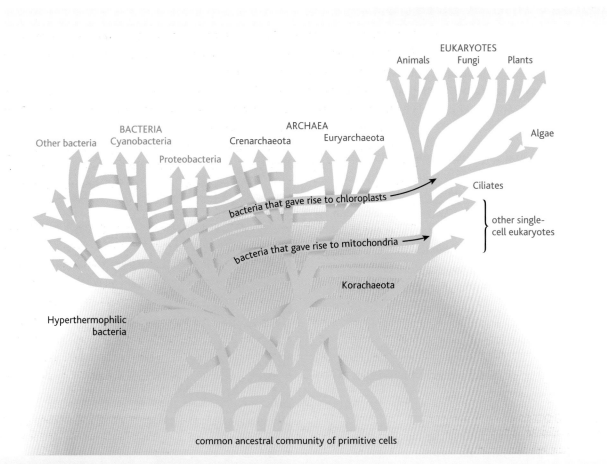

EUKARYOTES
Animals   Fungi   Plants

BACTERIA
Other bacteria   Cyanobacteria

ARCHAEA
Crenarchaeota   Euryarchaeota

Proteobacteria

Algae

bacteria that gave rise to chloroplasts

Ciliates

other single-cell eukaryotes

bacteria that gave rise to mitochondria

Korachaeota

Hyperthermophilic bacteria

common ancestral community of primitive cells

fig. 4.3.4 **A tree of life devised by Ford Doolittle to explain the evolution of different organisms.**

In contrast Professors Norman Pace and Ford Doolittle from Canada are adamant that the three-domain structure has been thoroughly supported by genomic evidence and there can be no really informed dissent to the theory. Most biologists agree with them.

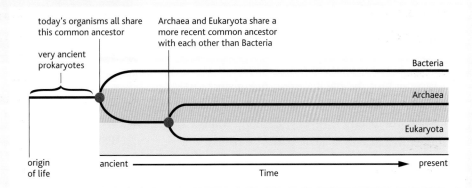

today's organisms all share this common ancestor

Archaea and Eukaryota share a more recent common ancestor with each other than Bacteria

very ancient prokaryotes

Bacteria

Archaea

Eukaryota

origin of life   ancient ——————→ present

Time

fig. 4.3.5 **The three domains of the living world – most but not all biologists think that all three domains shared a common prokaryote ancestor.**

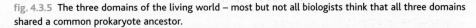

Questions

1   How is molecular phylogeny used to show genetic diversity and similarities between organisms?

2   Use the information on these pages along with other details you may find on the internet to help you explain the difficulties of drawing conclusions from evidence like this.

3   What factors may affect which theory becomes accepted by the scientific community?

Ecology and adaptation

Organisms do not exist in a vacuum. The various species are all part of a complex system of interactions between the physical world and other living organisms which we call **ecology**.

Each species exists in a particular **ecological niche**. This important concept is quite difficult to define. It describes the role of the organism in the community – a bit like a job description or a way of life. You can consider different aspects of a niche, such as the food niche or the habitat niche. Some niches are very large and general, eg organisms that eat grass; others are very small and specific, eg organisms that clean the teeth of other, larger organisms.

Useful terms

There are a number of useful terms in the science of ecology.

- An **ecosystem** is an environment that includes all the living organisms interacting together, the nutrients cycling through the system and the physical and chemical environment in which the organisms are living. An ecosystem consists of a network of habitats and the communities of organisms associated with them.

- A **habitat** is the place where an organism lives. You can think of the habitat of an organism as its address. Examples include a freshwater pond, a deciduous woodland or a rocky seashore. Many organisms live in only a small part of a habitat – a single tree or a rock pool for example. These are known as **microhabitats**.

- A **community** is all the populations of living organisms living in a habitat at any one time. For example, in a habitat such as a rock pool there will be populations of different seaweeds, sea anemones, shrimps, shellfish and small fish such as gobies and crabs, as well as other organisms.

fig. 4.3.6 The community of organisms seen here includes a wide variety of living organisms, from plankton to fish.

- A **population** is a group of organisms of the same species, living and breeding together in a particular niche in a habitat. The three-spined sticklebacks in a pond or the skin mites in your mattress are examples.

Adaptation to niches

A successful species is adapted well to its niche, meaning that individuals in that species have characteristics that increase their chances of survival and reproduction, and therefore of passing those characteristics on to the next generation. These adaptations may be of many different kinds, including anatomical, physiological and behavioural. **Anatomical adaptations** include the thick layer of blubber in seals and whales, and the sticky hairs on the sundew plant which enable it to capture insects ready to digest.

Physiological adaptations include differences in biochemical pathways or enzymes and therefore in the way the body works. For example, diving mammals can stay under water for far longer than non-diving mammals without drowning. Once they are under water their heart rate drops dramatically, so that the blood is pumped around their body less often and the oxygen in their blood is not used as rapidly (see **fig. 4.3.7**). The main body muscles can work more effectively using anaerobic respiration than those of land-living mammals, so the oxygen-carrying blood is directed to the brain and the heart where it is still needed. This is known as the mammalian diving response.

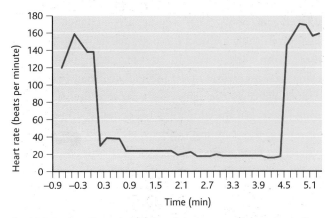

fig. 4.3.7 Bradycardia or the slowing of the heart rate in a seal as it dives.

Behavioural adaptations can also help organisms to survive better, particularly in difficult conditions. For example, many insects and reptiles orientate themselves to get the maximum sunlight on their bodies when the air temperature is relatively low. This allows them to warm up and move fast enough to feed and escape predators. When they get hot, they change their orientation to minimise their exposure to the sun, or shelter from it. Social behaviour such as hunting as a team or huddling together for warmth can improve the survival chances of both individuals and a group of organisms. Migrating to avoid harsh conditions, courtship rituals and using tools are other examples of behavioural adaptations.

David Grémillet and his team looked at the relative importance of physiological and behavioural adjustments in the great cormorant (*Phalacrocorax carbo*) in two contrasting environments – Normandy where the water temperature is 12 °C, and Greenland with a water temperature of 5 °C. Cormorants are not well insulated by fat and have poorly waterproofed feathers so they are easily affected by cold.

fig. 4.3.8 Cormorants rely on diving for fish – but they are not well insulated. Behavioural adaptations allow them to survive in cold conditions in spite of poor physiological adaptations.

The team found big differences in the feeding behaviour of birds breeding in the two regions. The birds living in Greenland spent 70% less time in the water than those in Normandy. They spent far less time swimming on the surface of the water between dives, and also returned to the land more often. The total daily energy intake of cormorants was similar in both areas but prey capture rates in Greenland were 150% higher than those in Normandy because the changes in their behaviour resulted in far greater efficiency at finding food. Behavioural adaptations were more important than physiological ones to their survival in the colder niche.

So natural selection leads to adaptations which give individuals an advantage in a particular niche. If conditions change, those adaptations may not be as successful, and the selection pressure will change. This may lead to change in the species, or **evolution**.

Questions

1 Explain the importance of the niche concept in understanding the adaptations of organisms.

2 Looking at the data from **fig. 4.3.7**, answer the following questions:
   a What do the negative numbers on the x-axis represent?
   b How long did the recorded dive last?
   c What was the percentage depression of the heart rate? Based on this, how long would you predict that the dive might have lasted without the bradycardia?

3 How do the behavioural adaptations of cormorants in Greenland help them to survive in their cold-water, fish-eating niche?

Natural selection at gene level

The niche an organism inhabits has a big effect on the genetic make-up of the population. Natural selection results in organisms that are adapted to fit a particular niche. Although we may describe natural selection acting at the morphological level, it is often acting at the level of the gene.

Mutations and natural selection

Mutations can cause small changes in genes and they are the source of variation on which natural selection acts. Mutations can increase the size of the **gene pool** of a population, by increasing the number of different alleles. The relative frequency with which a particular allele is found in a population is known as the allele frequency. Allele frequency is one way of measuring biodiversity (see page 228). Alleles in a population will be affected by natural selection. A mutation in a gene may result in a change in the physical appearance of an organism, in its physiology or even in its pattern of behaviour. If this change is advantageous, so that the individual will be more likely to survive and reproduce, then the frequency of those advantageous alleles within the population will increase. If the mutation is disadvantageous, natural selection will usually result in its removal from the gene pool. Sometimes the mutation is neutral, neither increasing nor decreasing the success of the individual. In this case, it will remain in the gene pool by chance.

For example, warfarin is a chemical which prevents the blood from clotting. It has been used as rat poison since about the 1950s – the rats die of internal bleeding.

fig. 4.3.9 **The presence of the allele that gives rats resistance to warfarin is obvious only once they are exposed to the poison.**

When warfarin was introduced, some rats already carried a harmless mutation which by luck gave them resistance to the poison. Not surprisingly, the powerful selection factor of the poison resulted in a rapid increase in the frequency of the resistance allele. Soon the majority of rats were resistant to warfarin and new, more powerful poisons had to be developed.

The effect of small populations

Large populations containing many individuals usually have large gene pools. This is because the chance of losing an allele from the gene pool by 'bad luck' is much less. For example, if an allele occurs in 10% of the population and that population consists of 10 individuals, then only one individual will carry the allele. Let's say that the allele is for running faster, but the individual breaks a leg. If a predator comes along, the 'running faster' allele won't save the individual and the allele will be lost from the population, despite being a favourable allele. However, in a population of 5000 individuals, 500 will carry the allele and the likelihood of all of those organisms being destroyed is remote. So there is a bigger chance of a potentially useful allele being maintained in the larger population.
This is one reason why large, genetically diverse populations are needed to maintain biodiversity, as you will see in chapter 4.4.

Similarly, when a small number of individuals leave the main population and set up a separate new population, genetic diversity is again easily lost. The alleles they carry may be a random selection of the gene pool, but probably will not include all the alleles at the same frequencies as in the original population. Any unusual genes in the founder members of this new population may become amplified as the population grows. This is known as the **founder effect** (**fig. 4.3.10**).

One example of this is shown by the Amish, an American religious sect which exists in three isolated communities. One of the groups has a high frequency of a very rare genetic disorder known as Ellis–van Creveld syndrome.

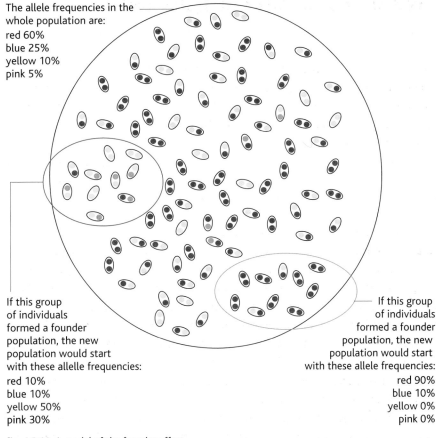

The allele frequencies in the whole population are:
red 60%
blue 25%
yellow 10%
pink 5%

If this group of individuals formed a founder population, the new population would start with these allelle frequencies:
red 10%
blue 10%
yellow 50%
pink 30%

If this group of individuals formed a founder population, the new population would start with these allele frequencies:
red 90%
blue 10%
yellow 0%
pink 0%

fig. 4.3.10 **A model of the founder effect.**

fig. 4.3.11 **Ellis–van Creveld syndrome is extremely rare in the population as a whole but relatively common in one isolated Amish population. This is an example of the founder effect, which can bring about a dramatic change in gene frequencies in a population.**

Mr and Mrs Samuel King emigrated to Pennsylvania in 1744 as members of a small group of 200 people who founded an Amish community. They had no idea of the legacy they took with them. One of the couple was heterozygous for Ellis–van Creveld syndrome. This rare genetic disease results in a type of dwarfism – the limbs are shortened, there may be extra digits and as a result of associated problems the affected person usually dies in the first year of life. The Samuel Kings produced many children, who in turn were also particularly prolific. This raised the frequency of the allele within the population way above its usual level. Purely by chance at least one other of the founding group also possessed this recessive gene in the heterozygous form. As a result of interbreeding in this isolated Amish community 1 in 14 individuals in the population now carry the gene. This results in a distressingly high number of affected births – in 1964 there were 43 cases of Ellis–van Creveld syndrome in a population group of 8000. More cases of Ellis–van Creveld syndrome have been found in this one small population than in the whole of the rest of the world.

Questions

1 Why do you think the resistance allele increased so rapidly in the rat population after the introduction of warfarin?

2 Why do small populations of organisms often have low genetic diversity, and what is the advantage of large genetic diversity?

Evolution in action

Individuals that are not well adapted may not survive to reproduce, or may produce fewer offspring than those that are better adapted, so their characteristics will become less common in the population. This is what Charles Darwin meant by the term 'survival of the fittest' and is what we now call **natural selection**. If the niche changes due to changes in the environment, other characteristics may make an individual more successful. Natural selection will favour the survival of individuals with those different characteristics, and we say the **selection pressure** has changed. Changes in selection pressure result in changes (evolution) within the species. Depending on how different the individuals are, they may even be considered a new species.

Adapting to change

Malpeque Bay on Prince Edward Island in Canada is home to massive oyster populations. In 1915 the oyster fishermen of Malpeque Bay began to notice that amongst their usually healthy catches there were a few diseased oysters that were small and flabby with pus-filled blisters. Before long the oysterbeds had been all but wiped out by this new Malpeque disease. However, a small number of the millions of offspring produced by each oyster in a year carried an allele giving them resistance to the disease. Because only individuals that had this allele were able to survive and reproduce, the frequency of this gene in the population increased rapidly. By 1935 a small oyster harvest was again possible and by 1940 the beds were as prolific as ever, but with a rather different gene pool – now containing a high frequency of disease-resistance alleles (see **fig. 4.3.12**).

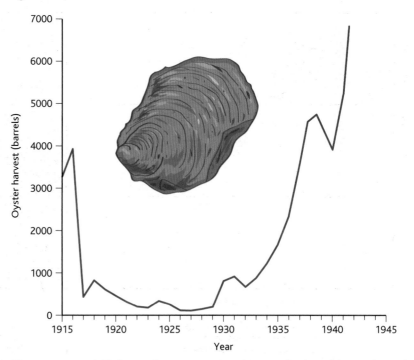

fig. 4.3.12 Oyster yields from Malpeque Bay, 1915–40. Disease devastated the populations but as a result of the increased selection of the disease-resistance allele within the population, large healthy oyster beds returned.

Selection for change – or stability

The oysters of Malpeque Bay are a good example of **directional selection**, 'classic' natural selection showing a change from one phenotypic property to a new one more advantageous in the circumstances (**fig. 4.3.13**). There are many examples of this type of evolution in progress.

Directional selection occurs anywhere that environmental pressure is applied to a population. It is frequently seen in populations of insects and plants that are regarded as pests and sprayed with chemical insecticides or herbicides. The chemicals may have a devastating effect initially, but directional selection ensures that within a relatively few generations resistant individuals become more common within the population.

The introduction of the rabbit disease myxomatosis into Britain in 1953 almost wiped out the rabbit population over the next ten years or so, but rabbits are now common once more. Many of them carry an allele that renders them immune to the ravages of myxomatosis – the frequency of that allele in the rabbit gene pool has increased enormously.

Diversifying selection is another variety of selection. The difference from directional selection is that the outcome is an increase in the diversity of the population rather than a trend in one particular direction (see **fig. 4.3.13** and later in this chapter). It occurs when conditions are very diverse and small subpopulations evolve different phenotypes suited to their very particular surroundings.

Balancing selection takes place when, as a result of natural selection, variety is maintained by keeping an allele within the population even though it might seem to be disadvantageous. An example from chapter 2.2 is the thalassaemia allele affecting human haemoglobin. Although the homozygous form of the allele is usually lethal, the heterozygous form gives protection against malaria and so the allele remains at a relatively high frequency within the population. This is known as **heterozygote advantage** or **hybrid vigour**.

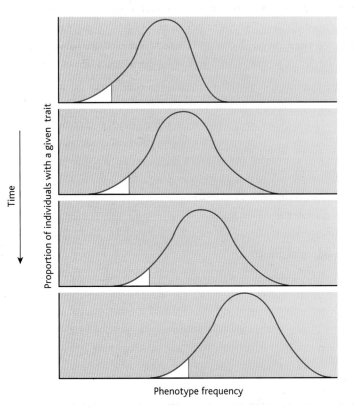

Directional selection leads to a change in the phenotypes of a population in a particular direction, making them better suited to their environment.

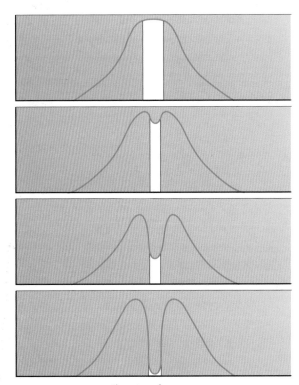

Diversifying selection increases the variety within a population – if the selection pressure is strong two non-overlapping populations may result which could end up as two new species.

fig. 4.3.13 Natural selection works in different ways in different circumstances, but in each case it results in populations that are well adapted to their niche. Individuals that fall within the white region on each graph are being selected against and leave fewer offspring.

HSW The story of the peppered moth

Our model of the way natural selection works is based in part on the observations of Bernard Kettlewell at Oxford University in the 1950s, who studied the peppered moth *Biston betularia*. Kettlewell proposed that the moth had undergone natural selection in response to environmental changes.

Evidence for industrial melanism

The normal, or typical, form of *Biston betularia* is a creamy speckled moth found in British woodlands. In the eighteenth century, relatively rare black specimens resulting from a random dominant mutation (known as the carbonaria form or melanics) were very popular with butterfly collectors of the day. This physiological change was a result of a change in the biochemistry of the pigmentation pathways. The dark coloration made these individuals easily visible against the pale bark of the trees, both for human collectors and for birds looking for a meal. This selection pressure meant that the frequency of the dark allele in the population remained low.

Then in the mid-nineteenth century the Industrial Revolution changed the face of much of Britain and Europe. The soot and smoke from the factory chimneys darkened the bark of the trees and the surfaces of buildings. As a result the melanic form of *Biston betularia* was at a selective advantage and the frequency of the allele within the population began to increase as more and more of the light-coloured moths fell prey to predators. This process became known as **industrial melanism**.

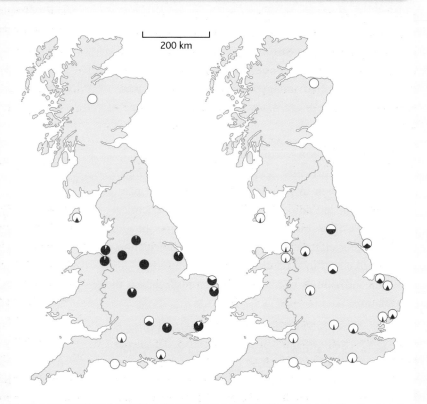

fig. 4.3.14 **The map on the left shows the proportions of melanic moths in the population from Kettlewell's 1956 data. The map on the right shows data on moths collected in similar areas by Bruce Grant and his colleagues in 1996.**

Kettlewell set up several experiments to examine the selection of the two forms. In Dorset, an unpolluted area with clean trees, he released equal numbers of light and dark moths; 12.5% of the light moths were subsequently recaptured, but only 6% of the dark ones were retrieved. In another series of observations birds were seen to eat 26 light moths and 164 dark ones as they rested on a light tree trunk. In Birmingham, at the time a very polluted industrial area, 40% of the dark moths were recaptured but only 19% of the light ones. Of the moths picked off the blackened trees from equal releases, 43 were light and 15 were dark. Kettlewell concluded that the change in frequency of melanic moths was due to the selection pressure of predation.

Reversing the trend

Anti-pollution legislation was passed in the 1960s. In the twenty-first century there is considerably less heavy industry in Britain, and the remaining industries produce much less pollution. As a result, trees and buildings have become cleaner and paler again, lichens have grown back on tree trunks and the selection pressure has moved back in favour of the paler moth (see **fig. 4.3.14**). The frequency of the typical or pale allele in the population has increased again.

Differences of opinion

After being accepted for a long time, the elegant story of *Biston betularia* has been challenged in recent years. Some scientists questioned whether the effect was in place all over the country. There were arguments about where on the trees the moths rest – Kettlewell chose positions on the trunks different from the normal resting sites – and other details. One author, Mike Majerus, looked for alternative explanations of the changes in the moths. He was interested in other influences that might have had an impact as well. However, he concluded:

> In my view the huge wealth of data collected since Kettlewell's initial predation papers does not undermine the basic qualitative deductions from that work. Differential bird predation of the *typical* and *carbonaria* forms, in habitats affected by industrial pollution to different degrees, is the primary influence on evolution of melanism in the peppered moth.

Other authors interpreted his results as evidence against industrial melanism and reacted as if Kettlewell's thesis was disproved. One scientist, Jerry Coyne, wrote that: 'My own reaction resembles the dismay attending the discovery ... that it was my father and not Santa who brought the presents on Christmas Eve.' This work was soon being used to disprove the whole idea of evolution. Arguing against bird predation being an agent of natural selection, Ted Sargent and others said in the journal *Evolutionary Biology* in 1998 that 'there is little persuasive evidence, in the form of rigorous and replicated observations and experiments to support this explanation at the present time'.

Then other scientists pointed out that there is no body of evidence against industrial melanism. In fact the evidence overwhelmingly supports Kettlewell's hypothesis. In recent years again, more and more scientists are standing up in support of Kettlewell's work and the model of natural selection demonstrated by industrial melanism.

Jim Mallet at University College London has looked into the evidence for and against industrial melanism. Most of the work on industrial melanism was carried out on the peppered moth, but there is evidence that similar melanism in polluted areas has occurred in over 70 different species of British moths. For example, the melanic form of the marbled beauty moth *Cryphia domestica* was dominant in London in the 1970s and 1980s. However, since the 1990s the normal pale form has reappeared in strength.

The fundamental story – that birds find it harder to see light moths on clean trees, and dark moths on dirty trees, and that this affects the gene pool and causes natural selection – seems hard to deny even if the debate still rumbles on.

fig. 4.3.15 **The camouflage effects of the two main forms of *Biston betularia* on the bark of a clean tree and a polluted tree.**

Questions

1 How can changes in a niche or habitat lead to changes in a species?

2 Using the data from **fig. 4.3.12**, how long did it take for Malpeque disease to virtually wipe out the oyster population in the bay, and how long did it take for the resistant allele to become sufficiently dominant for the population to recover? Why is this an example of the adaptation of an organism to its niche?

3 Suggest how the adaptation of diving in mammals could have evolved.

4 What arguments could you put forward for and against the fact that the position of the moths on the trees in Kettlewell's experiments is important in determining the validity of his model?

5 Using this section, the support material and any other resources you have, investigate the evidence for and against industrial melanism based on *Biston betularia*.

Increasing biodiversity

So far you have seen how changes in niches can result in different selection pressures and in change within a species. If that change is great enough, we may consider eventually that a new species has evolved from the old one. However, sometimes one species can evolve into more than one other species.

Darwin's finches

The classic example of how the availability of a range of different niches can provide different selection pressures and result in the evolution of several species from one is the story of Darwin's finches. These birds were discovered by the great nineteenth-century naturalist Charles Darwin on his voyage on *HMS Beagle*. On the Galapagos Islands near the equator there are a number of feeding niches for birds, eg small seeds, large nuts and insects living in rotten bark. The original finches that arrived on the islands were of a single species. No one is quite sure how they got there – the islands are 500 miles from land – but a small flock was probably carried there by a storm or a hurricane.

Within the birds that arrived at the islands, there would have been variation in alleles and characteristics, and different niches on the islands would have favoured individuals with different variations. For example, a bird with a slightly smaller, stronger bill would get more food by eating mainly seeds. This would enable it to thrive, reproduce and pass on its beak characteristics to its offspring. Over generations, natural selection resulted in individuals with small strong beaks ideally adapted to eating seeds. Similarly, a finch with a longer, thinner beak might well be more successful probing dead wood for insects and so again feed relatively exclusively in that way. By exploiting different niches the finches avoided competing for the same relatively scarce food resources. As a result, 14 different species of finch have evolved on the Galapagos Islands over several million years from one common ancestral species (fig. 4.3.16).

Because food was such an important selection pressure, it was important to mate with a finch with a similarly shaped beak to pass on the advantageous characteristic. Mating with a finch that had a differently shaped beak would produce a variety of offspring that were less well adapted to feeding, so there would also be a selective pressure on choosing the right kind of mate. So other phenotypic and behavioural changes that made choosing the right mate easier were also selected for. This is why the different species look different. Although the finches specialise and feed on particular types of food, and vary considerably in size and appearance, DNA analysis has shown that genetically they are remarkably similar.

fig. 4.3.16 Over several million years 14 species of finches evolved from the original ancestor species. The anatomy of the beaks adapted for different feeding niches of 12 of them can be seen here.

Selection for features that give reproductive success is known as **sexual selection**. In many species of birds, mammals and other animals, there are clear anatomical adaptations to help in attracting a mate. The tail of the peacock, the mane of a lion and the antlers of a stag are all anatomical adaptations that enable the male animal to attract the female and so pass on his genes.

HSW Sexual selection in action

Male African long-tailed widow birds have very long tails which appear to have very little use except in the mating season to attract mates. If our ideas of natural selection are correct, long tails should be selected for. To investigate these ideas, male birds were captured and their tails artificially lengthened (by gluing extra feathers on) or shortened (by cutting them). Once released, their reproductive success was measured by the number of nests with eggs and/or young in each male's territory.

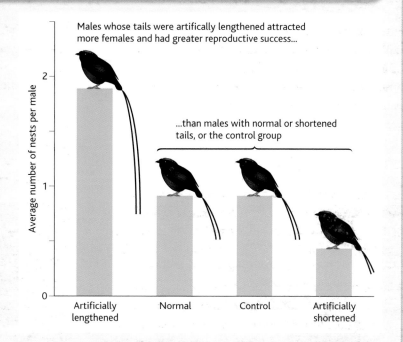

fig. 4.3.17 The selection for an anatomical feature that gives an advantage can be seen in African long-tailed widow birds.

Isolating mechanisms

For different species to evolve from one original species, different populations of the species usually have to become isolated from each other, so that mating, and therefore gene flow, between them is restricted. There are a number of ways in which this may happen.

- **Geographical isolation**: a physical barrier such as a river or a mountain range separates individuals from an original population.

- **Ecological isolation**: two populations inhabit the same region but develop preferences for different parts of the habitat.

- **Seasonal isolation**: the timing of flowering or sexual receptiveness in some parts of a population drifts away from the norm for the group. This can eventually lead to the two groups reproducing several months apart.

- **Behavioural isolation**: changes occur in the courtship ritual, display or mating pattern so that some animals do not recognise others as being potential mates. This might be due to a mutation that changes the colour or pattern of markings.

- **Mechanical isolation**: a mutation may occur that changes the genitalia of animals, making it physically possible for them to mate successfully with only some members of the group, or it changes the relationship between the stigma and stamens in flowers, making pollination between some individuals unsuccessful.

Allopatric speciation takes place when populations are physically separated in some way. **Sympatric speciation** takes place when two populations are geographically still close to each other.

Questions

1 Look at the skulls of the Darwin's finches. They look very different yet are genetically quite similar. Explain how this may have come about.

2 How do the results shown in fig. 4.3.17 support the current model of natural selection and evolution? What ethical issues might be raised with an investigation of this type?

3 Give examples of isolating mechanisms that lead to:
 a allopatric speciation b sympatric speciation.

Endemism

It should come as no surprise to you now that isolated islands frequently contain species that are found nowhere else. The availability of niches to species that first colonise the island, the different selection pressures of those niches compared with the 'home' environment and the founder effect of a limited gene pool all combine to result in the evolution of new species that occur only in that small area. This results in **endemism**.

Endemism describes the situation where a species of organism is found only within a particular area. The organism is said to be **endemic** to the region. It happens because the organism evolved within that region and hasn't migrated out to other areas. Migration is obviously limited by geographical boundaries such as water or mountains. So islands are more likely to have endemic species than parts of large continents, although endemism is certainly not confined to islands alone.

Examples of endemism

Endemic species of Madagascar

Madagascar, a large island off the coast of East Africa, provides good examples of endemism. Almost all of the species found there are endemic to the island. These range from the amazing giant baobab trees to ring-tailed lemurs, and from the bizarre elephant's foot plant to the small, prolifically breeding mammal, the yellow-streaked tenrec. The only species there that are not endemic are ones that have been taken to the island by people in relatively recent times.

fig. 4.3.18 The ring-tailed lemur is just one of the unique species that are endemic to Madagascar.

The isolated islands of Hawaii

The Hawaiian island populations show clearly how living organisms adapt to a particular niche or role in the community. The islands are very isolated – 4000 km from the nearest continental land mass and 1600 km from the nearest other islands. They also have a great deal of biodiversity in terms of species numbers – 1000 species of native flowers, 10 000 species of insects, 1000 species of land snails and around 100 species of birds. But before people introduced them, there were no reptiles and only one species of mammal – a bat. Analysis of the DNA of the native populations shows that they are very closely related, even though some of them look very different. All those insect species seem to have evolved from only around 400 original species, while there appear to have been only seven founder species of land birds. So in these isolated circumstances, a small group of founder organisms have adapted and evolved to take advantage of the different ecological niches that were available to them.

Places where endemism is common often have a rich biodiversity in terms of species numbers but relatively low genetic diversity. This is one reason why areas with many endemic populations are very vulnerable to the introduction of disease.

fig. 4.3.19 Hawaiian silverswords come in many different shapes and sizes. However, DNA evidence suggests they all evolved on the islands from a single ancestral species.

Australian marsupials and monotremes

Australia is well known for its unusual fauna and flora. Perhaps most unusual are two groups of mammals, the **marsupials** (which protect their young in pouches) and the even rarer egg-laying **monotremes**. In the rest of the world, the **placental mammals** dominate.

Until about 5.5 million years ago Australia was joined to the rest of the world, when the only mammals were marsupials and monotremes (**fig. 4.3.20**). After Australia separated from the other continents, the marsupials in particular evolved to fill an enormous range of niches, from the large herbivorous kangaroo with its wide-ranging niche to the koala with its eucalyptus tree niche, and others with different carnivorous niches, eg the quoll and the Tasmanian devil. Placental mammals evolved on other continents and mostly replaced the marsupials and monotremes that once lived there, but didn't reach Australia until humans carried them across the dividing ocean.

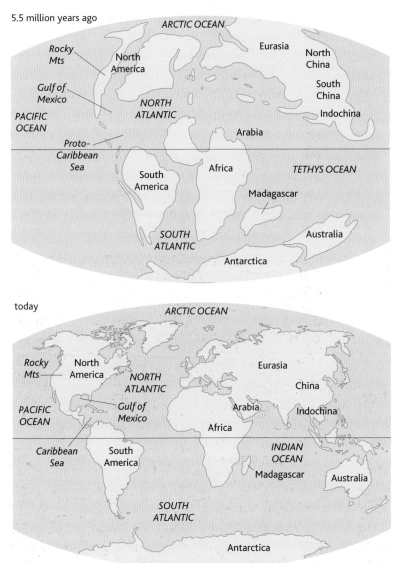

fig. 4.3.20 When the Australian land mass separated around 5.5 million years ago, the founder species that were isolated from the rest of the world had many different niches to fill.

Endemism and other isolating mechanisms

Endemism can also occur where there is plenty of food and animals don't need to travel far to find all they need to survive. Such conditions may be present in estuaries and rainforests. Here isolating mechanisms can easily separate populations, and complex interactions can develop between different species to produce new specialised niches. For example, in rainforests there are many plant species that can only be pollinated by one kind of insect, while in temperate forests such as in the UK, many species of bees act as general pollinators because there is less food.

A vulnerable resource?

As you have seen, islands tend to have a rich variety of plants and animals not seen anywhere else. However, island ecosystems are often relatively small and so they are very vulnerable to interference and damage by human beings. Of the animals that have become extinct in the last 400 years, 75% have been island species. An analysis of modern species under threat of extinction shows that again island species are at great risk – in the Pacific islands alone around 110 species of birds face possible extinction. In the final chapter of this book you are going to be looking in more detail at possible causes of extinction and the ways in which biodiversity can be measured, monitored and conserved around the world.

Questions

1 Suggest why endemic populations often have low genetic diversity.

2 There are many concerns about the introduction of species into isolated ecosystems such as islands. Use what you have learnt so far to predict why such introductions might be a problem.

4.4 The importance of biodiversity

Biodiversity

fig. 4.4.1 Lowland heath is home to all 12 of the native British species of amphibians and reptiles – this alone makes it an important source of biodiversity in the UK.

Biological diversity is decreasing at an alarming rate. Most scientists are agreed that this is not a good thing, even if they disagree about the causes. But what is biodiversity? Why should we preserve it – and how can we do this?

Most people have some idea what the term biodiversity means, but defining it clearly is not easy. The Convention on Biological Diversity, the largest international organisation working on the subject, has described it like this.

• Biodiversity comprises every form of life, from the smallest microbe to the largest animal, the genes that give them their specific characteristics and the ecosystems of which they are part. This includes diversity within species, between species and of ecosystems.

The number of different species is a useful basic measure, but the concept of biodiversity is much more far-reaching than this. The differences between individuals in a species, between populations of the same type of organism, between communities and between ecosystems are all examples of biodiversity.

Why is biodiversity important?

Does it really matter if there are fewer species of snails or beetles in the world, if an unknown plant species ceases to exist or if the genetic variation between the members of a rare population gets less and less? All the evidence suggests that it does.

In your work at GCSE and earlier, you learned how all the organisms in an ecosystem are interdependent, and how they can affect the physical conditions around them. These ecosystems are also interlinked on a larger scale across the Earth. If biodiversity is reduced in one area, the natural balance may be destroyed elsewhere. Healthy biodiversity allows large-scale ecosystems to function and self-regulate. The air and water of the planet are purified by the action of a wide range of organisms. Waste is decomposed and rendered non-toxic by many organisms, including bacteria and fungi, eg microorganisms in soil and water convert toxic ammonia into nitrate ions which are then taken up and used by plants.

Photosynthesis by plants plays an important part in stabilising the atmosphere and the world climate. Plants absorb vast amounts of water from the soil which then evaporates into the atmosphere through transpiration, as you saw in chapter 4.1. This helps to determine where rain will fall. Plant roots also hold the soil together, affecting how water runs off the soil surface and reducing the risk of flooding. Plant pollination, seed dispersal, soil fertility and nutrient recycling in systems such as the nitrogen cycle are vital for natural ecosystems as well as farming, and they depend on thriving biodiversity.

Biodiversity also provides the genetic diversity that has allowed us to develop the production of crops, livestock, fisheries and forests, and enables further improvement by cross-breeding and genetic engineering. This will help us to cope with problems arising from climate change and disease. Biodiversity also provides the potential of plants to produce chemicals that are important in many areas of human life.

Are some species more important than others?

When the media highlight extinctions and loss of biodiversity they usually cover large charismatic animals such as pandas, elephants, tigers and whales. However, animals or plants lower down the food chain are often vital in preserving biodiversity. Understanding the complex feeding relationships of animals can help us protect whole ecosystems.

For example, the members of the fig family are the staple food for hundreds of different species in many different countries – so important that scientists sometimes refer to figs as 'jungle burgers'. Animals from tiny insects to birds and large mammals feed on everything from the bark and leaves to the flowers and fruits. Figs also have very specific pollinators. There are several dozen different species of fig tree in Costa Rica, and each one is pollinated by a different type of fig wasp which has evolved to pollinate only that type of fig. So without the fig wasps the fig trees would die out, and without the fig trees many other species would be affected. The figs and the pollinator wasps are closely linked, and problems for either of them would result in a knock-on effect with a great loss of biodiversity. The seemingly humble fig wasps have a vital role.

Certain species, known as **keystone species**, seem to have a particularly important impact on the environment in which they live, even if they are not the most obvious species in the area.

fig. 4.4.2 **The sea otter (***Enhydra lutris***) is a keystone species in kelp forests – this one animal species maintains the biodiversity of the whole ecosystem.**

For example, sea otters play a major role in the survival of giant kelp forests in the ocean in some parts of the world (**fig. 4.4.2**). Kelp forests provide a home for a wide range of other species, but kelp itself is the main food of purple and red sea urchins. When there are lots of sea urchins free from predators, they roam the sea floor and eat the kelp as it starts to grow. This tends to keep the kelp very short and stops forests developing – which in turn reduces biodiversity. However, sea otters are major predators of sea urchins. If sea otters move into an area they eat many urchins, and the rest tend to spend their time in rock crevices to avoid the predators. This allows the kelp to grow – and it can grow many centimetres in a day. As the kelp forests form, bits of kelp break off and fall to the bottom to provide food for the urchins in their crevices. The sea otters thrive on hunting in the kelp for sea urchins, along with many fish and invertebrates that live among the kelp fronds. The sea otters are a keystone species here. But as large predators they are vulnerable. Their numbers are relatively small so disease or human hunters can easily wipe them out, and this in turn results in loss of the kelp forest and reduced biodiversity.

In Scotland the keystone species in the Caledonian Forest ecosystem is the Scots pine. In this case the keystone species is the largest and longest-living organism in the forest. A huge number of different species are dependent on these trees, from bacteria and fungi through to crossbills and red squirrels (see fig. 4.4.3).

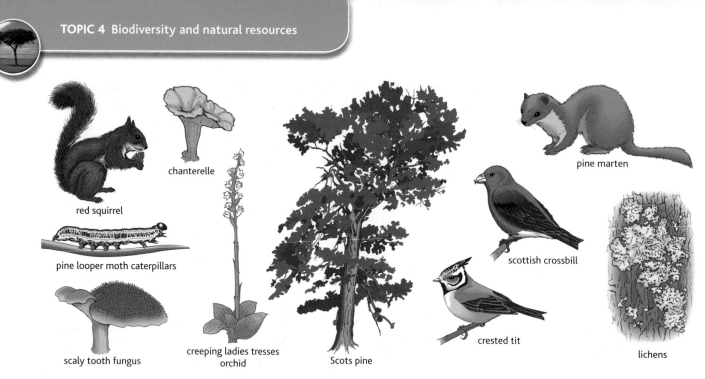

fig. 4.4.3 The Scots pine is a keystone species in the Caledonian Forest. These are just a few of the species that depend on these massive trees.

Are some places more important than others?

In terms of numbers of species, around the world biodiversity varies enormously. The wet tropics are generally the areas of highest biodiversity. For example, in the Amazon rainforests in 0.1 of a hectare (less than four tennis courts) you would expect to find 150–280 tree species – almost every tree is a different species! Imagine the other plants and animals associated with each type of tree and you can begin to appreciate the species richness of these areas. As you move away from the wet tropics, the species diversity tends to fall. In temperate rainforests, tree species richness drops to

20–25 species in the same size of sample area. Further north again in the boreal forest, eg in Scandinavia and northern Canada, it falls to 1–3 species. To highlight this, Norman Myers identified a number of **biodiversity hotspots** of unusual biodiversity and endemism. They occupy only 15.7% of the Earth's land surface but are home to 77% of the Earth's terrestrial vertebrate species, for example. Unfortunately, as you will see later in this chapter, these areas often coincide with areas where people live and need resources. Similar areas of particular species richness in tropical areas are seen for marine and freshwater species and among all the different types of organisms.

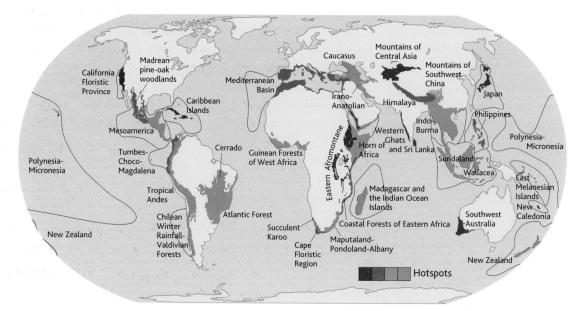

fig. 4.4.4 Known biodiversity hotspots around the world.

Species richness is not the only important factor in a biodiversity hotspot. Another important criterion is the number of endemic species in an area – species that are found nowhere else. If you look at fig. 4.4.5 you can see that the areas of greatest biodiversity are not always the same as the areas with the biggest number of unique species. This is why it is so difficult to prioritise areas for conservation.

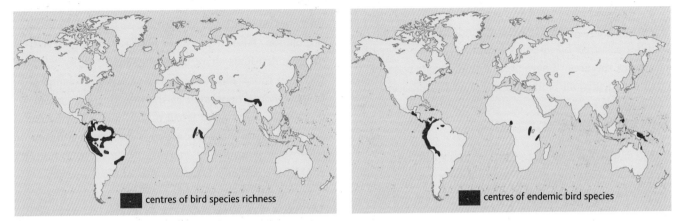

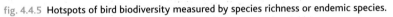

fig. 4.4.5 Hotspots of bird biodiversity measured by species richness or endemic species.

There have been many ideas about why some areas have particularly rich biodiversity – in fact around 125 different theories have been published! Most have been eliminated because they don't apply to all organisms or they are not supported by the evidence. A very stable ecosystem allows many complex relationships to develop between species. High levels of productivity (when photosynthesis rates are very high) can support more niches. A paper published in the *Biologist* in 2007 by Len Gillman and Shane Wright suggests that evidence is building that species diversity is linked to productivity through a speeding up of the evolutionary process. In other words, in areas where organisms grow and reproduce rapidly, more mutations occur which introduce more variety, enabling organisms to adapt to particular niches and evolve to form new species.

The risks to biodiversity are not evenly spread around the world. Certain areas are much more vulnerable to damage and loss, particularly small isolated ecosystems such as islands, rainforests, coral reefs, bogs and wetlands. Many of these areas are also biodiversity hotspots, so if they are damaged many species will be lost. Every time a species becomes extinct, the world's biodiversity decreases.

Questions

1 The term biodiversity is often used in the media simply to indicate the number of species of living organisms. Why does this give a limited picture?

2 Draw a flow chart to show the impact of sea otters on kelp forests and explain why they are regarded as keystone species.

3 Explain how high biological productivity and rapid mutation rates can explain how areas of high biodiversity such as the tropical rainforests come about.

4 Using fig. 4.4.5 explain how the areas of high bird biodiversity and high bird endemism differ, and why this might be.

Measuring biodiversity

Unless people understand the level of biodiversity in an area, it is impossible for them to recognise how important its loss might be, and so decide what to do about it. How can biodiversity be measured?

When considering an ecosystem, biodiversity is often simply measured by **species richness**, the number of species in the area. There are many different ways of quantifying this, all with advantages and limitations.

What do we measure?

It isn't always necessary to observe every different type of organism in an area to build up a picture of the health of the ecosystem. In many habitats certain species are particularly susceptible to change and these can be used as **indicator species** (or **bioindicators**) – changes in these species reveal changes in the overall balance of the ecosystem. For example, salmon are found only in rivers that are relatively pollution-free with high levels of oxygen in the water. So salmon in rivers are used as an indicator of water quality and lack of pollution – in other words, the health of the habitat. There has been great excitement in recent years as the water in the Thames has been cleaned up enough to enable salmon to live in it again.

Lizards are very common in sub-Saharan Africa and in many other hot or warm regions of the world. They eat insects and are in turn eaten by birds and small mammals, so they form an important part of many food chains and webs. They have a slow metabolic rate and are sensitive to many pesticides. So lizards have been shown to be useful indicators of the effect of pesticide spraying. The effect on the lizards gives information about the impact of spraying on insects, birds and mammals in the food chains and any knock-on effect on pollination rates of plants.

fig. 4.4.6 Lizards can be useful bioindicators of the effect on a whole ecosystem of human activities such as pesticide spraying.

As well as the number of different species, the size of the populations of different species can also give us an idea of the biodiversity. If there aren't enough organisms in a population, they may find it difficult to find a mate and there won't be a sustainable breeding population. Also, if the gene pool is reduced so that many individuals share the same alleles, faulty traits show up more and there is less variety. So changes in the environment such as the introduction of a new disease can wipe out a small population. The population sizes of keystone species are also important to monitor, because any significant increase or decrease in number could have a major effect on the whole food web.

When to measure biodiversity?

Biodiversity is not constant. For example, the animal species in an area can vary with the time of day. Many bat species flying on a warm summer's evening in the UK won't be visible the next morning! What is more, in the temperate and alpine areas of the world there are distinct seasons. This means that the picture of biodiversity in an area will change considerably through the year (**fig. 4.4.7**). If the number of plant species in the same area of a UK woodland floor or meadow is measured during the summer and the winter, the model of biodiversity will be very different. Similarly, wetland feeding sites for migrating birds are alive with aquatic and wading birds in the winter months, but during the summer they are relatively empty.

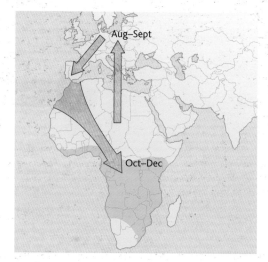

fig. 4.4.7 Willow warblers split their time between southern Africa, Germany and Spain. Measure biodiversity at the wrong time in any of these countries and you might miss the warblers completely.

HSW Measuring species richness

When measuring species richness you need to assess both the number of species in the area and the size of their populations. You also need to identify them correctly.

How would you compare the biodiversity of two grass meadows? Counting every plant would take forever and be almost impossible. Instead you can **sample** several small areas of each field. To make your results as reliable as possible it is best to sample as many separate areas as you can. Samples need to be chosen at random to stop you choosing areas that look particularly interesting, which would affect the results.

The simplest way to sample an area is to use a **quadrat** (see **fig. 4.4.8**). A quadrat is a square frame which you lay on the ground to identify the sample area. This method is particularly useful for plants and for animals that don't move much. Quadrats come in different sizes, but we often use ones with sides 1 m long which mark an area of a square metre (1 m²). This technique is known as **quantitative sampling**. It is a very important tool, but there are limitations because of factors such as the area you can sample, the randomness or otherwise of the sampling sites and the decisions you make about including or excluding organisms.

fig. 4.4.8 When you are using quadrats like this to measure biodiversity, you have to decide before you start if organisms partly covered by the quadrat count as 'in' or 'out'. It doesn't matter which, as long as you decide and stick to it!

There are many techniques for measuring the biodiversity of populations of organisms in other habitats. Each method has advantages and disadvantages when assessing biodiversity.

Light traps are often used to capture night-flying insects; they are very effective, as insects such as moths are strongly attracted to the light. Unfortunately you don't know where the insects have travelled from. **Capture/ recapture techniques** allow a clearer picture to be built up of which animals live in an area, or at least have it as part of their regular territory. This involves capturing animals on one occasion, marking them in a way that does not affect their survival chances and releasing them. Subsequent captures are examined to find out how many marked animals have been recaptured, and these data can be used to estimate population numbers in the area.

The animals in tree canopies are often collected by beating the branches and collecting what falls out. However, this method misses many populations that live at the tops of the trees, including the birds. For example, a rare weevil population in Richmond Park, UK, was ignored for years because it was missed by the collecting techniques being used.

fig. 4.4.9 The main limitation of fogging a tree as a collecting technique is that some organisms are safely hidden in the bark!

Some researchers spray an entire tree with anaesthetic gas or smoke (fogging), and then collect all the stunned organisms that fall out. Even this has its limitations – bark dwellers are immune because they are protected from the gas by the bark.

An effective but bizarre way of estimating population size is to use 'roadkill' numbers – the more squashed badgers or hedgehogs there are along a stretch of road, the higher the numbers in the local population. This might seem counter-intuitive, but only a proportion of a population ends up as roadkill, so lots of dead animals indicates lots of live ones as well! However, this method is not straightforward – Dr Jochen Langbein, who is surveying British deer

populations through road traffic accidents, has to take into account many factors. The number of cars on the roads of an area will have just as much impact on roadkills as the population of the animals. So although the largest proportion of the British deer population is found in the northern areas, numbers killed can be as high in the south of England just because there are more cars on the road.

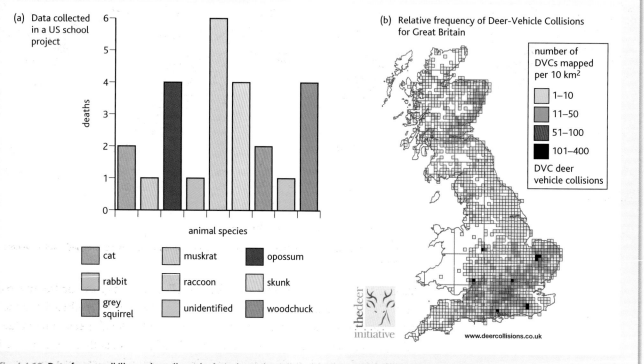

fig. 4.4.10 Data from roadkills can be collected relatively easily and can give scientists an idea of both the species diversity and the population numbers in an area as you can see from these two studies in the US and the UK.

Measuring genetic biodiversity

The genetic variety within a population is also an important measure of biological health and wellbeing – without variety a population is vulnerable. Modern technology has made it possible to build up a clear model of **genetic diversity** within a population by analysing the DNA and comparing particular regions for similarities and differences.

For example, scientists have discovered that cheetahs have very little genetic diversity. These beautiful members of the cat family, the fastest land animals on Earth, are not only low in numbers but have so little genetic biodiversity that they are in danger of being wiped out by a single disease or a small change in their environment. Scientists think that they must have been almost destroyed 10 000–20 000 years ago. As a result the modern populations are related to just a few founder members. Now that numbers are dwindling as their habitat disappears, there are serious worries about the survival of the species.

Models of the molecular phylogenetic relationships (see chapter 3.3) between related organisms based on DNA and other evidence has proved to be a very useful tool for measuring biodiversity. For example, scientists at the Natural History Museum have built up contrasting maps of biodiversity based on both numbers of species and DNA similarities. The ones shown in **fig. 4.4.11** show bee populations – the DNA map changes the most biodiverse area for bees from Ecuador to Kashmir. This type of study can have huge importance for conservation work. If you are trying to conserve biodiversity with limited funding – and funding is always limited – you need to be confident that you are choosing the area with the highest biodiversity.

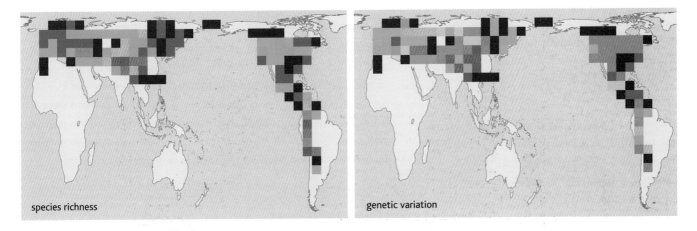

species richness

genetic variation

max ▓▓▓▓▓▓▓▓▓▓▓▓ min

fig. 4.4.11 **These maps show the biodiversity of bees measured as species richness and as genetic variation.**

Maps like these can be generated for overall biodiversity or for the diversity of particular groups of animals and plants. They can be produced for the whole world, for individual countries or for smaller local areas. The value of this type of data is that it can be used to highlight areas that need protection. It can also be used, with regular updating, to provide a way of monitoring changes in biodiversity in any particular area or worldwide.

Questions

1 Explain why a quadrat is a more useful tool for measuring the biodiversity of plants than of many animals.

2 What details would you want to know about the collection of data for **fig. 4.4.10 (a)** and **(b)** before you tried to interpret the figures?

3 What conclusions can you draw about the British deer population from the data on vehicle collisions in **fig. 4.4.10 (b)**?

4 Why is it important to measure both species richness and genetic diversity to give a full picture of biodiversity? How do the two maps in **fig. 4.4.11** give you more information than either of them alone?

The nature of extinction

Extinction means the permanent loss of all members of a species. Well-known examples of extinct species include the dinosaurs, sabre-toothed tigers and the dodo. Extinction is a common event – of the billions of species that have evolved since life appeared on Earth, only an estimated 2 million are still in existence today. Individual species become extinct because something changes in their habitat, such as changes in temperature affecting food supplies, a rise in sea level causing flooding, or the arrival of a new predator.

Scientists have calculated the 'natural' or 'background' extinction rate, based on the fossil record. They estimate that this is between 1 and 100 species each year. The current extinction rate is more like 27 000 species a year! This is mainly the result of human activity.

Before species become extinct, both population sizes and genetic diversity fall. Scientists try to evaluate what is happening to species to try and prevent them from becoming extinct. Species are classified as follows:

- **critically endangered** – facing an extremely high risk of extinction in the wild
- **endangered** – facing a very high risk of extinction in the wild
- **vulnerable** – facing a high risk of extinction in the wild.

The natural rate of extinction varies between different groups of organisms. Mammals on average become extinct more rarely (one every 200 years). Worryingly, 89 species of mammals have become extinct in the last 400 years – many more than would be predicted. What is more, another 169 mammal species are on the critically endangered list. Based on current trends, some scientists have predicted that up to 30% of the species on Earth could become extinct in the next 100 years. If that is true, the consequences for biodiversity would be severe.

Human causes of extinction

Human activities cause many of the extinctions being recorded at present. Habitat destruction is a major issue. People need land for building houses, roads and industrial premises, and for growing food. As the human population rises, the amount of land we use increases as well. This is a problem all over the world (see **fig. 4.4.12**).

Human activity also pollutes seas, rivers and lakes with sewage and with chemicals from industries and farms, eg fertilisers and pesticides. Pollution is a particular problem in closed-water systems such as the Great Lakes bordered by the US and Canada. Here the water has been shown to contain over 360 chemical compounds that have been introduced by human pollution. Seven out of the ten most valued fish species in the lakes have almost vanished from the ecosystem.

fig. 4.4.12 The site of the 2012 Olympic development in London. This area was home to bats, the rare black redstart, kingfishers and common lizards before the Olympics moved in. Brownfield developments (using land that was already built on) save using up the countryside, but they can have a bigger environmental impact than you might expect.

People eating plants and animals can sometimes drive them to extinction, as in the case of the dodo which lived on Mauritius (**fig. 4.4.13**). The islands were uninhabited until the late seventeenth century so these flightless birds had no fear of humans. Within 200 years of the first people landing on the island, the dodo was hunted out of existence by sailors eager for fresh meat.

fig. 4.4.13 The dodo was a large flightless bird, and provided easy fresh meat for sailors on ship's rations.

Species also become extinct as a result of other animals, plants or diseases introduced by people. For example, the prickly pear cactus, *Opuntia*, was introduced into Australia in the 1830s. By the early twentieth century the cactus had taken over 62 000 km² of land, considerably reducing the natural biodiversity of endemic species. In 1924 caterpillars of *Cactoblastis cactorum* (the cactus moth) were introduced to Queensland from South America. They devour the cactus and eat nothing else. Moth numbers soared rapidly, reducing *Opuntia* to a few small isolated patches and allowing the natural Australian flora to return. The moth was later introduced to some of the Caribbean islands, to control invading *Opuntia*. Unfortunately some moths have been transferred accidentally to the US mainland, where they are beginning to cause serious problems in the natural cactus populations of Florida (**fig. 4.4.14**). Here the cacti are part of the natural biodiversity, so the moths are destructive rather than helpful.

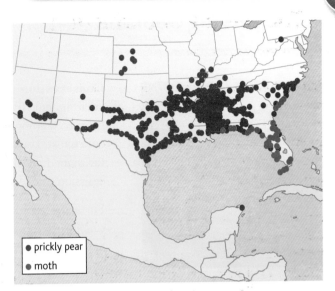

- prickly pear
- moth

fig. 4.4.14 In Australia the moth *Cactoblastis cactorum* played a useful role in restoring biodiversity. In the US, it threatens biodiversity by attacking the natural cactus flora.

There are fears that climate change as a result of human activity will also cause extinctions. So far no one really knows what the impact will be – it may just move the ranges of species rather than causing extinctions. However, extinction is inevitable when a species has nowhere else to move to (eg alpine plants in mountainous areas of the UK). Climate change may also interfere with food webs within an environment. For example, some breeding populations of sea birds on the North Sea coast depend on sand eels to feed their chicks. But the temperature of the North Sea has risen by about 1°C so sand eel populations are now peaking earlier and are no longer synchronised with the arrival of the chicks. This may cause extinctions – or the birds may breed earlier too as temperatures go up. There is still much to find out.

Questions

1 Consider the estimates of natural and human-induced extinction rates and how they may have been calculated. How reliable do you think the estimates are, and how much reliance should we put on these values when comparing them?

2 Why does the current extinction rate threaten biodiversity?

3 Look at **fig. 4.4.14**. Why are scientists concerned about the US cactus population? What measures might be taken to prevent the loss of biodiversity in the US cactus states?

Conserving habitats

fig. 4.4.15 **One of only a handful of Marine Nature Reserves in the UK is Lundy Island, it is also a habitat for a huge variety of sea birds.**

Conservation means keeping and protecting a living and changing environment. It is an active process involving an enormous range of projects from reclaiming land after industrial use to helping set up sustainable agriculture systems in the developed world, from the protection of a single threatened species to global legislation on pollution levels and greenhouse gas emissions.

For a species to be conserved, it needs a long-term habitat. Often it is not single species but large areas of countryside that are threatened – whole habitats and even ecosystems. Governments need to be involved to protect large areas such as these.

In the UK land is protected from development and allowed to sustain a rich diversity of animals and plants in a number of different ways, In England there are over 200 National Nature Reserves (NNRs), covering around 90 000 hectares of countryside, and **Sites of Special Scientific Interest** (**SSSIs**) which cover another 7% of land. These SSSIs are usually quite small areas that contain a particularly rich diversity of life, or a group of endangered plants or animals which thrive there. SSSIs play an important role in the conservation of many habitats in the British countryside.

The biggest units of protected land in the UK are the 14 National Parks which represent about 8% of the most beautiful countryside in Britain. These areas are all home to farming communities, villages and towns – but they are managed with a view to the best possible compromise between economic demands and the conservation of the countryside and the wildlife.

Similarly, National Parks are set up around the world to protect and conserve native species. For example, the Etosha National Park in the Kalahari Basin in Namibia is home to many rare species, including chimpanzees, and is the site of many animals' migration routes.

fig. 4.4.16 **These are rare mountain gorillas in the Volcanoes National Park in Rwanda.**

When land is set aside for conservation there are inevitably conflicts between the needs of people living there and the needs of the animals and plants being conserved. To maintain and conserve the area it costs money, which could be spent on health and education instead. People need the land to live on and to earn money, but their activities can lead to pollution, erosion and many other problems. What is the greater need – the long-term biodiversity of the local ecosystem or the welfare of the local people? And who makes the decisions about which gets priority treatment?

Sustainability

Habitats and ecosystems can be conserved with less conflict by sustainable methods of land use. For example, illegal logging operations in rainforests use 'slash and burn' – cutting down all the trees and burning the remains in the ground afterwards – to sell the wood and then use the soil for farming. However, if trees are harvested selectively and replanting is carried out, biodiversity can be maintained while people continue to use the forest for income.

fig. 4.4.17 Goods made with timber from sustainable forestry programmes like the FSC scheme help consumers choose products that are less damaging to biodiversity.

Sustainable agriculture includes farming methods which minimise damage to the environment and avoid monoculture. These are becoming increasingly important around the world. This does not mean organic farming in the strict sense used in the UK. It means using organic fertilisers where possible, using artificial fertilisers as little as possible, using biological pest control and as little chemical spraying as possible, maintaining hedgerows and leaving verges around fields. Large-scale farming is vital to provide the food we need. Often sustainable methods such as using biological pest control can increase yields and improve profits as a cheaper long-term option than using chemicals.

Prioritising effort is very important. People and politicians have limited sympathy for conservation and so it is important to target spending accurately. Around the world there is a growing understanding of the need for sustainable agriculture and sustainable tourism to conserve biodiversity, whilst still providing the food and income that people need. Research continues into how food and other resources can be produced in a way that minimises loss of biodiversity – and even increases it again. Similarly tourism can be developed in a way that is sustainable, does minimal damage to the environment, provides jobs and money for local people and at the same time conserves the environment and thus maintains biodiversity.

Maintaining biodiversity is a major issue for the twenty-first century. Success or failure will affect the whole planet, and the potential consequences of failure could be devastating for us all. In a global economy, we can all play a vital role in maintaining biodiversity by the choices that we make now.

Questions

1 Preservation means preserving something exactly as it is now. How does conservation differ from preservation?

2 Two different plans are put forward for a piece of local wasteland – to build a small shopping area with a car park or to make a small wildlife reserve with a nature trail and field study hut. Discuss possible advantages and disadvantages of both schemes for local people.

Conserving species

Saving animal species

It is often not possible to conserve animal species in the wild because the conditions that have put them under threat of extinction continue. Zoos and wildlife parks used to exist just for public interest, but today they play an important role in animal conservation. In **captive breeding programmes** individuals of an endangered species are bred in zoos and parks in an attempt to save the species from extinction. Usually the ultimate aim is to reintroduce the captive-bred animals into the wild to restore the original populations. Reintroduction can be more successful in national parks or other protected areas. Species that have been saved this way include the Californian condors in the US, and captive breeding programmes for the white rhino and Przewalski's horse in the UK and elsewhere have led to successful reintroductions of these animals into protected areas in their own countries (**fig. 4.4.18**).

fig. 4.4.18 Przewalski's horses (a truly wild horse species) have been reintroduced successfully to a National Park in their home range in Mongolia as a result of captive breeding programmes at Whipsnade Zoo in the UK and elsewhere.

There are several problems with captive breeding and reintroduction:

- There isn't enough space or resources in zoos and parks for all the endangered species.

- It is often difficult to provide the right conditions for breeding, even if scientists know what those conditions are. For example, the giant panda is notoriously difficult to breed even when conditions are ideal.

- Unless the original reason for the species being pushed to the edge of extinction is removed, reintroduction to the wild will be unsuccessful.

- Animals that have been bred in captivity may have great problems in adjusting to unsupported life in the wild.

- When the population is small, the gene pool is reduced and this can cause serious problems. Zoos try to overcome this by keeping detailed records of the genetic data of their breeding individuals. Sperm can be swapped with other zoos (for artificial insemination) to maximise genetic variation in the offspring.

- Reintroduction programmes can be very expensive and time-consuming and they may fail.

For some of the most endangered species, captive breeding may include **cross-species cloning** – cloning animals using closely related species as surrogate mothers. In 2001 scientists cloned a guar, an endangered species of ox. DNA from the cells of a male animal which had died 8 years earlier was fused with cows' ova from which the nucleus had been removed. Of the 692 ova used, only one produced a healthy embryo, which was implanted in an ordinary cow. The guar bull calf was born strong and healthy but he died within 48 hours from a common gut infection

Some people believe this technology can be used to bring back species that have recently become extinct. Conversely, there are scientists who feel that this cutting-edge research is a waste of valuable resources, at least until the conditions which drove the organism to extinction have been addressed.

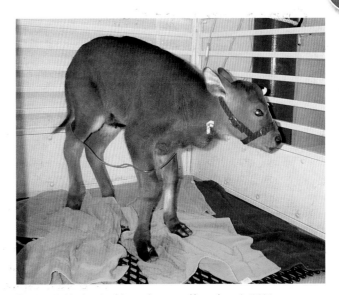

fig. 4.4.19 Noah, a healthy male guar calf was born in 2001. Unfortunately he died of an infection which scientists do not think was linked to the cloning process. Is this the way forward to saving endangered species?

Saving plant species

It has been estimated that 25% of the world's flowering plant species could disappear within the next 50 years. There are thought to be about 242 000 species of flowering plants now, so this would mean 60 500 species disappearing in less than one human lifetime!

Plants are of vital importance to all our lives (see chapter 4.2). The genetic material of these extinct species would be lost forever, which would be a disaster not only for the plants but also possibly for human survival. Cross-breeding back to original wild plants, or using them to supply genes for engineering, are ways in which the long-term health of our crop plants can be maintained.

Saving the seeds

Botanic gardens (zoos for plants!) maintain collections of many of the world's most interesting and unusual plants. In the 1960s, with a view to conservation, the Royal Botanic Gardens at Kew set up a seed bank which is now home to the seeds of over 4000 wild flowering plants (fig. 4.4.20). The Millennium Seed Bank had two main aims. The first was to collect and conserve the seeds of the entire UK native flora by the year 2000. Of the 1442 native plants over 300 are already threatened with extinction. The second aim is to conserve the seeds of an additional 10% of the flora of the whole world by the year 2010, concentrating particularly on the drylands – arid, subarid and subhumid regions – which are experiencing some of the most rapid loss of habitat.

A seed bank can preserve many plants in a state of effective suspended animation. Live seeds are collected from the wild, removed from the fruits and cleaned. They are screened using X-rays to make sure that they contain fully developed embryos. Then they are dried, put into jars and stored at between −20 and −40°C where many will survive and remain capable of germinating for up to 200 years. In general, the lifespan of a seed doubles for every 5°C drop in temperature or 2% fall in relative humidity. Some of the seeds stored may even germinate in several thousand years' time.

fig. 4.4.20 **Stored at low temperatures, seeds like these will still be able to germinate in hundreds of years – by which time hopefully their habitats will have been restored and conserved!**

Most plants make huge numbers of seeds, so they can be collected without damaging the natural population. Seeds are usually small, so large numbers of them can be stored fairly cheaply in a small space, and they contain all the genetic material of the plant so that they are a record of the genetic make-up of the species as well as a potential new plant for the future. Other countries have set up similar projects and there are now over 1000 seed banks around the world.

About 80% of the known species of plants could be stored in this way. However, seeds of some species do not store well. Unfortunately these include many crop plants such as mango, rubber, oak, avocado, cacao and coconut. These plants have to be conserved differently. They may be grown where they are found naturally, in field gene banks such as plantations, orchards and arboretums or as tissue cultures (see chapter 3.3). The species is grown on, year after year. One problem is that field gene banks take up a lot of room and a lot of work, eg the world potato collection at the International Potato Centre (CIP) in Peru contains around 4100 different clones of potato, all of which have to be planted annually! Using tissue cultures to conserve plants and growing plants on as needed takes up a lot less space and time, and allows more variety to be conserved.

Education and scientific research

Zoos and botanic gardens are centres of scientific research as well as general education. Most zoos, for example, work closely with university departments in research and many have education departments that are widely used by schools to raise awareness of conservation issues.

Education is an important part of the conservation process. Schools now cover the problems of falling biodiversity and conservation in some detail, but the wider world needs to recognise the problems, and this isn't always easy. Politicians have their own agendas and they are serving many different groups of people. Sometimes the needs of conservation are in direct conflict with the demands of industry or agriculture. Education can help to raise awareness of all the available opportunities for conservation.

The media have a role to play in highlighting the problems of falling biodiversity. Television, radio, magazines, books and the internet can all help to raise awareness of what is happening in the world around us and what we can – and can't – do to help. The messages given out on the media may occasionally be oversimplified, doom-laden or over-optimistic, but they help to educate the general public about the need for conservation.

HSW Linking conservation and economics

Economic and political factors affect the success of conservation projects. Conservation schemes that are also economically attractive have a greater chance of succeeding.

A group of scientists led by Taylor Ricketts at Stanford University in California carried out research into the role played by bees which pollinate the coffee plantations in Costa Rica. The bees live in and depend on the tropical forests of the region. Coffee plantation owners tend to cut down forest to make space to plant coffee bushes, destroying a habitat of rich biodiversity and replacing it with a monoculture. The scientists developed a hypothesis that the presence of the forest and the bees that live in it would actually increase the yield of coffee. The team investigated an area where the coffee plantations were mixed with patches of forest. They showed that, in line with their theory, coffee production was highest nearest the forest. Twelve experimental sites were set up near (50 m from) forest patches, at an intermediate distance (800 m) and far (1600 m) from the natural forest. The researchers hand-pollinated some coffee bushes, so they knew they had been pollinated. This gave them a control. Then they compared the relative fruit set and seed masses of the coffee beans in the different experimental areas left to the bees. The bees increased the yields of coffee beans by about 20%. The benefit to the plantation owner was about £30 000 a year! This provided a clear economic incentive for coffee plantation owners to maintain forest habitats rather than cutting them down.

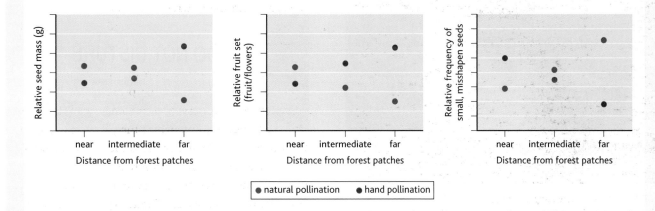

● natural pollination   ● hand pollination

fig. 4.4.21 This investigation showed that coffee plants closer to the forest produced far more coffee beans, which were bigger and healthier, because the bees acted as pollinators. This gave growers a financial incentive to preserve the forest and its biodiversity.

Questions

1. Discuss the advantages and disadvantages of cross-species cloning as a method of preserving biodiversity.

2. Why is it much easier to conserve plant biodiversity than animal biodiversity?

3. What do the data from fig. 4.4.21 show about the relationship between the proportion of seeds set, the mass of the coffee beans produced and proximity to the forest?

4. How does this evidence support the hypothesis being tested by the scientists from Stanford University and why is it important for the future of biodiversity in Costa Rica?

5. Discuss the main advantages and disadvantages of captive breeding and reintroduction programmes.

HSW Making choices – the story of Guam

The story of the brown tree snakes of Guam highlights many of the factors which affect biodiversity and conservation efforts around the world.

In the late 1940s shipments of timber arrived on the island of Guam from New Guinea with some brown tree snakes (**fig. 4.4.22**) as stowaways in the ships. Guam had only one tiny indigenous snake. The newcomer, which can grow to almost 3 m long, feeds on birds and lizards and hunts at night. In Guam the brown tree snakes found forests full of birds and lizards endemic to the island, and no predators to attack snakes. The endemic birds had no defences against the night-time predator, which took adults as they slept, as well as young and eggs from the nests.

fig. 4.4.22 **The brown tree snake.**

No one realised what was happening for almost 30 years, until the snake population had grown large enough to affect bird numbers noticeably. The snake wasn't properly identified by scientists until 1983 and by then only about 160 hectares of woodland were uninhabited by the snakes. Once scientists recognised the role of the snake, they started captive breeding programmes of the threatened endemic birds. But the damage to the island biodiversity had been done. Today, seven species of Guam birds are extinct. Two exist only in zoos where there are intensive breeding programmes to try and maintain the species. The remaining nine species are rare or uncommon. Five of the native lizard species have disappeared as well, and the only small mammals are introduced house mice and rats.

The biggest concern now is that the brown tree snake will spread from Guam to neighbouring islands and cause similar problems there. Biosecurity is in place around the airfields and port. There are electric barriers to stop the snakes, and special sniffer dogs trained to sniff them out in cargo bays.

People are trapping the snakes – in fact about 6000 a year are removed in this way, but there are always plenty more. One female can lay up to 24 eggs in two clutches each year and recent estimates put the population at over 2 million!

fig. 4.4.23 **Biosecurity is tight on Guam in an attempt to prevent the spread of the brown tree snake to other islands.**

Tackling the problems of Guam

In a situation like Guam there is a limit to what scientists can do. The brown tree snake has been extensively studied and it is remarkably well adapted for its niche on Guam. Everyone accepts the need to aim for control rather than eradication.

A predator for the snake could be introduced, but any snake predator could attack other island species too. Scientists at the US National Zoological Park are investigating diseases specific to the brown tree snake that could be used, just as myxomatosis was used to control rabbits. There are two viruses that cause disease in the snakes and kill over 50% of those infected. The viruses can't survive in animals with temperatures above 35°C, so birds and mammals, including humans, are not at risk.

The alternative approach is to try and conserve those endemic bird species that remain. Two species survive only in zoos. The captive breeding programme for the Guam rail has been very successful, and reintroduction to the island would be possible if the snakes could be kept under control. A breeding population has been established on the (so far) snake-free island of Rota, 31 miles away from Guam.

fig. 4.4.24 **The Guam rail survives – just – as a result of a captive breeding programme.**

Breeding the Micronesian kingfisher has been much more difficult. Scientists knew very little about the birds before the captive breeding programme began. They didn't know what the adult birds ate in the wild, so they fed the captive birds newborn mice and they did very well. They also found that the birds need a particular type of log to use as a nest site. When a few birds finally nested and hatched eggs successfully, the babies disappeared. Unfortunately the parents seemed to mistake their pink, bald babies for newborn mice and ate them! From then, the scientists fed the adults small lizards which more accurately mimicked their natural diet, and the cannibalism decreased. At the start of the programme 29 birds were captured from the wild. By 2007 there were almost 100 Micronesian kingfishers in captive breeding programmes – not yet enough to try establishing a wild population.

Much time and money has been spent attempting to solve these problems of Guam. Some people argue that the ecosystem should be left to evolve without the intervention of science. Eventually a new biodiversity would develop, based around tree snakes and organisms which can survive their predatory habits. Then the money and research efforts could be used elsewhere on biodiversity issues that affect a wider range of species of larger areas of the world, or on medical research, for example.

Wherever biodiversity is under threat and conservation is important, issues such as these are raised. Science cannot answer these questions – people and society have to make their choices and live with the consequences.

fig. 4.4.25 **Without a captive breeding programme, the Micronesian kingfisher would also be extinct.**

Questions

1 Explain in terms of niche and adaptation why the brown tree snake has had a devastating effect on biodiversity in Guam.

2 Evaluate the role of scientific research, captive breeding and reintroduction programmes in restoring the endemic bird populations of Guam. Summarise the difficulties faced as well as the progress made in each case.

3 Develop arguments both justifying and disagreeing with the money and time spent on:

   a biosecurity measures

   b control measures for the brown tree snake

   c captive breeding programmes for endangered bird species on the island of Guam.

Examzone: Topic 4 practice questions

1 Read the following passage about the palisade cells of a leaf. Copy the passage, completing the spaces with the most appropriate word or words.

The palisade cell is typical of plant cells in that it has three structures, ..., ... and .... none of which is present in animal cells. In common with animal cells, plant cells (such as palisade cells) have membrane-bound organelles which are not present in .... cells. In a leaf, palisade cells are grouped together as a layer just below the epidermis forming a .... the function of which is to carry out photosynthesis.

(Total 5 marks)

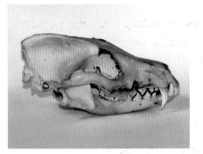

2 The fox (*Vulpes vulpes*) is a common mammal living in both rural (country) and urban (town) areas of the United Kingdom. Foxes eat a variety of foods including berries, rabbits, small birds and rodents such as rats and mice.

a The photograph below shows the skull of the fox. Describe two features of the teeth of this fox that are an adaptation for feeding on small mammals. **(2)**

b A number of studies have investigated the differences between the diets of rural foxes and foxes living in urban areas. The results of one study are shown below.

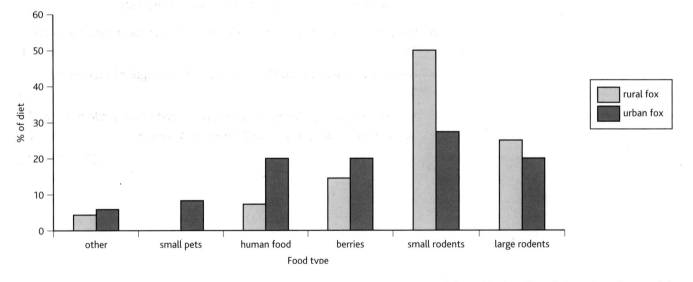

i Compare the diet of the rural fox with the diet of the urban fox. **(2)**

ii Human food was found to have a high content of carbohydrates and fats. Suggest how this could affect the time spent looking for food and the quantity of food eaten by the urban foxes. **(2)**

c It has been suggested that the teeth of the urban fox are changing as their diet changes. Describe how the rural and urban foxes could evolve into separate species. **(4)**

(Total 10 marks)

3 Garlic is known to contain an antimicrobial substance called allicin.

    a Suggest an advantage to a plant in producing antimicrobial substances in its cells. (1)

    b The presence of antimicrobial substances in garlic can be demonstrated by grinding garlic in ethanol to produce an extract. A sample of this extract is then applied to a small disc of filter paper.

        i Describe how you would demonstrate that this disc contained an antimicrobial substance. (4)

        ii A newly discovered rainforest plant is thought to contain a powerful antimicrobial substance. Explain how you would compare the effectiveness of this new substance with allicin in garlic. (3)

    c Suggest how this newly discovered plant substance might have a useful application which does not involve genetic modification. (1)

                                                        (Total 9 marks)

4 Many plants are grown for their fibres for use in making cloth. For example, nettle fibres can be used for this purpose. In the future nettle fibres may be developed in the United Kingdom as an alternative to imported cotton plant fibres.

A student investigated the strength of plant fibres to find out if they could be made into cloth. The following method was used.

Step 1. Harvest nettle plants and remove leaves from stems
Step 2. Place stems in buckets and cover with water. Leave for ten days
Step 3. Remove stems and wash fibres under running water

    a Suggest and explain the changes that take place in the nettle stems during Step 2. (2)

    b Describe a reliable method for measuring the strength of these plant fibres. (4)

    c Suggest why, in the United Kingdom, cloth made from nettle fibres could be made more cheaply than cloth made from cotton. (2)

                                                        (Total 8 marks)

Index

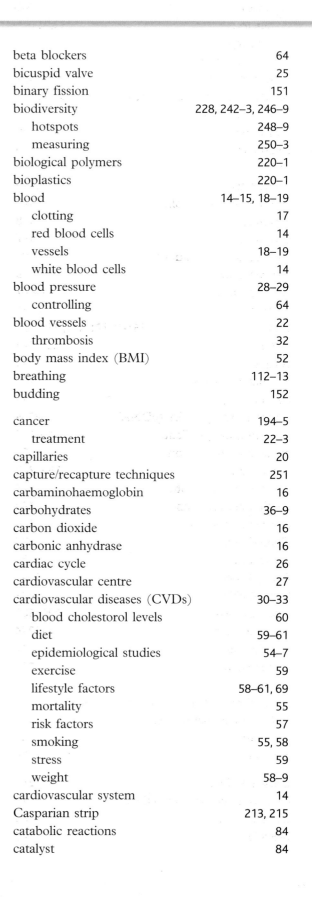

Pearson Education Limited
Edinburgh Gate
Harlow
Essex CM20 2JE

© Pearson Education Limited 2008

The right of Ann Fullick to be identified as the author of this work
has been asserted by her in accordance with the Copyright, Designs
and Patents Act of 1988.

First published 2008

Fifth impression 2011

ISBN 978-1-4058-9632-0

Development editor and external project management Sue Kearsey
Design by 320 Design Ltd
Illustration by Oxford Designers and Illustrators
Picture research by Charlotte Lippmann, Kay Altwegg, Sarah Purtill
Index by Ian Crane
Printed in China   (GCC/05)

Acknowledgements
The publishers are grateful to the following for their collaboration in
reviewing this book:
Martin Furness-Smith, Susan Millins, John Sears, James Williams
and Mark Winterbottom.

Photo acknowledgements
The publishers are grateful to the following for their permission to
reproduce copyright photographs:

(Key: b-bottom; c-centre; l-left; r-right; t-top)

Advertising Archives: 68br; **Alamy Images:** AGStockUSA, Inc.
218; B.A.E. Inc. 67; blickwinkel 215, 250; Brandon Cole Marine
Photography 247; Bubbles Photolibrary 59t, 126; camera lucida
environment 251l; Choups 258; Ashley Cooper 246; DAVID
NOBLE PHOTOGRAPHY 237; Luca DiCecco 59; Chuck Franklin
219; Helene Rogers 156; Holt Studios International Ltd 62; Janine
Wiedel Photolibrary 12, 193; Medical-on-Line 225r; Mike Abrahams
192; Andrew Moss 257; louise murray 228br; Nic Hamilton
Photographic 264; Peter Arnold, Inc. 200t; PHOTOTAKE Inc.
20tr; PHOTOTAKE Inc. / Tomasz Szul 139; Picture Partners 154c;
Chris Rout 51br; Andre Seale 234; Ken Welsh 98; **Ardea:** Bill
Coster 241t; John Mason 241b; **CF Trust:** 121; Trevor Clifford:
Pearson Education 13, 64; **Corbis:** Bettmann 94bl; You Sung-Ho/
Reuters 179; **FLPA Images of Nature:** CHRISTO BAARS/FOTO
NATURA 244br; Dembinsky Photo Ass. 222; GERRY ELLIS/
Minden Pictures 262l; David T. Grewcock 236; Frans Lanting 242;
MARK MOFFETT/Minden Pictures 251r; Phil McLean 228t; Nigel
Cattlin 93, 214b, 214c, 214t; Fritz Polking 40, 256b; Gary K Smith
223; **Food Features:** 50; **John Frost Newspapers:** 226;
GeoScience Features Picture Library: M. Hirons 148; **Getty
Images:** David McNew 182c; Image Source 43; J.D. Pooley 182t;
Robert Harding World Imagery: G Richardson 256t;
iStockphoto: 181; Johns Hopkins Institute for Cell Engineering:
176; **Maggie Bartless, NHGRI/www.genome.gov:** 23; **MRC
Human Genetics Unit:** 125bl; **Natural History Museum
Picture Library:** 229l, 229r; **Nature Picture Library:** Bristol
City Museum 255; Pete Oxford 244l; Andrew Parkinson 235; Dan
Rees 263t; Shattil & Rozinski 186; **NHPA Ltd / Photoshot
Holdings:** Anthony Bannister 228c; Karl Switak 171; Lutra 228bl;
PA Photos: PA ARCHIVE IMAGES 254; Peter Byrne 153; **Panos
Pictures:** Giacomo Pirozzi 51bl; **Photodisc:** Modern Technologies
8tl; Professional Science 8tc; **Photolibrary.com:** Foodpix 216;
Phototake, Inc: ISM 200b; Ken Wagner 203; **PunchStock:**
Corbis 8tr; **Robert Lanza MD, Advanced Cell Technology:** 259;

Royal Botanic Garden Kew: 260; **Science Photo Library Ltd:**
25, 32, 76, 142tl, 202; AJ PHOTO 125br; Asa Thoresen 144bl;
ATHENAIS, ISM 31; Biomedical Imaging Unit, Southhampton
General Hospitasl 142tr; BIOPHOTO ASSOCIATES 11;
A. BARRINGTON BROWN 77t; CHUCK BROWN 19t; BSIP,
VILLAREAL 56; BSIP,Sercomi 172b; DR JEREMY BURGESS
39t, 204; Scott Camazine 213; Claude Nuridsany & Marie Perennou
172c; CNRI 39c, 90, 147; David Scharf 151, 162; Don W. Fawcett
138r; EQUINOX GRAPHICS 87; DR. TIM EVANS 45; Eye of
Science 100, 114, 167, 206b; PASCAL GOETGHELUCK 194,
194l; STEVE GSCHMEISSNER 17; HEALTH PROTECTION
AGENCY 123; Herve Conge, ISM 208; IAN HOOTON 29;
Hubert Raguet/Eurelios 225l; HYBRID MEDICAL ANIMATION
22; I. ANDERSSON, OXFORD MOLECULAR BIOPHYSICS
LABORATORY 47; Innerspace Imaging 144bc, 144br; J C Revy
118tl, 118tr; John Reader 134; RUSSELL KIGHTLEY 74; JAMES
KING-HOLMES 119b, 119t; Klaus Guldbrandsen 177, 180; DR
KARI LOUNATMAA 103; DR P. MARAZZI 83; MAXIMILIAN
STOCK LTD 34, 68bl; Mehau Kulyk 8t, 154; PETER MENZEL
46, 262r; Anthony Mercieca 263b; Micahel Abbey 185; Dr Gopal
Murti 136br, 143; NIBSC 102; Omikron 173; ALFRED PASIEKA
84; D. PHILLIPS 20tl; NOBLE PROCTOR 244tr; Professors P.
Motta & T. Naguro 138l, 150; Paul Rapson 221; Science Source
77b; Scimat 187; Scott Sinkler 170; SOVEREIGN, ISM 33; Andrew
Syred 207, 209l; SHEILA TERRY 159; DR. KEITH WHEELER
206t; HATTIE YOUNG 120; **STILL Pictures The Whole Earth
Photo Library:** Biosphoto/Huguet Pierre 220; Roland Birke 10;
Peter Arnold, Inc./ Ed Reschke 136tr; **UKAEA:** 209r; **Wellcome
Trust Medical Photographic Library:** M. I. Walker 27

All other images © Pearson Education

Cover photo © Gregory Ochocki / Science Photo Library

Text acknowledgements
We are grateful to the following for permission to reproduce
copyright material:

Blackwell Publishing Ltd for figure 1.4.5 on page 58 adapted from
"Heritability of death from coronary heart disease: a 36-year
follow-up of 20,966 Swedish twins" by S. Zdravkovic, A. Wienke,
N. L. Pedersen, M. E. Marenberg, A. I. Yashin & U. De Faire
published in *Journal of Internal Medicine* 1 September 2002 copyright
© Blackwell Publishing Ltd, 2002; International Task Force for
Prevention of Coronary Heart Disease for figures 1.4.10 & 1.4.11
on page 61 adapted from *Kit 4: PROCAM (Münster Heart Study):
Determinants of Mortality in the PROCAM Study* by P. Cullen,
H. Schulte and G. Assmann, 1997; 96:2128-2136 reproduced with
permission; Internet Scientific Publications LLC for figure 1.4.16 on
page 65 from "Enas A Enas, Annamalai Senthilkumar: Coronary
Artery Disease in Asian Indians: An Update and Review" published
in *The Internet Journal of Cardiology*, 2001 Vol 1, 2 copyright ©
Internet Scientific Publications LLC; Unilever plc for graph 1.4.17
on page 66 "impact of flora Pro-Active on blood cholesterol levels"
www.proactivescience.com, based on international studies and
produced by Unilever Australia, reproduced with permission;
Scientific American, Inc for figure 4.3.4 on page 233 from "Uprooting
the Tree of Life" by W. Ford Doolittle, February 2000 copyright ©
2000 by Scientific American, Inc. All rights reserved; Paul D. Colombo,
EduTel Communications Inc for figure 4.4.10a on page 252 "Rt2
Roadkill '97" graphic/data contributed by Paul D. Colombo, EduTel
Communications Inc published on http://roadkill.edutel.com,
reproduced with permission; Dr J. Langbein on behalf of The Deer
Initiative National Deer Collisions Project for figure 4.4.10b on page 252
Deer on our Roads – Counting the Cost published on www.deercollisions.
co.uk reproduced with permission; Natural History Musuem for
figure 4.4.11 on page 253 'Biodiversity and WorldMap' published
on www.nhm.ac.uk copyright © Natural History Museum, London;
GeoResources Institute for figure 4.4.14 provided by the GeoResources
Institute, Mississippi State University, USA, through the Cactus
Moth Detection and Monitoring Network project at URL: http://www.
gri.msstate.edu/cactus_moth, with funding provided by US Geological
Survey and the National Biological Information Infrastructure
(NBII); and PNAS and Taylor H. Ricketts for figure 4.4.21 on
page 261 from *Economic value of tropical forest to coffee production* by
Taylor H. Ricketts, Gretchen C. Daily, Paul R. Ehrlich and Charles
D. Michener published on 24 August 2004, Vol. 101, No. 34,
12579-12582 copyright © 2004 National Academy of Sciences,
U.S.A.

Every effort has been made to trace the copyright holders and we
apologise in advance for any unintentional omissions. We would be
please to insert the appropriate acknowledgement in any subsequent
edition of this publication.

Single User Licence Agreement:

Edexcel AS Biology Students' Book with FREE ActiveBook CD-ROM

Warning:

This is a legally binding agreement between You (the user or purchasing institution) and Pearson Education Limited of Edinburgh Gate, Harlow, Essex, CM20 2JE, United Kingdom ('PEL').

By retaining this Licence, any software media or accompanying written materials or carrying out any of the permitted activities You are agreeing to be bound by the terms and conditions of this Licence. If You do not agree to the terms and conditions of this Licence, do not continue to use the Edexcel AS Biology Students' Book with FREE ActiveBook CD-ROM and promptly return the entire publication (this Licence and all software, written materials, packaging and any other component received with it) with Your sales receipt to Your supplier for a full refund.

Intellectual Property Rights:

This Edexcel AS Biology Students' Book with FREE ActiveBook CD-ROM consists of copyright software and data. All intellectual property rights, including the copyright is owned by PEL or its licensors and shall remain vested in them at all times. You only own the disk on which the software is supplied. If You do not continue to do only what You are allowed to do as contained in this Licence you will be in breach of the Licence and PEL shall have the right to terminate this Licence by written notice and take action to recover from you any damages suffered by PEL as a result of your breach.

The PEL name, PEL logo, Edexcel name, Edexcel logo and all other trademarks appearing on the software and Edexcel AS Biology Students' Book with FREE ActiveBook CD-ROM are trademarks of PEL.

You shall not utilise any such trademarks for any purpose whatsoever other than as they appear on the software and Edexcel AS Biology Students' Book with FREE ActiveBook CD-ROM.

Yes, You can:

1 use this Edexcel AS Biology Students' Book with FREE ActiveBook CD-ROM on Your own personal computer as a single individual user.

You may make a copy of the Edexcel AS Biology Students' Book with FREE ActiveBook CD-ROM in machine readable form for backup purposes only.

The backup copy must include all copyright information contained in the original.

No, You cannot:

1 copy this Edexcel AS Biology Students' Book with FREE ActiveBook CD-ROM (other than making one copy for back-up purposes as set out in the Yes, You can table above);

2 alter, disassemble, or modify this Edexcel AS Biology Students' Book with FREE ActiveBook CD- ROM, or in any way reverse engineer, decompile or create a derivative product from the contents of the database or any software included in it:

3 include any materials or software data from the Edexcel AS Biology Students' Book with FREE ActiveBook CD-ROM in any other product or software materials;

4 rent, hire, lend, sub-licence or sell the Edexcel AS Biology Students' Book with FREE ActiveBook CD-ROM;

5 copy any part of the documentation except where specifically indicated otherwise;

6 use the software in any way not specified above without the prior written consent of PEL;

7 Subject the software, Edexcel AS Biology Students' Book with FREE ActiveBook CD-ROM or any PEL content to any derogatory treatment or use them in such a way that would bring PEL into disrepute or cause PEL to incur liability to any third party.

Grant of Licence:

PEL grants You, provided You only do what is allowed under the 'Yes, You can' table above, and do nothing under the 'No, You cannot' table above, a non-exclusive, non-transferable Licence to use this Edexcel AS Biology Students' Book with FREE ActiveBook CD-ROM.